# A HANDBOOK OF BIOTECHNOLOGY

## Part - I

# A HANDBOOK OF BIOTECHNOLOGY

## Part - I

*Edited by*
**Neelam Yadav**

**ANMOL PUBLICATIONS PVT. LTD.**
NEW DELHI - 110002 (INDIA)

**ANMOL PUBLICATIONS PVT. LTD.**
4374/4B, Ansari Road, Daryaganj,
New Delhi - 110 002
Phone : 23278000, 23261597, 23255577

*A Handbook of Biotechnology — Part I*



First Edition, 2003

ISBN 81-261-1483-5

[Responsibility for the facts stated, opinions expressed, conclusions reached and plagiarism, if any, in these volumes is entirely that of the Editor. The publishers bear no responsibility for them, whatsoever.]

PRINTED IN INDIA

Published by J.L. Kumar for Anmol Publications Pvt. Ltd.,
New Delhi - 110002 and Printed at 'Tarun' Offset Press, Delhi.

# Contents

# Preface

Biotechnology is a branch of science that combines the study and inter-action of biology and technology. The complex relationship between science and technology has benefited the mankind, since prehistoric times. Those were times when Biotechnology was used and practised in processes like fermentation by common people, without knowing their complex mechanisms. However, it was in the twentieth century and particularly its latter half that Biotechnology witnessed a revolutionary change. Many researches were conducted and newer horizons reached during this period. Newer products and drugs are being produced on a large scale now. This has benifited a huge section of society in particular, and the whole mankind in general.

This work is an effort to provide the students of Biotechnology with its latest elements, uses and applications. Tissue culture techniques, gene technology, involving recombinant DNA, gene cloning and PCR, hybridoma, protein engineering and enzymes etc. are discussed in some detail in this book. One great use of Biotechnology is in identifying criminals, with the help of the samples of semen, blood, urine, tears, saliva, perspiration and hair roots etc. This aspect also forms part of this book.

Metabolic engineering, agricultural research industrial microbiology, intellectual property rights and patents, environmental applications and implications are some other topics taken up here. Intellectual property and environmental aspects are in particular of great relevance to India, today. Towards the end of twentieth century, India has seen hectic activity in this field. Efforts are on to create better infrastructure and trained manpower.

Good literature on any subject helps a lot to all those, involved with the subject, be they students, researchers industrialists or others in any way connected with the discipline. This effort aims at providing exactly that to those, associated with the discipline of Biotechnology. It requires a lot of devotion, hard work, research and compilation on the part of the individual or individuals involved in creation of such a work.

This book has been written in simple language, still being a book on technology, it is full of technical terms. The text matter of the book is supported by relevant data in the form of tables. Several figures are also given to enrich the subject matter.

The subject of Biotechnology has gradually become very important. Its importance can be gauged from the fact that a National Biotechnology Board was constituted by the Government under the Department of Science and Technology on the recommendations of Scientific Advisory Committee of the Parliament in 1982. Biotechnology plays an important role in employment, productivity, trade, economics and quality of life. As such, literature on the subject is of great importance and relevance to lots of people.

Hopefully, this effort would prove to be purposeful for and acceptable by those, concerned.

**— *Editor***

# SECTION - 1

# GENERAL

# ONE

# History and Background

From the prehistoric times, man has been engaged in commercializing activities such as leavening the bread, wine making and brewing, food preservation and several fermentation processes including soy sauce and vinegar production. In India particularly, many ayurvedic (the most ancient system of medication and disease treatment) medicines have been prepared over the years with self generated alcohol. Undoubtedly, interest in the subject has grown over the last ten years. Besides the traditional Universities and research organizations, a large number of private entrepreneurial biotechnology companies have sprung up in several countries. Although biotechnological operations started long ago, the recent advent of genetic engineering and recombinant DNA techniques have really boosted up the biotechnological advancements exploited primarily for industrial and public needs. In the present chapter, an attempt has been made to bringforth a comparative definition and structure of Science and Technology in relation to Biological Sciences and finally give a clear and concrete objective view of "What is Biotechnology."

**Technology and Science**

Science refers to a body of knowledge ascertained by observation, experimentation, systematic and critical testing. It is a self correcting system dependent upon the critical evaluation of facts. Logistically, the property of science resides in the fact that it is recorded in such a way that what one generation does and learns, the next generation can also learn and use. Sometimes, scientific facts are lost and re-appear in due course with greater fervent. Thus, though the Egyptians and Babylonians learned to measure time by the Sun and Stars and by the flow of water, nobody introduced any important new principle for the determination of time for three thousand years. After that vast lapse, an unknown person invented a mechanical clock only in 1250 A.D.

Technology, on the other hand, refers to the application of Science to industry or anywhere else. Technology has been in the news for the last two decades, because of its application to several public and private sectors. Consequently, all sorts of technologies—chemical technology, glass technology, food technology, fibre technology, paint and wax technology, Jute technology, soap technology, laser technology, oil technology and several others—have sprung up recently. In India particularly, recognition of the potential significance and application of technology has given rise to several Institutes, to train students interested in different aspects.

## Technology and Engineering

Before the end of the Middle Ages, Man's knowledge of the world and its industrial techniques had outgrown the discoveries of ancient science. Scientific investigations were dealing with the most fundamental aspects of nature, matter and energy. However, these fundamental aspects needed manouevering and modification before they could be applied. Consequently, the applications of science gradually took to Engineering which is more concerned with modelling and making designs within the realm of science for public uses, including machinery and electricals. It also deals with public works such as roads, railways, sewers, bridges, harbours, canals, aircraft and ship building besides constructing and managing military fortifications. In recent years, new disciplines such as genetic engineering and genetic technology have also sprung up. These are more or less techniques to manipulate the structural and organizational make up of the living cell, instead of inanimate objects.

## Fundamental Biosciences

Biology is the science of life; of all living creatures, from viruses to man himself. The earth is populated by millions of different types of living beings. Each lives differently, but all share some common structural and chemical components resulting in organized life. Investigations into the unifying features of living objects will only indicate what biologists know about life and how they describe various structural, compositional, functional and behavioural aspects of life processes. Unifying theories explain isolated facts. Science seeks theory to organize an accumulation of poorly-understood facts. One of the greatest unifying theories of biology is that all, or nearly all forms of life have a common and fundamental basic structure. This fact is not obvious; for instance, an elephant and a tree do not seem to resemble each other. Yet both are fundamentally alike in being composed of cells. This one aspect of fundamental Bioscience is categorization of the structure of living organism. If nearly all forms of life have a common basic structure (such as cells), then do they have common functions? This is another aspect of fundamental Bioscience. This biological question is also related to the chemistry, and thus understanding the chemistry of living organisms is another aspect of Bioscience. At one stage, it also takes into consideration the argument that life can be explained in terms of the same general laws applicable to all matter. The point is highly debatable and fundamental, because living organisms are composed of atoms, common in the nonliving world, too. But these atoms are much more intricately organized, in living beings, than in nonliving matter. This forms the major aspect of study in Fundamental Biosciences. It accounts for the relationship between life and matter either through vitalism (that life is made possible by some force that is neither chemical nor physical) or mechanism (that life can be explained entirely in chemical and physical terms). Living organisms today are the temporary manifestations of a lineage of life that extends back to several billion years. Individuals die, but life continues in their offspring. Reproduction and heredity form another aspect of fundamental biosciences. As a corollary, fundamental Biosciences deal with the study of living systems in a wider perspective (Fig.).

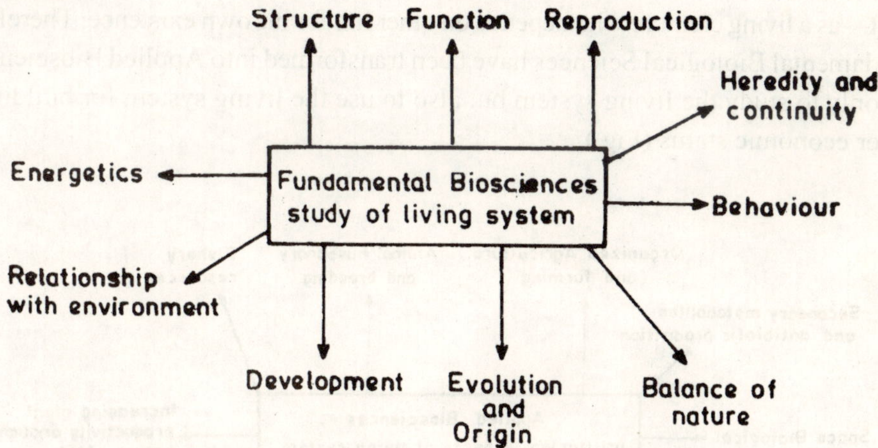


**Fig.** The spectrum of studies in fundamental Biosciences

## Applied Biosciences

While curiosity persists, information obtained has to be processed, to modernize instructional and fundamental material in Biology. It leads to the fullest possible consideration of new perspectives in Biology, growing from the astonishing increase in scientific knowledge. At the current rate of scientific advance, there is about six times as much significant biological knowledge today as in 1930, and about twenty times as much as in 1900. By the 21st century, at this rate, there will be hundred times as much Biology to 'cover' in the introductory course as at the beginning of this century. Obviously, one must be highly selective in one's choice of what scientific facts, concepts and principles to present, so that the most important developments of new biology, together with the most profound insights of the classical biology (or better older biology), may be included. This highly selective approach means, however, that many feasible view points and approaches may claim equal validity and equally high merit. Consequently, the biologists may get convinced that although authoritative fundamental facts, concepts and principles must be studied, an economically beneficial and utility-oriented study catering to the needs of growing society is also required.

For example, primitive man generally lived in balance with his surroundings. He was an integral part of the biological community. Modern or civilized man, especially in the current century, has approached nature with an arrogance that could be his own undoing. By poor agricultural practices, he has destroyed vast areas of productive soil; he has devastated forests, only to discover his water supply diminishing and his soil supply eroding. He has fouled rivers, lakes and even oceans, thereby making their water unsuitable for his needs. But today, he is beginning to behave as he must—as a living creature who depends on other life for his own existence. Therefore, Fundamental Biological Sciences have been transformed into Applied Biosciences, not only to study the living system but also to use the living system for building a better economic status (Fig.).

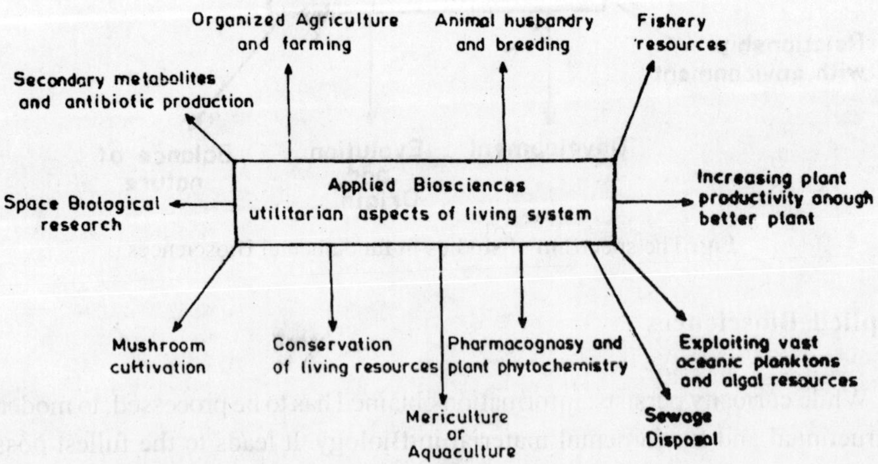


**Fig.** The specturm of studies in applied biosciences

**Advent of Biotechnology**

With the advances in Bioscience (both fundamental and Applied) as well as the development of individual disciplines such as Biochemistry, Microbiology, Genetics and Molecular Biology, and also Engineering and other technologies, man's thinking has become more lucid and almost directed towards harvesting economic benefits. The relationship between these subjects has become extremely close and overlaps with diverging applications, just like mixing parts of different things to get something that tastes and looks new. Consequently, the birth of Biotechnology took place—a link between the biological sciences, physical scineces and Technological achievements. It may even be referred to as the clever science of biology.

The word 'biotechnology'—a package term for 'biological technology'came into use in the mid 1970s, gradually superceding the more ambiguous 'bioengineering'. This was (and still is) variously used, to describe both 'biomedical engineering' which refers to the design and manufacture of such products as heart valves, artificial hips and body scanners, and 'biochemical engineering', which refers to chemical engineering processes using biological substances or organisms (such as fermenter design and control and product recovery and purification). Such linguistic changes and developments, though confusing, are natural and unavoidable; it should be remembered that the word 'biology' was only introduced around 1800 as a substitute for physiology or zoology. Even definitions of technology vary from the simple "applied science" to "the scientific study of the practical or industrial arts". So perhaps it is not surprising that not everyone can agree about biotechnology.

**The Definition**

One of the most recent quoted definitions of biotechnology is as follows: "the application of scientific and engineering principles to the processing of materials by biological agents to provide goods and services". The wording is carefully chosen. Agents includes a wide range of biological substances such as enzymes, and whole cells or multicellular organisms; the word services covers processes such as waste and water treatment. Another frequently quoted definition is: "The application of biological organisms, systems or processes to manufacturing and service industry." Both these definitions are rather vague about the nature of the organisms or agents involved and, therefore, do not settle the question of whether agriculture itself is biotechnology. It can be said with a greater degree of certainty, however, that Agriculture is not considered a biotechnological process, although genetic manipulation of plant is considered so. It can be argued that systematic farming of plants and animals for food and fuel falls within both these definitions but, in practice, most people do not include conventional agriculture within biotechnology. Other definitions have restricted biotechnology to the use of microorganisms and cultured cells. Even this gives rise to problems, because certain food processes, such as baking, brewing and cheesemaking are clearly still included within this definition of biotechnology. These methods were developed long before any knowledge of the involvement of microorganisms existed.

Here an attempt has been made to describe the information sources and processes which are available in the field of biotechnology. Biotechnology is a multi-disciplinary applied science, which has received considerable attention particularly during the last five years. The governments of most developed and developing countries including India have allocated large sums of money to support research and industrial development in biotechnology. More than 300 specialist biotechnology companies have been formed world-wide, several of which have made very successful public share offerings. Besides, the proliferation of articles, books, journals, patents and newsletters about biotechnology has been enormous. Many people have, therefore, heard of biotechnology but few seem to know how to find out more about it.

In mid 1983, biotechnology was definitely an 'in' subject; a 'buzzword' even, increasingly used to create a 'high technology' image by companies and universities in the hope of attracting attention and money. One aspect of biotechnology, genetic engineering, was also becoming a topic of considerable public interest and debate because of its long-term possibilities in altering and improving the genetic heritage of agricultural plants, farm animals and even human beings. This potential raises a number of important moral, ethical and ecological questions apart from posing a formidable scientific challenge.

However, there is much more to biotechnology than genetic engineering and most genetic engineering is at present concerned with the manipulation of bacteria and yeasts rather than higher organisms. Biotechnology can, with some difficulty, be defined and treated as a single subject area, albeit one with very grey edges.

The demand for biotechnologists in a wide range of manufacturing and service industries is increasing and their specialised expertise in biotechnological processes that reduce costs compared to traditional methods.

In the chemical industry, the transformation of substances by biocatalysis offers new routes to products that are difficult or expensive to produce by chemical transformations, and can give rise to new products such as biopolymers. Activities range from improving or engineering raw materials such as oil seeds; processing waste into fuel or potential intermediates such as industrial feedstock; generating new antibiotics or producing alternative structural materials to plastics.

The food processing industry has a constant demand for acceptable additives such as colourants, sweeteners and sauces, including soy sauce, or for bulk protein, novel catalysts and preservatives. The incorporation of novel fungi into milk also increases the monetary value of products such as yoghurt and cheese.

Agricultural applications range from the improvement of plant and animal stock to pest control and animal foodstuff. Waste treatment is essentially a large scale fermentation industry requiring new catalysts, improved reactors and has constant application to the problems of treating toxic wastes. The energy industries require systems to provide new fuel sources and improved efficiency of energy recovery from existing ones. The conditions under which biocatalysts function make their application an important energy conservation activity in their own right as biological fuel cells.

A substantial part of the activity of the new biotechnology companies is related to health care. Not only are vaccines and drugs such as interferon and hormones made available, more early but the whole area of diagnostics has expanded from home-based pregnancy tests to sophisticated continuous monitoring of patients by biosensors in hospitals.

## Through Ages

The utility and manipulation of several cell constituents to encourage a special fermentation process is a practice that dates back hundreds of years and most types of fermented foods available today were widely used and consumed during the Roman Times.

## Fermenting Alcohol and Leavening Bread

Processes such as baking, brewing and cheese making have been in use for many years and were developed long before man knew of the existence of the microorganisms involved. For convenience, today many people exclude these traditional processes from what seems to comprise biotechnology.

Louis Pasteur studied at the French School Ecole Normale and achieved notoriety for his accomplishments in organic chemistry. He discovered that tartaric acid, an organic compound, formed two types of crystals and succeeded in separating these microscopically. He became interested in discovering why the local wines were souring. The popular belief was that the fermentation of grape juice to wine was a natural chemical process involving the breakdown of the protein albumin. Even before Pasteur began his work on the origin of microbial life, three other workers (de La Toure from France and Schwann and Kutzing from Germany) had already proposed that the products of fermentation, primarily ethyl alcohol and carbon dioxide were produced by microbes. This was highly criticized by the leading Chemists of that time (including Berzelius, Liebig and Wohler) who did not believe a microbe could be responsible for fermentation. They strictly observed fermentation as a chemical reaction and as a team of successful organic chemists, they were able to uphold this view. In a series of classic experiments, Pasteur first demonstrated that alcohol could be produced from grape-yeast mixtures but not from albumin-yeast mixtures. This showed that albumin was unnecessary for fermentation reactions. Using heat, he then destroyed the yeast, and alcohol failed to form in the grape juice. However, when yeasts were returned to the flasks, fermentation took place and wine was formed. Acceptance of Pasteur's techniques terminated the problem of sour wine. Also, the chemical theory of fermentation was now seriously being doubted. Louis Pasteur also discovered that each type of fermentation is mediated by a specific microorganism which gains energy during the process. The latter observation led to the accidental discovery of cell-free metabolism in 1897 by Edward Buchner of Germany. Buchner found that an extract of macerated yeast freed of intact cells by filtration, retained the ability to convert sugar into alcohol. The pressure of World War I, the British Naval blockade and the consequent production of Glycerol by Germany led the Germans to quickly develop Buchner's discovery and produce 1,000 tons of glycerol per month by Industrial fermentation. Similarly, Britain had to develop acetone-butanol fermentations (using bacterium Clostridium acetobutylicum) by Chaim Weizmann (who later became the First President of Israel). Although the fermentation production of acetone-butanol faded in the following years, it is returning once again (See Chapter 3). Louis Pasteur is credited with the historic discovery of the process of fermentation and could certainly be remembered as the founder of Industrial Biotechnology.

Yeast has been commonly used as a leavening agent in baking during ancient times by the Jews, Egyptians, Greeks and Romans. At that time leavened bread was made by mixing some leftover dough from the previous batch of bread with fresh dough. In recent years, however, pure cultures of selected strains of Saccharomyces cerevisiae are mixed with the bread dough to bring derived changes in texture and flavour.

**Cheese in Ancient Greece**

Milk was probably one of the first agricultural products. Since milk is quickly infected by bacteria, which sour it by converting the sugar (lactose) into lactic acid, it is likely that cheese was one of the first fermented foods. Over the years, this spontaneous process was gradually exploited and developed to make cheese and similar products.

Cheese making is a milestone in the history of food processing and technology. Earlier writings reveal that it was consumed as early as 2000 B.C. The Greeks considered it to be of divine origin and it was often part of their sacrifice to the Gods. Roman soldiers carried cheese with them to war fields, and by the Middle Ages, Cheese had become a staple food of the Middle Eastern diet. Europeans adopted it on the return of the crusaders and soon its production came under the protection of the Church. In fact, cheese production is one of the first major examples of advanced technology in which a process has contributed to increasing the nutritive value of the final product through the increased production of essential nutrients or the synthesis of nutrients not present in the original food. When the English settlers came to America they brought a number of cows and cottage and cheddar cheeses were soon common in the home. Roquefort cheese is still made in the Southwest France in a village of the same name, as it was centuries ago. Ewe's milk is cured in limestone caves maintained at a constant temperature and humidity.

A popular legend on the origin of cheese speaks about a Bedouin traveller who filled his goat skin pouch with milk and set off on his journey. At midday, he stopped to eat, and found that milk was filled with large lumps. He, however, was too hungry to eat the lumps and found these surprisingly good.

Technically, cheese the dairy product derived from microorganisms is produced from the Casein of the milk. Initially, the protein is curdled by the acid produced by the streptococci in the milk. The process is accelerated by adding rennin, an enzyme usually obtained from the stomach lining of the calf. The acid curd is separated from the whey that is used to make whey solids from which processed cheese is prepared.

## Japanese Enzymes

The Japanese have, for years, been producing several enzymes through solidstate cultures of several microbes on a number of cereals such as barley, wheat, soybeans and rice. This process in Japan is generally known as Koji method. Several species of Aspergillus, Rhizopus and Monascus are used. This Koji is primarily used as a source of enzymes in the manufacture of shoyu, miso and sake (wine from rice).

In this method, the substrate is soaked in water, drained and heat sterilized followed by cooling. It is then inoculated with the appropriate mold. This substrate is then placed in trays or in large shallow tanks in a room at an appropriate temperature or alternatively in any ordinary room with natural temperature fluctuations. The mold is allowed to grow for 3-4 days with occasional turning of the material either mechanically or by hand. During this "ripening process" the substrate is used as a source of several enzymes. Quite often, the substrate is inoculated simultaneously with more than one strain of mold.

The method is an ingenious one for the natural production of several enzymes. In India too, a similar method has been utilized by soaking edible cereals in water and eating the bean sprouts since several enzymes are produced during the soaking. Extracting enzymes useful for various purposes is now a commercial exploitation on a large scale. However, the Koji method is a reflection of the ancient biotechnological process for enzyme production which is now scaling up at the Industrial level. Even now, the Koji method is used to manufacture microbial remnant to make curd in the cheese industry. It has also been used successfully to produce colouring agents and in the production of feed made from liquid animal waste mixed with maize.

## Indonesian Tempeh

Tempeh is a very common vegetarian dish, rich in vitamin B-12; it is prepared from soybeans which are soaked and fermented. This is fairly common in the East Indies and other tropical countries, especially Indonesia. Generally, fungal fermentation with Rhizopus sp. is preceded by a bacterial acid fermentation when soybean is soaked. Fermentation is completed when the beans are knitted into a compact cake by the mold mycelium. The cake can then be sliced and fried in deep fat or cut into chunks and used in soups as a meat replacement. The fresh tempeh has a very pleasant, dough-like flavour and is often dipped in shrimp, fish or soy sauces before and after deep frying.

Commercial tempeh is very rich in vitamin B-12, produced by an unidentified bacterium which grows on the soybeans simultaneously with the mold. However, vitamin B-12 is not synthesized if the tempeh is made by inoculating with pure mold. Tempeh is indeed a very fine vegetarian blend of both vitamin B-12 and protein.

In Indonesia, soybeans are fermented in packets made from wilted banana leaves, with the mold inoculum obtained from the earlier batch of up cooled tempeh, Fig 2.1. A similar product, ontjom, also known as Oncom, is prepared in Indonesia, but using peanut press cake (residue left after extracting oil) as the substrate and Neurospora sitophila as the fermenting fungus. The fermentation technology is similar to that for tempeh and the end product is a pinkish-textured meat substitute with an almond or mince meat flavour. Tempeh has been made using various cereals such as wheat, sometimes in combination with soybeans to make a product that tastes like bread or popcorn. Besides, tempeh has also been shown to contain antibiotic substances active against certain bacteria.

Whole, clean soybeans

↓

Soak overnight (or 24 hours) to hydrate soybeans and allow bacterial fermentation and acidification

↓

Dehull by hand or by passing through machine to loosen hulls
Remove hulls by floatation on water

↓

Boil cotyledons for 60 minutes

↓

Drain and cool and allow surface moisture to evaporate
Inoculate with tempeh mold

↓

Ferment small packets of inoculated soyabeans wrapped in banana leaves or in clean shallow covered pans

↓

Incubate at a temperature of 30°C-35°C until soyabeans are completely covered with mold mycelium (generally 24 to 36 hours)

↓

Tempeh cakes can be sold in the market or used in home by slicing thin strips and deep frying or cutting into chunks and cooking in soups

↓

**Fig.** Flow Sheet: Indonesian Tempeh Fermentation

Tempeh is not a new or fashion product. It has been in use in Indonesia for hundreds of years, and does represent the remnant of one of the fine biotechnological operations in the East Indies. The tempeh Industry is now fast developing and efforts are being made now to produce tempeh in small packages for distribution at low cost.

## Indian Idli

Idli is a very common and age old South Indian dish which is now eaten all over India, just like Pizza in the west. It is acidic steamed bread which is very rich in protein. The dough is a blend of ground rice and dehulled black gram (mung beans) or sometimes even lentils. Both the cereals are soaked separately in the ratio of 3 : 1 for 12-24 hours. Then the soaked rice and black gram are ground separately with a mortar and pestle. Generally, rice is coarsely ground and black grain is finely ground. It is mixed up with salt as desired and the batter is incubated overnight. During this process, the batter becomes acidic and it is leavened by carbon dioxide, primarily produced by Leuconostoc mesenteroides. The acidity is attributed to the presence of Streptococcus faecalis.

The leavened and acidified batter is cooked with steam in special metal cups and the batter on steaming is puffed up and becomes spongy and soft. It is common dish at breakfast or even during major meal time during the day.

Many people replace mung beans with 'urd' (a special cereal commonly used in India) depending upon their taste and price. High salt content has excellent preservative action and is valuable as a condiment.

## African Porridges

In several parts of Africa, naturally fermented porridges forms the staple food. West African gari is prepared from an acid-fermented cassava porridge. During the fermentation (at 35°C), any cyanide-containing sugars present are hydrolyzed, removing the cyanide. Thus, cassava becomes acidic and the characteristic flavour of gari appears. This fermentation generally takes 3-4 days. The major fermenting organism is corynebacterium manihot which hydrolyzes the starch exothermally and produces lactic acid and formic acid. As the product becomes acidic, a yeast-like fungus Geotrichum candidum develops which oxidizes the acid and develops the gari flavour. Generally, the fluid is squeezed out from the ferment and the starchy residue is either used directly as fugu or dried in a basket over the fire resulting in gelatinized granules. African also use millet, maize and sorghum for fermentation and to produce these acidic porridges. These low cost technologies have been used since time immemorial in various parts of the Africa.

# SECTION - 2

# GENE MANIPULATION

# TWO

# Molecular Manipulation

## I. Isolation

Isolation and molecular manipulation of specific genes is one of the most important areas of research in biotechnology. These genes, therefore, should be either isolated or artificially synthesized before they are manipulated and used for transformation leading to the production of transgenic animals and plants. The isolated genes are also sometimes sequenced for a better understanding of its structure, that needs to be manipulated. We, therefore, discuss the gene technology available for isolation, sequencing and synthesis of specific genes in this chapter, 'Isolation of Genes'

During the last two decades, significant progress has been made in the techniques for isolation of a variety of genes, including those for (i) ribosomal RNA; (ii) specific protein products; (iii) phenotypic traits with unknown product and those for (iv) regulatory functions e.g. promoter genes, etc. Different techniques have been used for the isolation of these different types of genes and will be briefly described in this section.

### Past attempts for isolation of ribosomal RNA genes in Xenopus

As we know, ribosomes consist of ribosomal RNA (rRNA) and proteins Ribosomal RNA (rRNA) can be of four *sues* namely, 28S, 165, 5.85 and 5S (S stands for Svedberg units, a measure of speed with which a molecule sediments on centrifugation). This ribosomal RNA makes 80 per cent of cellular RNA and is synthesized on ribosomal genes, which could be isolated. Isolation of ribosomal genes was considered convenient clue to their following three characteristic features : (i) availability of homogeneous rRNA; (ii) differences between ribosomal RNA genes and other genes due to relatively high G + C content in rRNA ( rRNA has 45-

GO';'o G + C while the remaining RNA has only 40% G + C) ; (iii) ribosomal genes are present in multiple copies, their number within a cell sometimes approaching several thousands.

In view of the above, ribosomal RNA genes were isolated for the first time (in 1965) in an amphibian named Xenopus, by Hugh Wallace and Max L. Birnstiel, working at the University of Edinburgh (U.K.). Following steps were involved in this isolation : (i) rRNA was isolated from ribosomes of Xenopus and made radioactively labelled due to its replication in a medium containing tritiated uridine; (ii) ribosomal DNA was isolated by density gradient centrifugation followed by its denaturation (G+C content of rDNA differs from that of bulk DNA and helps in its separation by centrifugation); (iii) single stranded DNA was fixed on a filter paper; (iv) labelled rRNA was added on filter paper carrying single stranded DNA; (v) DNA-RNA hybridization was allowed to take place; (vi) excess labelled RNA was washed : (vii) radioactivity was measured and duplex hybrids isolated, which on denaturation, gave single stranded DNA, which could be made double stranded.

Isolation of ribosomal DNA allowed the first characterization of ribosomal genes, which was followed by the isolation of 5S genes by Donald Brown. For a detailed account of isolation of ribosomal genes in Xenopus as originally done, readers are referred to an article Isolation of Genes in Scientific American (August, 1973).

**Isolation of genes coding for specific proteins**

Isolation of genes for specific proteins became possible only after the discovery of reverse transcriptase enzyme in 1970. This enzyme can he easily used for the synthesis of copy DNA on complementary DNA (cDNA) from mRNA. This complementary DNA can then be used for the isolation of the corresponding gene from genomic DNA. It is, therefore, obvious that for the isolation of a specific gene, techniques should first be available for the isolation of specific mRNA. For this purpose, antibodies are produced against a specific protein for which the gene is to he isolated. Therefore, isolation of a gene coding for a specific protein involves the following basic steps (Fig) ; (i) purification of the protein product of the gene; (ii) production of antibodies against thin protein product by immunizing animals like rabbit or mouse: (iii) precipitation of polysomes engaged in synthesizing specific protein with the help of the antibody raised in step (ii); (iv) isolation and purification of mRNA from the polysome fraction; (v) using mRNA for synthesizing cDNA with the help of reverse transcriptase; (vi) cloning of cDNA thus synthesised for preparation of a cDNA library; (vii) immunological and electrophoretic analysis of the translation products of cDNA clones to identify specific cDNA clone meant for the specific protein in mind; (viii) use of the specific cDNA probes selected in step (vii) for the identification and isolation of the gene from genomic DNA through screening a complete or partial genomic library (for preparation of cDNA and genomic DNA libraries, consult Chapter).

In step (vi) above for preparation of cDNA library, cDNA may be cloned in expression vectors (vectors in which gene can be expressed due to the presence of promoter sequences for RNA polymerase enzyme) like gt 11, which can accept gene insertions into ß galactosidase gene. The chimeric vector will then produce hybrid protein if correctly expressed. The hybrid protein can be identified using antibody as above. Plaques belonging to the chimeric vector carrying the desired gene can also be identified by the reaction of antibody attached to a radioactive protein. This procedure was used to clone a number of genes including a gene regulating zero synthesis in maize.

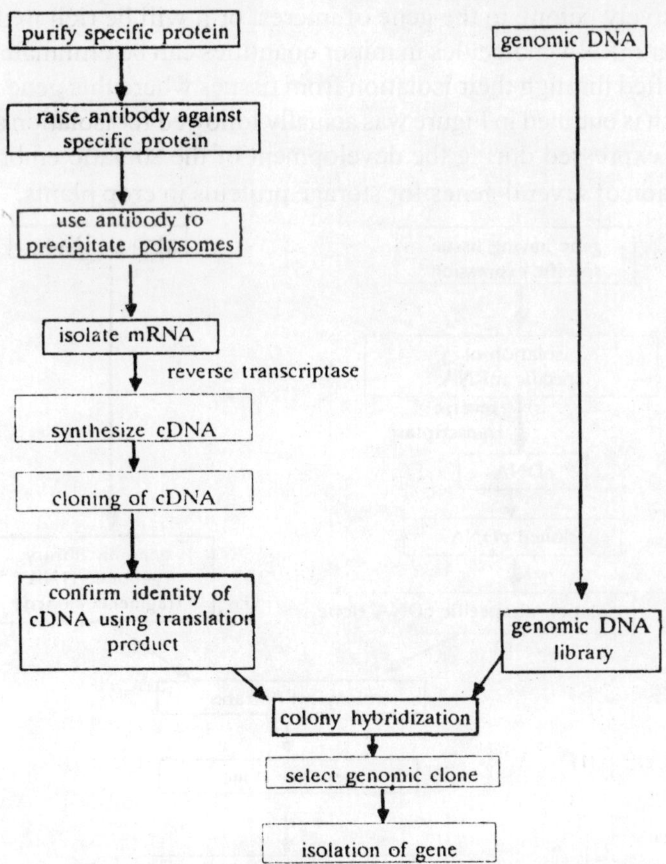


**Fig.** Isolation of a gene for a specific protein

The cDNA clones may also be used directly for gene manipulation and transformation experiments. However, when cDNA clone is used, it should be described as synthesis of gene rather than as isolation of gene, since cDNA has been artificialy synthesized, and not isolated from the organism. The above technique for isolation of gene has now been successfully utilized both in plants and animals. However, the first genes isolated were those which existed in multiple copies and were expressed at high levels in specific tissues, e.g. ovalbumin gene in chicks, globin and immunoglobulin genes in mouse, genes for storage proteins in cereals and legumes, amylase genes in barley, actin genes in some legumes, etc.

## Isolation of genes which are tissue specific in expression

It is much easier to isolate genes which are expressed in specific tissues. For instance, genes for storage proteins are expressed only in developing seeds, ovalbumin gene is expressed in oviduct or globin gene is expressed in erythrocytes. Such genes can be easily isolated because mRNA extracted from these specific tissues will either exclusively belong to the gene of interest or it will be rich in this species of mRNA. Other mRNA molecules in minor quantities can be eliminated, since these can be identified through their isolation from tissues where this gene is silent. This strategy which is outlined in Figure was actually followed for isolation and cloning of carrot genes expressed during the development of the somatic embryos, and also for the isolation of several genes for storage proteins in crop plants.

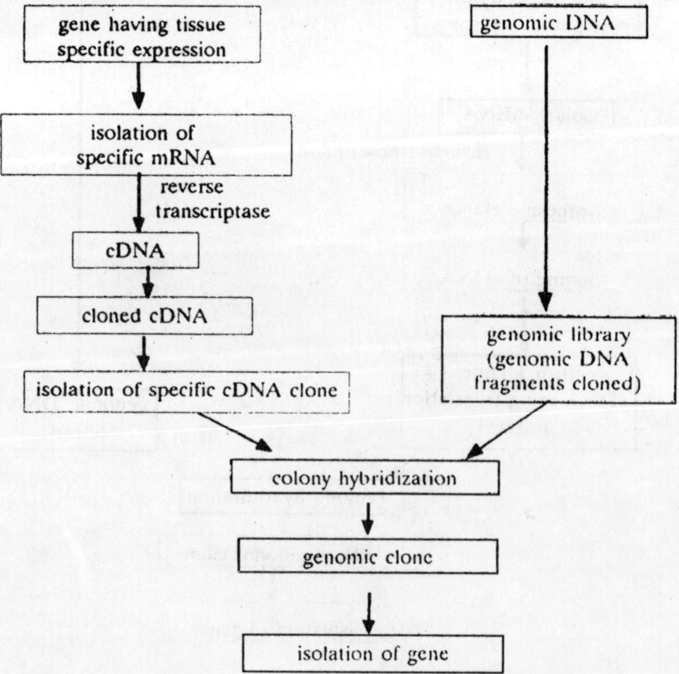


**Fig.** Isolation of a gene with tissue specific expression using its mRNA

## Isolation of genes using DNA or RNA probes

Specific molecular probes (whether DNA or RNA probes), if available can be used for isolation of specific genes. These probes may be available either from another species for the same gene or may be artificially synthesized using a part of the amino acid sequence of the protein product of the gene in question.

Use of heterologous probes—The probes obtained from one species and used for another species are called Ireteroloyorrs probes. These heterologous probes have been found to be effective in identifying gene clones during colony hybridization or plaque hybridization or on Southern blots (see Chapter). For instance, the gene for chalcone synthase (CHS) has been isolated from Antirrhinum majus and Petunia hybrida using heterologous cDNA probe from parsley. Similarly, heterologous Antirrhinum cDNA probe was used for isolation of CHS gene from barley and heterologous probes from maize were used for isolation of barley genes Wx (waxy gene) and AI (aleurone gene). Heterologous probes should ordinarily be used with cDNA library and not with the genomic library, since in the latter case, unrelated genes or pseudogenes (which do not express) may be isolated and cloned. These heterologous clones if available in expression vector (e.g. pGEM-4 bloc) can also be used or getting RNA probes, which have been found W Ire more sensitive and efficient.

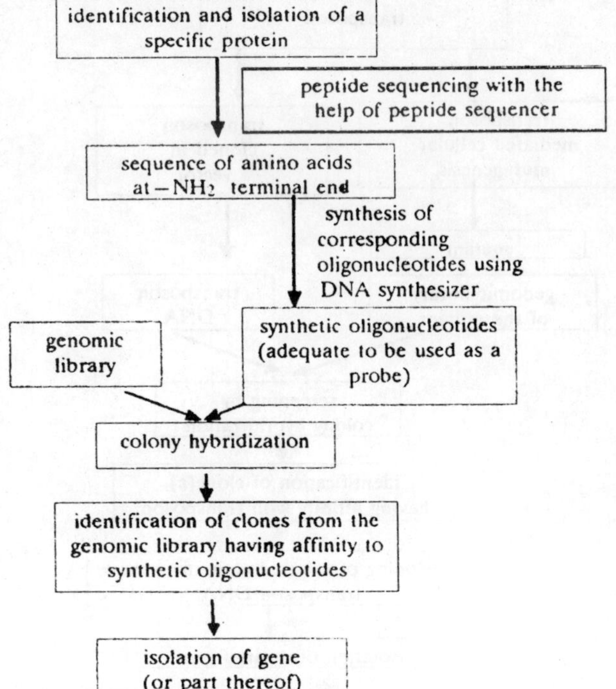


Fig. Isolation of a gene using the probe, artificially synthesized, on the basis of amino acid sequence of a part of the protein

However if protein, purified using the technique of two dimensional gel electrophoresis, is used for microscquencing of 5-15 consecutive amino acids, this information can be used for the synthesis of oligonucleotides (using automated DNA synthesizers). These oligonucleotides may then be directly utilized for screening of cDNA or genomic libraries for isolation of specific genes (Fig.) These can also be used for designing primers for the amplification of gene in DNA extracted from a tissue. A 32 kilodalton (KD) glyeoprotein (isolated from style) associated with the S, incompatibility allele of Nicotiana alata, was used for synthesis of an oligonucleotide leading to the isolation of a related gene.

## Isolation of genes coding for unknown products

In some cases, we are interested in isolation of n gene whose phenotypic effect is known, but the gene product has not been identified or (all not to be isolated). Such genes include those for morphological traits like dormancy, photoperiodicity, disease resistance, etc, his area of research, in which genetics is studied by isolating the gene first without knowing the gene product is often described as reverse genetics. The methods used for the isolation of these genes are different from those used for genes coding for known proteins. Some of these methods are described in this section.

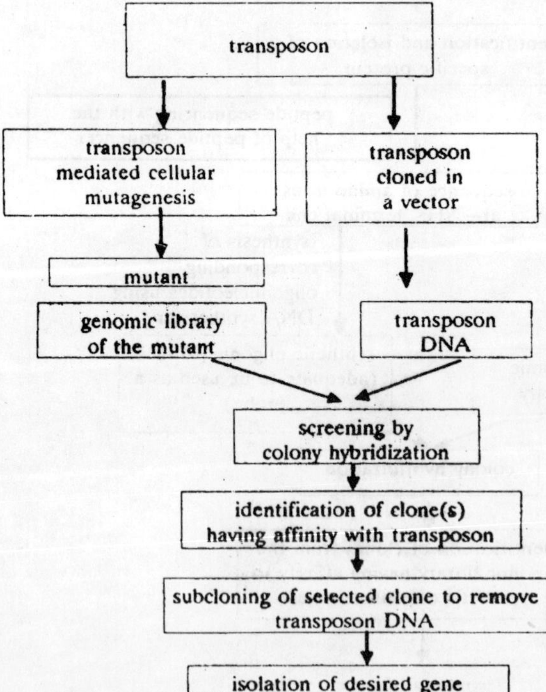


**Fig.** Use of transposon mutagenesis for tagging for identification and isolation of a gene

Use of transposable elements (transposon tagging)—The transposable elements (TEs), in some cases, have been effectively utilised for isolation of genes, when the gene product is unknown. In this case, a transposon works as a mutagen and therefore as a gene gag. Following steps which are illustrated in Figure, are involved in this procedure : (i) Clone a known gene with a scorable phenotypic effect. (ii) TE is transposed to this gene to get an unstable allele. (iii) This unstable allele is cloned and TE is isolated from this unstable allele (this is to select a TE which can produce unstable allele). (iv) This TE is transposed to a gene of interest with known phenotypic effect to produce unstable allele. (v) The DNA is extracted from this mutant. (vi) TE sequence is used as a probe to isolate and clone the mutant gene (carrying inserted TE), so that we can then isolate the gene of interest.

**Table**

Genes isolated using transposon tagging in plants

| *Gene* | *Gene function* | *Species* | *Transposon (Tn)* |
|---|---|---|---|
| A1 | Enzyme required for anthocyanin biosynthesis | Maize | Mtg En/Spm |
| Bz1 | Enzyme required fur anythocyanin biosynthesis | Maize | Ac |
| Bz2 | Enzyme required for anthocyanin biosynthesis | Maize | Ds; Mu |
| C1 | Regulatory gene | Manze | En/Spm |
| C2 | Enzyme required for anthocyanin biosynthesis | Maize | En/Spm |
| P | Regulatory gene | Maize | Ac |
| R | Regulatory gene | Maize | Ac |
| Vp1 | Regulatory gene | Maize | Mu |
| Knotted | Leaf development | Maim | Ds |
| Opaque-2 | Regulatory gene | Maize | Ac; En/Spm |
| hcf-106 | Chloroplast development | Maize | Mu |
| Pallida | Enzyme required for anthocyanin biosynthesis | Snapdragon | Tan 3 |
| deficiens | Regulatory gent | Snapdragon | Tan 7 |

In maize, TE like Ac/Ds, En/Spm and Mu l have been isolated using the genes Wx, C2 and Adh1. Similarly, TEs like Tam3 and Tam7 have been isolated from snapdragon (Antirrhinum majus). These TEs have been used for gene tagging experiments leading to isolation of genes. In maize, several genes like Bz1, P, Al, Cl and C2 have been isolated successfully using gene tagging method (Table)

For transposon tagging, often transposable elements endogenous to species like maize and snapdragon have been used. However, rarely transposons available from one plant species can be moved into the genome of another plant species, whose gene is to be isolated. For instance, Ac element of maize has been transferred to tobacco, where it can integrate into any locus permitting transposm tagging and gene isolation.

Mutant complementation—In this technique, DNA clones from the wild type strain are selected which should be able to complement the mutants, which are thus transformed into wild type. Once these DNA clones are available, the protoplasts derived from the mutant plant may be transformed using wild type clones and the transgenic plants are produced. Once this is done, the gene of interest can be isolated from DNA extracted from the transformed plants using wild type complementary clone as a probe.

Use of RFLP maps or chromosome walking for gene isolation. In recent years, RFLP (restriction fragment length polymorphism) maps have been prepared in maize, tomato, rice, etc. Linkage of RFLP loci with genes of interest are also being worked out as an aid to plant breeding. If RFLPs are known which are very close to the target gene (as may be done through RFLP maps or chromosome walking), then by locating the RFLPs on either side of the target gene, a long intervening DNA segment may be isolated. This fragment may be used for subcloning leading to the isolation of the desired gene. If RFLPs are examined in two near isogenic lines (NTLs) differing for a single gene, the difference in RFLP maps will help in locating the position of gene on molecular map which can then be utilized for isolation of the gene.

Use of chromosome jumping for gene isolation— as described earlier, the technique of chromosome, jumping for walking long chromosome distances thus bringing the available known sequence (related to the available probe) closer to the gene of interest. Once this has been achieved, the techniques described in the previous section may he conveniently used for isolating this gene for further study.

**Isolation of novel genes by Asis Datta at JNU, New Delhi**

In 1992, Prof Asis Datta of JNU, New Delhi was selected for the Birla Award for Science and Technology for cloning and characterization of the following two novel genes, relevant human health.

Gene for oxalate decarboxylase—In 1997, Prof Datta's laboratory reported for the first time the cloning and characterization of the oxalate decarboxylase gene, which degrades and will thus reduce the content of oxalate or oxalic acid, which is harmful in many ways including the following: (a) Several green leafy vegetables (e.g. -Amaranthus, spinach, etc.) being rich in oxalate, if ingested in large quantity are toxic, since oxalate chelates calcium and also destroys renal tissues in kideny; (b) The attack and spread of a fungns (Whetzelinirdsclerot iorium), which causes damage to crops like sunflower involves accumulation of oxalic acid in the infested tissue; (c) Oxalic acid is an essential substrate for the synthesis of the neurotoxin called ß-N-oxalyl I-L-a, ß diaminopropionic acid (ODAP), found in different parts of the plants of Lathyrus sativus (Khesari dal). Consumption of this legume causes neurolathyrism, a well known disease causing damage to leg muscles, etc. ODAP interferes with metobolism of glutamic acid, which is involved in transmission of nerve impulses in the brain, so that despite its rich protein content, L. sativus can not be used as a food source. In all the above cases (leafy vegetables, sunflower & L. Sativus), transfer of single gene for oxalate decarboxylase to produce transgenic plants will improve these plant species by reducing the oxalate content. This new gene will also allow the development of a 'diagnostic kit to measure oxalate in blood and urine', since prevalent methods for clinical tests are quite expensive.

Gene for a seed specific protein with nutritionally balanced composition of amino acids—Prof Datta's laboratory recently also reported, for the first time (in 1992) isolation of a gene from Amaranthus that encodes a seed specific protein rich in all essential amino acids including lysine and S-amino acids. Interestingly, the amino acid composition of this protein corresponds to that of WHO recommended protein standard for the optimum human nutrition. It is known that wed proteins in cereals as well as in legumes (peas, etc.) are deficient, the cereal proteins being deficient in lysine and pulses being deficient in sulphur containing amino acids. In view of this, the gene isolated from Amaranthus will be utilized in future for compensating the amino acid deficiencies of many seed proteins, once it is genetically engineered into the target crop plants.

## II. Sequencing

Once a gene or a DNA fragment has been cloned, its further study involves DNA sequencing. The technique of DNA sequencing was very laborious, till 1975, when a breakthrough was made in DNA sequencing methods. Two different methods are now routinely used for determination of DNA sequences.

### Maxam and Gilbert's chemical degradation method

In this method, which is illustrated in Figure, following steps are involved: (i) Label the 3' ends of DNA with $^{32}P$. (ii) Separate two strands, both labelled at 3' ends. (iii) Divide the mixture in our samples, each treated with a different reagent having the property of destroying either only G, or only C, or 'A and- G' or 'T and C". The concentration of reagent is so adjusted that 50% of target base is destroyed, so that fragments of different sizes having $^{32}P$ are produced. ( iv) Electrophorese each of the four samples in four different lanes of the gel. (v) Autoradiograph the gel and determine the sequence from positions of bands in four lanes as shown in Figure.

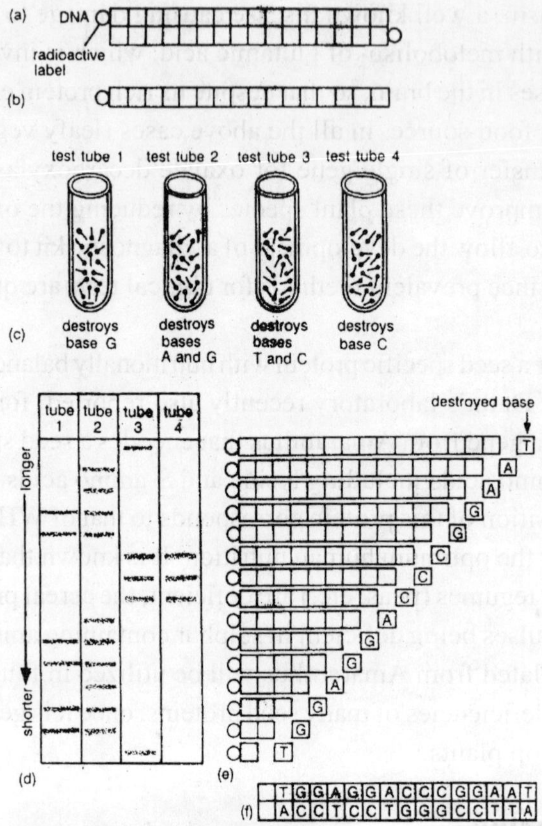


**Fig.** Different steps involved in Maxam and Gilbert's method of DNA sequencing

## Sanger's dideoxynucleotide synthetic method

Fred Sanger (who won Nobel Prize twice) had initially developed a method for DNA sequencing, which utilized DNA polymerase to extend DNA chain length. This was termed plus-minus method. Subsequently, he developed a more powerful method, utilizing single stranded DNA as template for DNA synthesis, in which 2', 3' dideoxynucleotides (Fig.) were incorporated leading to termination of DNA synthesis. These dideoxynucleotides are used as triphosphates (ddNTP) and can be incorporated in a growing chain, but they terminate synthesis Fig, since they can not form a phosphodiester bond with next incoming deoxynucleotide triphosphate (dNTP). Following steps are involved in Sanger's dideoxy method for DNA sequencing. (i) Four reaction tubes are set up, each containing single stranded DNA sample (cloned

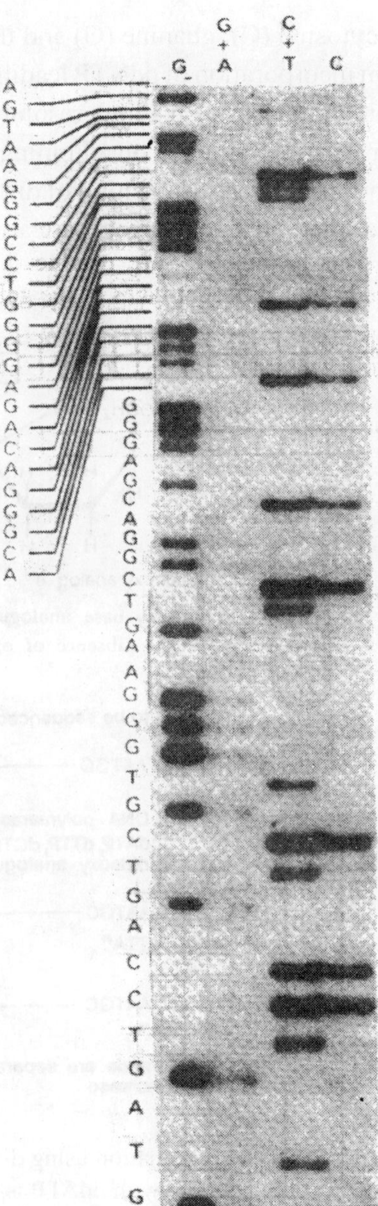


**Fig.** An autoradiograph of a Maxam and Gilbert's sequencing gel (the sequence read from the gel is shown to the left : modified from Ayala and Kiger- 'Modern Gentics', 1984)

in M13 phage) to be sequenced, all four dNTPs (radioactively labelled) and an enzyme for DNA synthesis (DNA polymerase I). Each tube also contains a small amount (much smaller amount relative to four dNTPs) of one of the four ddNTP, so that four tubes have each a different ddNTP, bringing about termination at a specific base- adenine (A), cytosine (C), guanine (G) and thymine (T). (ii) The fragments, generated by random incorporation of ddNTP leading to termination of reaction, are then separated by electrophoresis on a high resolution polyacrylamide gel. This is done for all the four reaction mixtures on adjoining lanes in the get. (iii) The gel is used for autoradiography so that the position of different bands in each lane can be visualized. (iv) The bands on autoradiogram can be used for getting the DNA sequence as shown in Figure.

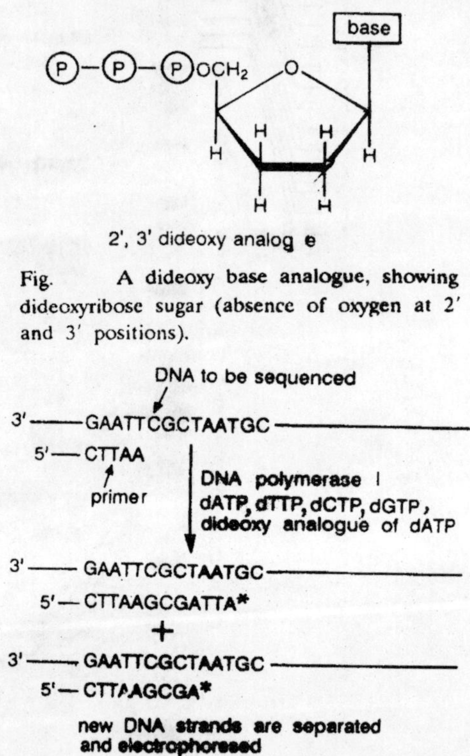


**Fig.** Stragey used in chain termination using dideoxy analogue of a base (only reaction with ddATP is shown)

A variant of the above dideoxy method was later developed, which has allowed the production of automatic sequencers. In this new approach, a different fluorescent dye is tagged to the oligonucleotide primer in each of the four reaction tubes (blue for A, red for C, etc.). The four reaction mixtures are pooled and electrophoresed together in a single polyacrylamide gel tube. A high sensitivity fluorescence detector, placed near the bottom of the tube, measures the amount of each fluorophore as a function of time. The sequence is determined from the temporal order of peaks corresponding to four different dyes (Fig.).

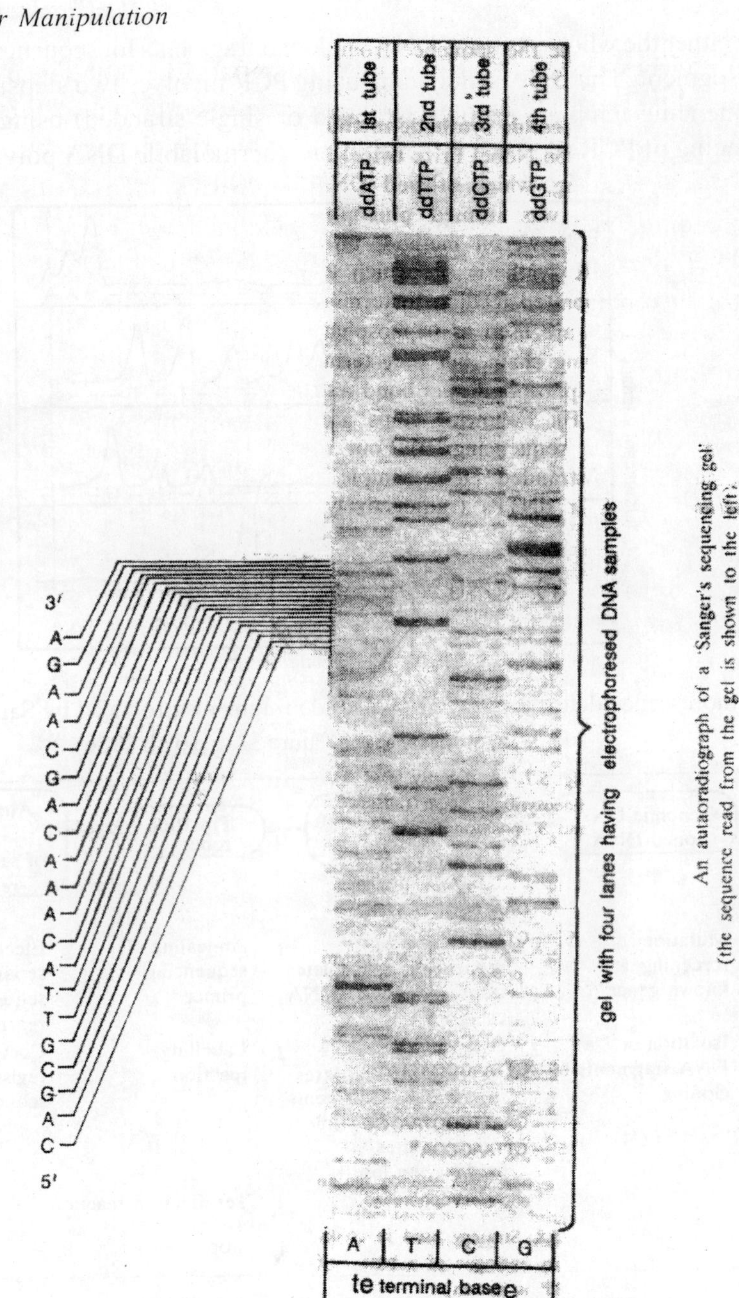


An autoaradiograph of a 'Sanger's sequencing gel' (the sequence read from the gel is shown to the left).

## Direct DNA sequencing using PCR (also called ligation mediated PCR=LMPCR)

Polymerase chain reaction (PCR) has also been used for sequencing the amplified DNA product. This method of DNA sequencing is faster and more reliable and can utilize either the whole genomic DNA or cloned fragments for sequencing a particular DNA segment. The DNA sequencing using PCR involves two steps: (i) generation of sequencing template (double stranded or single stranded) using PCR and (ii) sequencing of PCR products either with the thermolabile DNA polymerise or with

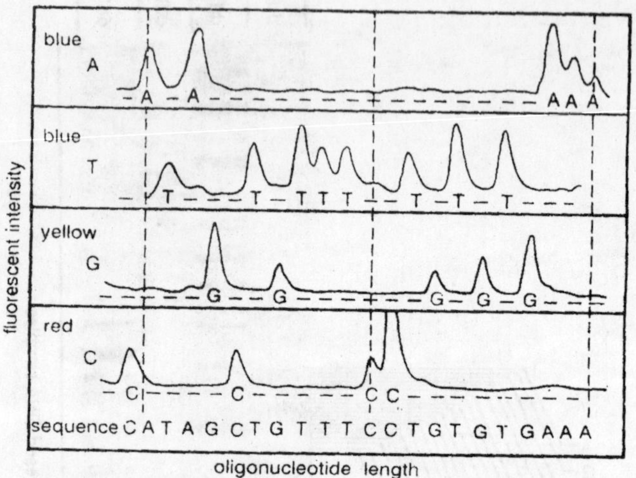


**Fig.** Fluorescence detection of oligonucleotide fragments produced by Sanger's dideoxy method (modified from Nature 321 : p 674: 1986).

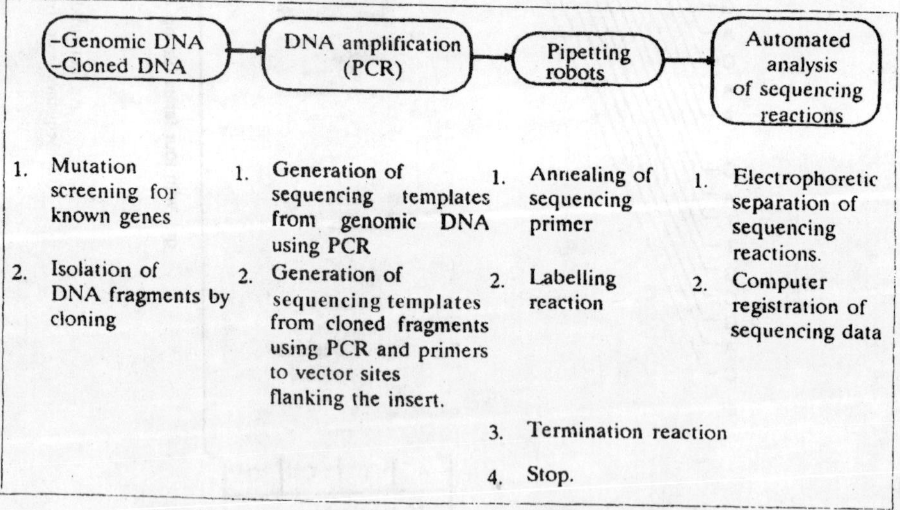


**Fig.** Use of PCR in automation of DNA sequencing

the thermostable Taq DNA polymerise. Thus the DNA sequencing method using PCR eliminates the need of cloning the DNA in single stranded DNA phage vector i.e. M13. Although double stranded DNA product of PCR can be utilized for sequencing, this may cause difficulty due to reassociation after denaturation thus preventing the sequencing primer from annealing to its complementary sequence to allow extension. Toy reduce this problem, either a variant of the standard method for sequencing double stranded DNA is employed or single stranded DNA templates are produced following asymmetric PCR earlier discussed in Chapter.

A number of thermolabile DNA polymerases have been used for sequencing of in vitro amplified DNA. Alternatively, Taq polymerase (thermostable enzyme) used for PCR, can also be used for sequencing reaction. In either case, Sanger's synthetic method involving incorporation of dideoxynucleotide (ddNTP) for chain termination, is used for sequencing. However, Maxim and Gilbert's chemical degradation method can also be used.

In Sanger's method, as usual, four mixtures are prepared, each using one of the four ddNTP. The sequencing primer is labelled with 3'P and the mixtures with amplified DNA, Taq polymerise and appropriate buffer are incubated at 70°C for 5 min. The reaction is stopped by addition of formamide stop solution in each tube and mixtures are run on polyacrylamide sequencing gel to obtain the ladders, which can be read by computer or manually. This method using PCR helped in automation of DNA sequencing (Fig.)

## III. Synthesis

There are two approaches available for the synthesis of genes: (i) When the detailed structure of a gene is available, this gene can be synthesized by a purely chemical method as done by H.G. Khorana for the synthesis of gene for a tRNA (reported for the first time in 1970). (ii) If the detailed nucleotide sequence of the gene is not available, one may utilize the 'RNA directed DNA polymerase' enzyme for the synthesis of the gene in question in the form of complementary DNA (cDNA) from the mRNA of the gene isolated in its pure form. As discussed above, cDNA can also be used for isolation of a gene either (i) from DNA extracted from living cells utilizing the technique of Southern blot hybridization, or from (ii) a genomic library.

### Chemical synthesis of tRNA genes

As outlined above, before one can start the chemical synthesis of a gene, the structure of gene should be known. The structure of a gene earlier could not be worked out by direct chemical analysis, since there were no means for isolating a gene (techniques for isolation of genes were developed later and were described earlier in this chapter). The structure of gene could, therefore, be inferred only from its product. For instance, if a gene is responsible for giving rise to a polypeptide chain and the structure of this chain is known, then from the genetic code dictionary, structure of the gene could be easily inferred. Such genes, were initially considered to be too long to be synthesized because an average gene contains about 1,500 base pairs. On the other hand, since tRNA molecules are fairly small in size (about 80 nucleotides), a gene responsible for giving rise to a tRNA molecule was within the reach for synthesis.

Synthesis of gene for yeast alanyl tRNA—As readers may know, structures of large number of tRNA molecules are now known. The first tRNA whose structure could be known was yeast alanyl tRNA. R.W. Holley who died in early 1993 and his coworkers (in 1965) gave the detailed structure of yeast alanyl tRNA. This information was immediately used by H.G. Khorana and his coworkers to deduce the structure of the gene for yeast alanyl tRNA. The structure of this gene would obviously be such that one of the two strands of DNA (gene) would be complementary to base sequence of yeast alanyl tRNA. The other strand would then automatically have the same sequence as in tRNA except that in place of uracil of tRNA, there would be thymine in DNA.

Synthesis of such a long chain (77 base pairs) of double stranded DNA was rather difficult in 1965 (not in 1990's) but Khorana and his coworkers had extensive experience of synthesizing; DNA of known base sequences. It was decided that such a long chain could not be synthesized by adding a single base each time. Therefore, it was decided that small oligo deoxyribonucleotides ranging in length from 5 to 20 nucleotides should first be synthesized. These segments would be single stranded and would cover the whole length of both the strands of DNA. These would then be joined to form double stranded DNA, 77 nucleoside pairs long. In actual synthesis of yeast alanyl tRNA gene, the following steps were involved.

(a) Synthesis of oligonucleotides—Fifteen oligonucleotides ranging from pentanucleotide (5 bases) to an icosanucleotide (20 bases) were synthesized. Synthesis was conducted through condensation between hydroxyl group at the 3' position of one nucleotide and phosphate group at 5' position of the second nucleotide. In order to bring about condensation, all other functional groups, not taking part in condensation were protected using specific protective groups. After protecting the groups, reaction between a nucleotide with protected 5' end and another nucleotide with protected 3' end proceed according to Figure. Subsequently, condensation was also brought about between groups of two, three or four nucleotides.

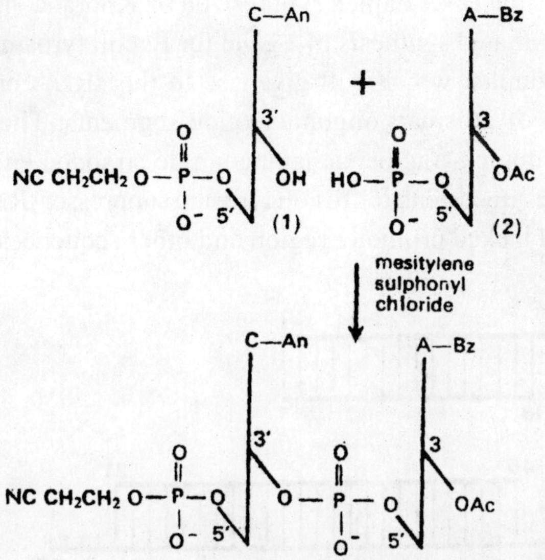

**Fig.** Condensation between two nucleotides with protected 5'OH and 3'OH groups in the sugars and protected amino groups in the nitrogenous bases (Ac = acetyl, An = anisoyl and Bz = benzoyl groups)

(b) Synthesis of three duplex fragments—With the help of 15 oligonucleotides (single stranded), three duplex fragments, each with a single stranded end, were prepared (Fig.). These three fragments consisted of the following : (i) the first 20 nucleotides, with the nucleotides 17-20 as single stranded; this fragment was called 'A'; (ii) the nucleotide residues 17-20 with single stranded regions 17-20 and 46-50; this segment was called 'B'; and (iii) the nucleotide residues 46-77 with single stranded region 46-50; this segment was called 'C'.

(c) Synthesis of gene from three duplex fragments—The three fragments 'A', 'B' and 'C' were joined to give complete gene in each of the two following ways; (i) In one scheme, 'A' was joined to 'B' taking advantage of the overlap in residues 17-20; 'C' was then added, with the overlap in the region 46-50 residues. The complete double stranded DNA with 77 base pairs representing the gene was thus ready. (ii) In the other scheme, 'B' was first added to 'C' and to this 'A' was added in the end to obtain the complete gene. This gene, which was synthesized in 1970, was hopefully used for subsequent work, since this gene could replicate and make its own copies.

Synthesis of gene for a true precursor tRNA—Before Khorana could complete the synthesis of gene for yeast alanyl tRNA in 1970, as outlined above, it became obvious that tRNA was not the direct product of transcription. Instead a precursor molecule is first synthesized which subsequently, after losing segments of RNA by cleavage, gives rise to tRNA. Obviously, therefore, the actual gene for yeast alanyl tRNA will be longer than the DNA duplex synthesized by Khorana. In view of this, Khorana subsequently initiated synthesis of a gene for E. coli tyrosine; suppressor tRNA precursor. DNA duplex which will give rise to this tRNA precursor, was synthesized in the form of 26 small oligonucleotide segments. These were then arranged into six DNA duplex fragments having single stranded ends. These six fragments gave rise to presumed gene for E. coli tyrosine suppressor tRNA precursor. This gene, however, still lacked promoter region and other sequences essential for processing.

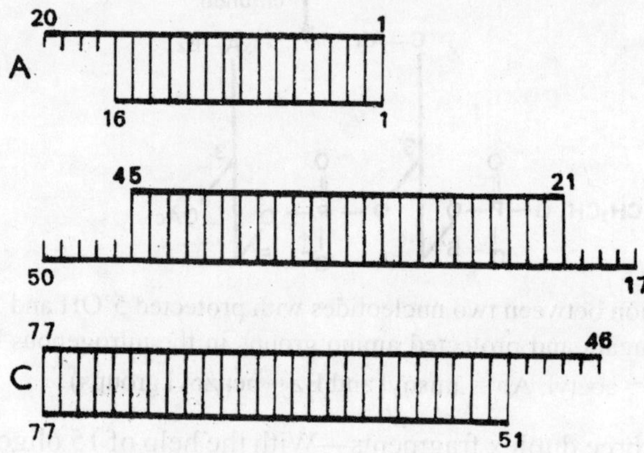


**Fig.** Three duplex fragments, with single stranded sticky ends, produced for the synthesis of yeast alanyl (RNA gene by H.G. Khorana)

Later, in 1979, Khorana reported completion of the total synthesis of a biologically functional tyrosine suppressor transfer RNA gene carrying all regulatory sequences. This gene was 207 base pairs long and included the following : (i) a 51 base pairs long DNA promoter region; (ii) a 126 base pairs long DNA corresponding to the precursor tRNA and (iii) a 25 base pairs long duplex DNA corresponding to 16 base pairs adjoining CCA end of tRNA and the remainder, a modified sequence including EcoRI endonuclease specific sequence. The details of the complete gene are shown in Figure. The complete synthetic gene was cloned in the vector bacteriophage (lambda virus) by gene cloning method which is taken up in separate chapter. On transformation of E.coli, the phage could multiply with the cloned gene.

## Total synthesis of a human leukocyte interferon gene

Interferon is a protein, which has a property of inhibiting viral infection and cell proliferation and thus can be used for treatment of viral infection and malignancies (cancer). These properties have generated great interest in human interferons (IFNs). At least three kinds of interferon genes are known in human genome: (i) leukocyte interferon genes (IFNa genes), (ii) fibroblast interferon gene (IFN-ß genes), and (iii) immune interferon gene (IFN-g) genes). Genes for different kinds of interferons differ and details of base sequences are not known for all of them.

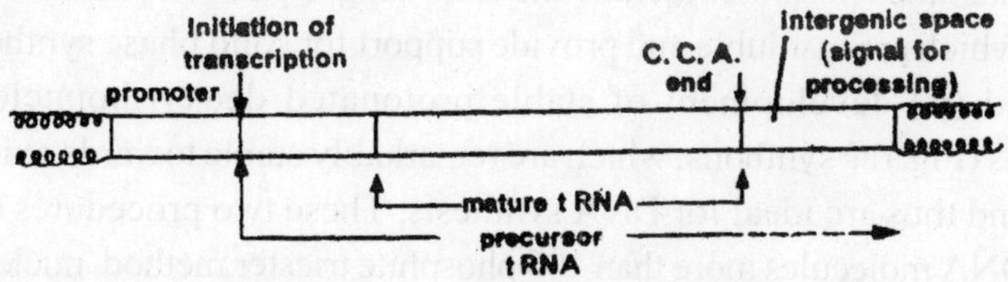


**Fig.** Different parts of the complete E. coli tryosine suppressor tRNA gene (redrawn from Science, Vol. 203; 1979)

In 1980, Weismann and coworkers published the nucleotide sequence of first IFN-a gene derived from cloned IFN-a cDNA. On the basis of this information, total synthesis of this human leukocyte interferon gene; 514 base pairs long, DNA was achieved and published in 1981 (Edge etal., 1981). Since there are 514 base pairs in the gene, it involved synthesis of DNA strands with 1028 nucleotides in predetermined way. This was the longest gene synthesized till 1982. The synthesis involved coupling of initial nucleotide onto a polyacrylamide resin to which further nucleotides could be added in pairs. Sixty six (66) oligonucleotide segments varying in size from 14 to 21 were first synthesized, which were then arranged in a predetermined way and joined. This gene has been incorporated into a plasmid and is capable of synthesizing a- interferon. Further details about the synthesis of this IFN-a gene are available in the original article published in Nature (Vol. 292, pages 756-761, 1981). The 514 base pairs code for 166 amino acids = 498 base pairs, and include the initiation and termination signals. There are no intron sequences present in this gene (introns are noncoding sequences, common in eukaryotic genes that are known as split genes).

The plasmids with this newly synthesized gene have been transferred into bacteria (E. coli), which are being utilized now for synthesis of interferon in industry. This has greatly reduced the cost of production of this drug (interferon) which was earlier sold at the rate of Rs. 16 million per 50 mg.

## Gene synthesis machines

Methods for synthesis of DNA molecules of known base sequences have improved dramatically during 1980s and 1990s. Therefore, genes can now be synthesized rapidly and in high yields. For instance, a gene for a tRNA took 20 man years effort during 1965- 70 (described earlier in this chapter), but the same gene can now be synthesized within a matter of few days with the help of Gene Synthesis Machines developed in 1980's. The key innovations which made the availability of gene synthesis machines possible, include the following: (i) development of silica based supports, which are insoluble and provide support for solid phase synthesis of DNA chains and (ii) development of stable protonated deoxyribonucleoside phosphoramidites (Fig.) as synthons, which are remarkably stable towards oxidation and hydrolysis and thus are ideal for DNA synthesis. These two procedures can be used to prepare DNA molecules more than 100 phosphite triester method. nucleotides long within a few days. Different steps involved in solid phase synthesis of a DNA chain are shown in Figure.

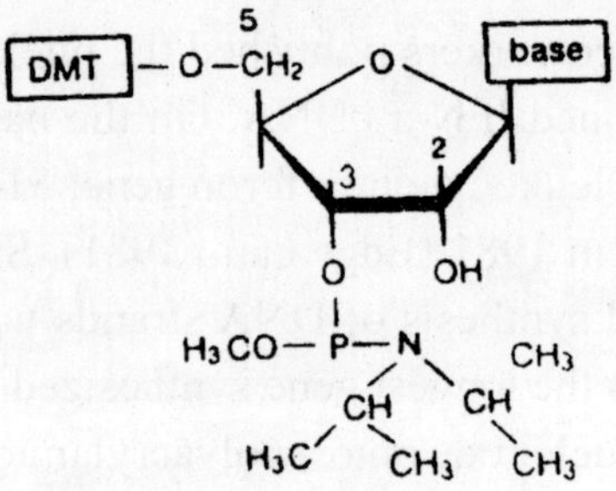

**Fig.** A protonated phosphoramidite (5' hydroxyl is blocked by DMT = dimenthoxytrityl group)

Several versions of gene machines are now available. These machines, under the control of a microprocessor, synthesize specific short sequences of single stranded DNA automatically. The desired sequence is entered on a keyboard and the microprocessor automatically opens the valves of the containers of successive nucleotides, reagents and solvents needed at each step, into a synthesizer column, which is packed with tiny silica beads. These beads provide support on which DNA molecules are assembled (Fig.).

## Synthesis of genes from mRNA

H. Temin and D. Baltimore, in 1990 discovered the presence of 'RNA directed DNA polymerase' enzyme which has the ability of synthesizing DNA on RNA template. This enabled molecular biologists to synthesize complementary DNA (cDNA) using mRNA as a template. If mRNA transcribed from a specific gene is made available in purified form, the complementary DNA (cDNA) synthesized with its help will represent the synthesized gene. To what extent this synthesized gene is a faithful copy of native gene will depend on the fidelity of copying mRNA

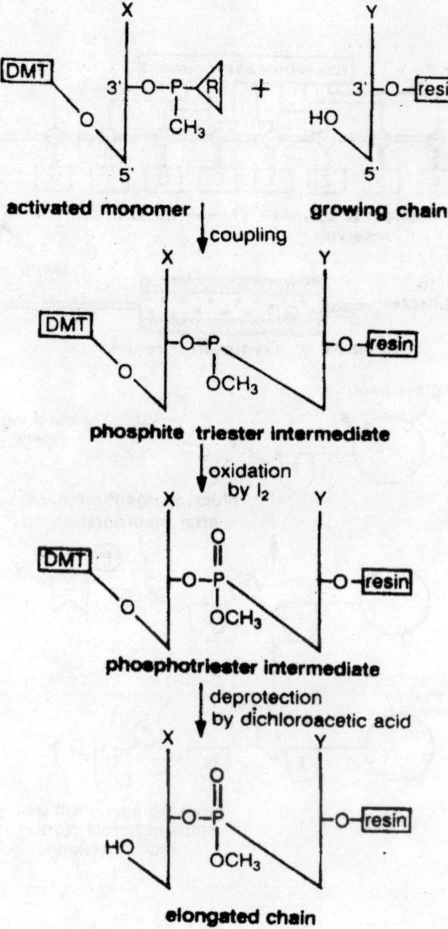


**Fig.** Different steps involved in solid phase syntehsis of a DNA chain by the phosphite triester method

and also on the stability of DNA thus synthesized. Moreover, since mRNA of a gene does not have the complete transcript of the gene in vivo, the synthesized gene will be smaller than the gene in vivo.

By copying eukaryotic purified mRNA, several genes have been artificially synthesized. Most important of these genes are the genes for sea urchinhistone proteins, ovalbumin gene in chicken and globin genes in mammals. The gene synthesized as cDNA from a P globin mRNA has been inserted into a plasmid in order to study its behaviour. As earlier pointed out, this gene will lack any additional regulatory sequences absent in mRNA but present in a globin gene in vivo. This

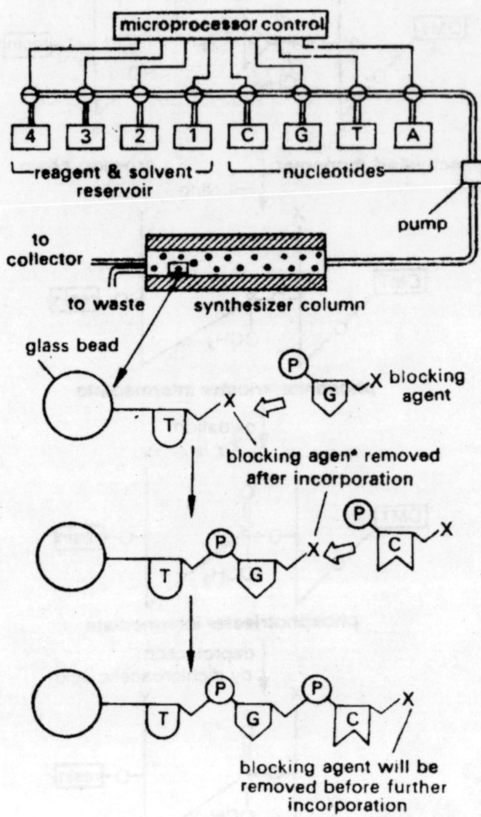


**Fig.** An outline of a gene synthesis machine, which synthesizes a gene automatically once the nucleotide sequences are entered in the microprocessor

type of gene will also lack intron sequences found in eukaryotic split genes. Another factor causing a difference between synthesized gene and the gene in vivo will depend on the sincerity with which the enzyme will synthesize cDNA, because in many cases cDNA was found to be smaller than the mRNA itself. These genes (cDNA) have wide-ranging applications in molecular biology.

# THREE

# DNA and Gene

DNA is the primary genetic material of almost all organisms (with some exceptions) is Deoxyribonucleic acid-DNA. DNA is the repository of the total information for the synthesis of all the enzymes, proteins and RNA molecules. There has been a terrific surge of interest in the chemistry and biology of this master molecule, both for the fundamental and applied aspects. Consequently, the birth of genetic engineering took place with what is known as recombinant DNA (rDNA) technology, which revolutionized the prospects of manoeuvring the genetic make up of a microbe, plant, animal and human being to suit the requirements of humanity. The last ten years have witnessed the birth of an industry based on a scientific discipline which is scarcely elder. Investigations in the field by Stanley Cohen, Paul Berg and Herbet Boyer were soon rewarded with the development of recombinant DNA, sequence analysis and rapid DNA synthesis. With this, the first demonstration that a human hormone could be produced in bacteria came in 1977 with the expression in E. coli of the peptide somatostatin. Rapid advances in a fiercely competitive environment led to the production of the first three rDNA products to enter human clinical testings: insulin (1978), Growth hormones (1979) and alpha interferon (1980). Molecular DNA techniques may ultimately have their most significant effect on agriculture. Recent advances have raised the possibility of the development of new plant germplasm through the introduction of any gene from any organism into plants. Although there can be several aspects of recombinant DNA, yet only such aspects are discussed in this Chapter as would support the modern biotechnological operations.

### Conventional Genetic Recombination

Genetic recombination refers to changes in the sequence or grouping of nucleotides. These changes may include making and breaking concatemers (a single giant chromosome containing two or more linearly repeated phase chromosomes), integration and excision, crossing over, gene conversion, separation of sister chromosomes in prokaryotic and eukaryotic cell divisions, segregation and independent segregation in meiosis, conjugation, cell fusion and many other modes of interchange of genetic material. A recombinant by definition is a product of recombination which might have taken place by any one of the methods. Whenever rearrangement leads to a combination of genes that is new, and relative to some pre-existing combination, the new group is called a recombinant. A 'recombinant envisages a permanent genetic change unlike localized denaturation or renaturation, folding or unfolding. Mutations could also arise by recombinations. However, recombinants or genetic recombination is not necessarily the result of mutations only. Genetic recombination is not under the direct control of an expenter, although the process can be facilitated. Besides, much depends upon the chance factor and time to achieve a particular and desired recombinant. Besides, the selection procedure to identify and isolate a recombinant is fairly cumbersome.

Recombinant DNA is a novel technique for controlled recombination which makes it possible to achieve the target in a short period and also to break barriers between totally incompatible genomes. It is a method of bringing together and joining pieces of DNA from unrelated organisms (say bacterium and human genome) achieving the aim and profit of general recombination.

Recombinant DNA technology now provides methods by which genes or other segments of large chromosomes can be isolated, replicated and studied by nucleic acid sequencing (which is otherwise very laborious if one has to search for a gene in the main DNA molecule at random) and electron microscopy. The various steps involved in preparation of recombinant DNA could be briefly summarized.

1. Selection of DNA of interest (commonly referred to as the foreign or target or passenger DNA).
2. A cloning vector or vehicle to carry the inserted piece of target DNA
3. Restriction endonucleases which makes internal cuts at specific sites on DNA.
4. DNA ligase-the enzyme that joins pieces of nucleotides/DNA together.
5. A host to replicate the vector containing foreign DNA. This may be a prokaryotic or eukaryotic cell.
6. Screening test for recombinants produced by insertion of DNA.

## Purified DNA and Restriction Enzymes

General methods for the isolation and purification of DNA are available. These involve gentle disruption (lysis) of cells containing the target DNA followed by separation of the DNA from all other molecules such as proteins, polysaccharides, and lipids. This is generally achieved by placing the complex mixture (cell lysate) in a centrifuge tube containing a concentrated solution of cesium chloride or sucrose.

This is centrifuged at a very high speed in an ultracentrifuge for several hours (about 10-12 hours). Consequently, the nucleic acids will form a specific band in the solution in the centrifuge tube, depending upon the density of DNA. This band can then be removed (by piercing the tube with specially designed syringes) and purified further by chemical methods.

The purified DNA is then subjected to the action of restriction endonucleases (Table). These enzymes recognise specific sequences in double stranded DNA and then catalyze the cleavage of the DNA at particular points. Each restriction enzyme recognizes a specific short sequence of bases and cleaves DNA at a particular site within that sequence. In the presence of a large population of identical DNA molecules, each molecule will be cut by a particular endonuclease at precisely the same site, yielding multiple identical molecular fragments. These subsets of fragment can be conveniently separated from one another, according to size, by gel electrophoresis. A sequence of at least four bases is generally required for recognition. The net result is the production of DNA fragments with defined ends. As seen in the Table, there are a large number of restriction enzymes each of which recognise different sequences and cleave DNA at specific sites in a controlled fashion. Some of the restricted enzymes (as shown by arrow heads) will cut the recognition sequence at different positions on the two DNA strands

**Table**

Target Sites and Products of Digestion of Some Restriction Enzyme

| ***Name*** | ***Source*** | ***Target Sequence*** | ***Product*** |
|---|---|---|---|
| *Eco* RI | E. coli RY13 | 5' G ↓ AATTC 3'<br>3 ~ CTTAA ↑ G 5' | 5 ' G ATTC<br>CTTAA G<br>5'5' |
| *Hindi* III | Hacmophilus influenzae R | 5' A ↓ AGCTT 3'<br>3' TTCGA ↑ A$_{5'}$ | A AGCTT<br>TTCGA A<br>5' |
| *Sma* I | Serratia marcescens Sb | 5' CCC ↓ GGG3'<br>3' GGG ↓ CCC$_{5'}$ | A AGCTT<br>TTCGA A |
| *Hae* II | Haernophilus aegyptius | 5' GG CC 3'<br>3' CC GG 5' | GGCC<br>CCGG |
| *Hae* III | " " | " | " |
| *Pst* I | Providencia stuartii 164 | 5' CGGCA ↓ G 3'<br>%' G ↑ ACGTC 5' | CTGCA 3' G<br>G ACGTC |
| *Sst* I | Streplomyces standford | 5' GAGCT ↓ C 3'<br>3' C ↑ TCGAS 5' | GAGCT 3' C<br>C TCGAG |
| *Hsu* I | Haemophilus suis | 5' A ↓ AGCTT 3'<br>3' TTCGA ↑ A 5' | A 5' AGCTT<br>TI CGA |

**Table**
Various Restriction Endonucleases and Source

| *Enzyme* | *Source* | *Target Sequence* |
|---|---|---|
| Accl | Acinetobacter calcoacetlcus | GT↓ (CA) (TG) AC |
| Acyl | Anaena cylindrica | GPu↓CGPyC |
| Alul | Arlhrobacter luteus | AG↓ CT |
| Aosl | Ananaena oscillaroides | TGC↓ GCA |
| Aosll | Aanaena oscillaroides | GPu ↓ CGPyC |
| ApyyI | Arlhrobacler pyrldinolis | CC↓ (AT) GG |
| Asul | Aanaena subcylindrica | G↓GNCC |
| AsuaII | Ananaena subcylindrica | TT↓ CGAA |
| Aval | Ananaena varlabilis | C↓PyCGPuG |
| Avall | Ananaena variabilis | C↓ G(AT) CC |
| Ball | Brevibacterlum albldum | TGG↓CCA |
| BamHI | Bacillus amyloliquefaciens H | G↓ GATCC |
| BamHII. | Bacillus anryloliquefaciens | GAA(N)½ $(N)_3$TTC |
| BamNx | Bacillus amyloliqueJaciens N | G↓ G(AT) CC |
| Bbel | Bifidobacterium brave | GGCGC↓C |
| Bell | Bacillus caldolytlcus | T↓GATCA |
| Bgll | Bacillus globlgii | $GCC(N)_4$↓NGGC |
| BGlII | Bacillus globigii | A↓GATCT |
| BluI | Brevibacierium luteum | C↓TCGAG |
| Bpel | Bordetella pretussis | A↓AGCTT |
| Bst1 | Bacillus sterothermophilus 1503-4R | G↓GATCC |
| BstEll | Bacillus sterolherrrrophilus ET | G↓GTNACC |
| BstNl. | Bacillus sterothermophilus N | CC↓ (AT)GG |
| BstPl | Bacillus slearothernnophilus | G↓ GTNACC |
| BsuRl | Bacillus subtilis R | GG↓CC |
| BVUl | Bacillus vulgatis | GPuGCPy↓C |
| Caull | Chloroflexus aurantiucus | CC↓(GC) GG |
| Clal | Caryoplranorr latum L | AT↓ CGAT |
| Clil | Caryop/:anon latum | GG↓CC |
| Ddel | Destrlfovibrio desulfuricans Norway strain | C↓ TNAG |
| Dpnl | Dip!ococcus pneumonlae | GA↓ TC |
| Ecal | Enlerobacler cloacae | G↓GTNACC |
| EcoRI | Escherichia colt RY13 | G↓AA7-CC |
| EcoRI | Escherlchia coli RY 13 | PuPuA↓TPyPy |
| EeoRII | Escherlcltia roll R245 | CC(AT) GG |
| FnuAI | Fusobacterium nucleatum A | G↓ANTC |
| FnuCI | Fusobacterium nucleatum C | ↓GATC |

*(Table Contd.)*

| *Enzyme* | *Source* | *Target Sequence* |
|---|---|---|
| FnuDI | Fusobacterium nucleatum D | GG↓CC |
| FnuDII | Fusobacterium nucleatum D | CG↓CG |
| FnuDIII | Fusobacterium nucleatum D | GCG↓C |
| FnuEI | Fusobacterium nucleatum E | ↓GATC |
| Fnu4HI | Fusobacterium nucleatum 4H | G↓CNGC |
| FspAI | Flavobacterlum species | G↓GTNACC |
| GdiI | Gluconobacter dloxyacetonicus | AGG↓CCT |
| GdiII | Cluconobacter dioxyacetonicus | Py↓GGCCG |
| HaeI | Haemophilus aegyptus | (AT) GG↓CC(AT) |
| HaeII | Haemophilus aegyptius | PuGCGC↓Py |
| HaeIII | Haemophilus aegyptlus | GG↓CC |
| HapII | Haemophilus aphrophilus | C↓CGG |
| HgaI | Haemophilus gallinarum | GACGC(N)$_5$↓<br>CTGCG(N)10↓ |
| HgiAI | Herpelosiphon gigateus Hpg5 | G(AT)GC(AT)↑C, |
| HgiBI | Harpelosiphon giganteus Hpg5 | G↓G(A)CC |
| HgiCI | Herpelosiphon giganteus HP99 | G↓GPyFuCC |
| HgiCII | Herpelosiphon giganteus Hpg9 | G↓G(A)CC |
| HgiCIII | Herpelosiphon giganteus Hpg9 | G↓TCGAC |
| HgiDI | Herpetosiphon giganteus Hpa2 | Gpu↓CGPyC |
| HgiDII | Herpetosiphon giganteus Hpa2 | G↓TCGAC |
| HgiEI | Herpetosiphon giganteus Hpg24 | G↓G(T)CC |
| HgiGi | Herpetosiphon giganteus Hpal | GPu↓CGPyC |
| Hhal | Haemophilus haemolyticus | GCG↓C |
| HincII | Haemophilus influenzae Rc | GTPy↓PuAC |
| HindIII | Haemophilus inlluenzae Rd | A↓AGCTT |
| Hinfl | Haemophilus influenzae Rf | G↓ANTC |
| Hpal | Haemophilus parainfluenzae | GTT↓AAC |
| Hpall | Haemophilus parainlluenzae | C↓CGG |
| Hphl | Haemophilus parahaemolyticus | GGTGA(N)$_8$↓<br>CCACT(N)$_{71}$ |
| Hsul | tfaemophilus suis | A↓AGCTT |
| Kpnl | Klebsiella pneumoniae | GGTAC↓C |
| Mbol | Moraxella bovis | ↓GATC |
| Mboll | Moraxella bovis | GAAGA(N)$_{8\downarrow}$<br>CTTCT(N)$_{7\uparrow}$ |
| Mlal | Mastigoladus laminosus | TT↓CGAA |
| Mnol | Moraxella nonliquefaciens | C↓CGG |
| Pstl | Providencia stuartii 164 | CTGCA↓G |
| Pvull | Proteus vulgaris | CAG↓CTG |

*(Table Contd.)*

| *Enzyme* | *Source* | *Target Sequence* |
|---|---|---|
| Rsal | Rhodopseutiomonas sphaeroides | GT↓AC |
| Rshl | Rhodopseudomonas sphaeroides | CGAT↓CG |
| Rrul | Rhodospirillum rubrum | AGT↓ACT |
| Rrull | Rhodospirillum rubrum | CC↓(AT)GG |
| Sacl | Streptomyces achronaogenes | GAGCT↓C |
| Sacll | Streptomyces achromogenes | CCGC'GG |
| Sall | Streptomyces albus G | G↓TCGAC |
| Sau3A | Staphylococcus aureus 3A | ↓GATC |
| Sau961 | Staphylococcus aureus PS96 | G↓GNCC |
| Sfal | Streptococcus faecalis var. zymogenes | GG↓CC |
| Slat | Streptomyces lavendulae | C↓TCGAG |
| Smal | Serratia marcescens Sb | CCC'GGG |
| Sphl | Streptomyces phaeochromogenes | GCATG↓C |
| Sstl | Streptomyces stanford | GAGCT↓C |
| Sstll | Streptomyces stanford | CCGC↓GG |
| Taql | Thermus aquaticus YTI | ↓CGA |
| Thai | Thermoplasma acidophilum | CG↓CG |
| TthHB8 | Thermus thermophilus I-IBS | T↓CGA |
| Tthllll | Thermus thermophilus III | TGACN'NNGTC |
| Tth11III | Thermus thermophilus III | $CAAPuCA(N)_{11}$↓ $GTTPyGT(N)_{9}$↑ |
| Xbal | Xanthomonas badrii | T↓CTAGA |
| XhoI | Xanthomonas holcicola | C↓TCGAG |
| XhoII | Xanthomonas holcicola | Pu↓GATCPy |
| Xmal | Xanthomonas malvacearum | C↓CCGGG |
| XmaIII | Xanthomonas malvacearum | C↓GGCCG |
| Xpal | Xanthomonas papavericola | C↓TCGAG |

(i.e. asymmetrically), where as others will cut the target sequence at the same position on both the strands. In the former case, the cleavage product will have protruding 5' or 3' termini (referred to as sticky or cohesive end) whereas in the latter case, the cleavage products are said to have blunt or flush ends.

Till recently, more than 250 restriction enzymes have been isolated and identified. These enzymes are named based on an abbreviation of the name of the organism from which the enzyme was isolated. According to the nomenclature employed, the first initial of the genus and the first two initials of the species form the basic name. When the enzyme is present in a specific strain, these three italicized letters may be followed by a strain designation. The third portion of the name is reserved for a

Roman numeral indicating the order of discovery of the enzyme on the particular strain. For example, Hae II is purified from the strain Haemophilus aegypticus and is one of three restriction enzymes present in the strain. Bam I is the restriction enzyme present in Bacillus amyloliquifaciens. Rru I is the restriction enzyme present in the photosynthetic bacterium Rhodospirillum rubrum. Restriction enzymes which have been purified from isolated and purified from different organisms might recognize and cleave at the same sequence in the DNA. These restriction enzymes are known as perfect isoschizomers (e.g. Hind III and Hsu I). Imperfect isoschizomers, on the other hand, may recognize the same sequence but cleave at different nucleotides within the sequence (e.g. Xma I and Sma I).

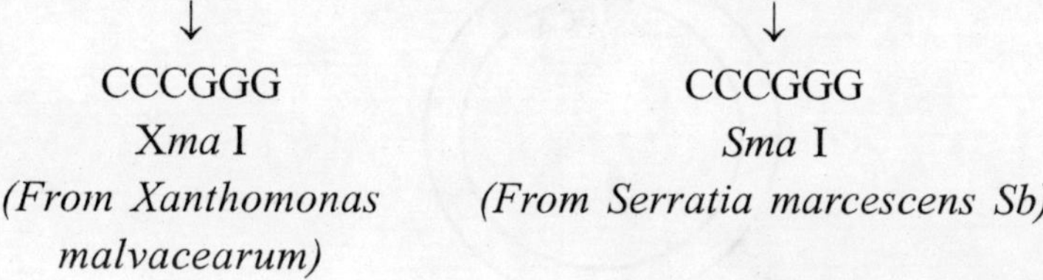


In general, restriction enzymes are classified as Type I and Type II. Type I are large multimeric proteins that are capable of both cleaving and modifying DNA and require ATP, $Mg^{2+}$ and SAM (S-adenosylmethionine). Type II does not require ATP and SAM. It, however, does require $Mg^{2+}$ and this cleaves DNA at specific sites and produces DNA fragments with cohesive terminii. Type II are most widely used in genetic engineering research. These were discovered fairly late in 1968 by Smith and his coworkers in the bacterium Haemophilus influenzae typed. The recognition sites for Type II are generally tetramer, pentamer or hexamer base pains in length.

Digestion with Hind III produces its protruding ends and a key feature of this is that each of the free ends is capable of base-pairing with another free end produced by the same enzyme. So the opposite ends of the same DNA fragment can base pair and, therefore, circularize (intramolecular) or the ends of different DNA fragments can associate (intermolecular) (Fig.). Such base pairing, however, is not very stable and can be disrupted by heat. The joining of complementary ends or blunt ends is made possible by DNA ligase.

**Vectors**

DNA from viruses or plasmids that replicate extrachromosomally in E. coli are commonly used as vectors for making recombinant DNA. These can be conveniently manipulated in vitro and reintroduced into the bacteria. The DNA to be inserted into the vector can be derived from E. coli or other bacteria, plants or animals. Small circular duplex plasmid DNAs from E. coli, especially modified derivatives of these plasmids, modified X-phage DNA and single-stranded M 13-phage DNA have all been very useful as vectors in recombinant DNA work. These can easily replicate in E. coli after insertion by transformation or transfection.

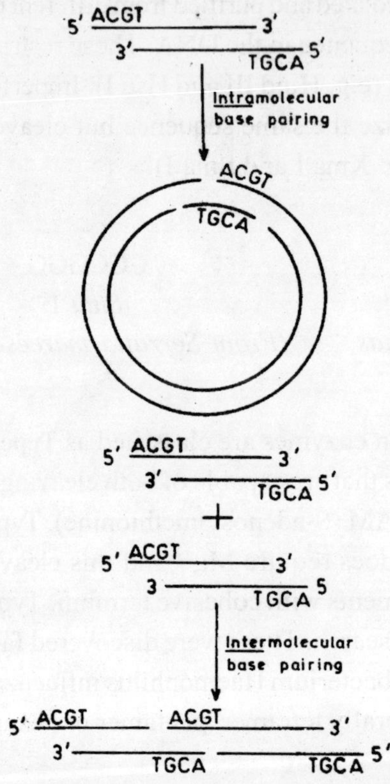


**Fig.** Circular (intramolecular) and associated (intermolecular) base pairing

Plasmids are found in almost all bacterial species. Plasmids are circular DNA molecules that account for 1-3 per cent of the bacterial cell genome and code for several of the proteins responsible for resistance against many heavy metals and antibiotics. The number of plasmids in a cell may vary from 1 to 100 and there is a great variety of size. A plasmid, like most other DNA molecules is a double helix each strand of which is a linear array of the four nucleotide bases as found in DNA and the bases are complimentary.

Since most DNA fragments are incapable of self-replication, an autonomously replicating segment of DNA must be used to accomplish replication of the desired DNA fragments. Most cloning vectors have thus been derived from extrachromosomal replicons such as bacteriophage, viruses and plasmids. Three most thoroughly studied plasmids include: (i) F and F plasmids; (ii) R plasmids—plasmids carrying genes for resistance to antibiotics and (iii) col plasmids-plasmids coding for colicins, proteins that kill sensitive E. coli cells. Plasmids are also knwon in bacteria that code for bacteriocins other than colicins. For example, plasmids are known that code for vibriocins; these are proteins that kill sensitive Vibrio cholerae cells.

For any sequence of DNA to be passed on from one cell division to the next, it must be replicated, so that, at division, a new copy of the gene is passed on to the daughter cell. Therefore, the sequence of a particular DNA needs to be inserted into a vector. Plasmids, as vectors, have several unique restriction endonuclease recognition sequences for cloning. In principle, it is possible to insert a target gene of interest into either a plasmid or phage vector; either is capable of carrying the target gene into a bacterial cell where it would be replicated. If the vector is a plasmid, both the plasmid DNA and the inserted target gene can be propagated indefinitely in the dividing bacterial cells. If the vector is a phage, its replication in the bacterial host would result in every phage carrying an identical copy of the target gene. When phage vectors are used, a population of cells is infected with the viruses and virus replication proceeds spontaneously. When plasmid vectors are used, a population of cells is bathed in the plasmid DNA containing the inserted DNA. Eventually, the phage DNA containing the target gene is inserted into the bacterial chromosome, where it will replicate as though it was a set of normal chromosomal genes. Then, these vectors alongwith the target genes are introduced into a bacterial host, after they have been joined. This is usually achieved with the enzyme DNA ligase. It is important for ligation that the vector DNA and the target DNA have been cut with the same restriction endonuelease (Fig.). Insertion of target DNA into the vector causes a large break in the integrity of one of the plasmid antibiotic resistance genes. This results in inactivation of that function (insertional inactivation) and so in the hybrid or recombinant plasmid, one of the antibiotic resistances will be lost i.e. any bacterial cell carrying this new recombinant plasmid will be sensitive to this particular antibiotic. The recombinant DNA so produced is introduced into bacterial cells by a method known as transformation (for plasmid DNA) and transfection (for phage DNA). Bacterial cell walls are not ordinarily permeable to such recombinant DNA, but a dilute solution of calcium chloride renders the bacteria permeable and a few of them take up the new plasmids. Once inside the cells, these vectors start replicating and on spreading the cells on to a nutrient agar plate, the individual transformed or transferred cells will begin to grow and divide as separate units. The replicated DNA in each case is passed on to daughter cells giving rise to clones (genetically homogeneous population of cells derived from a single cell).

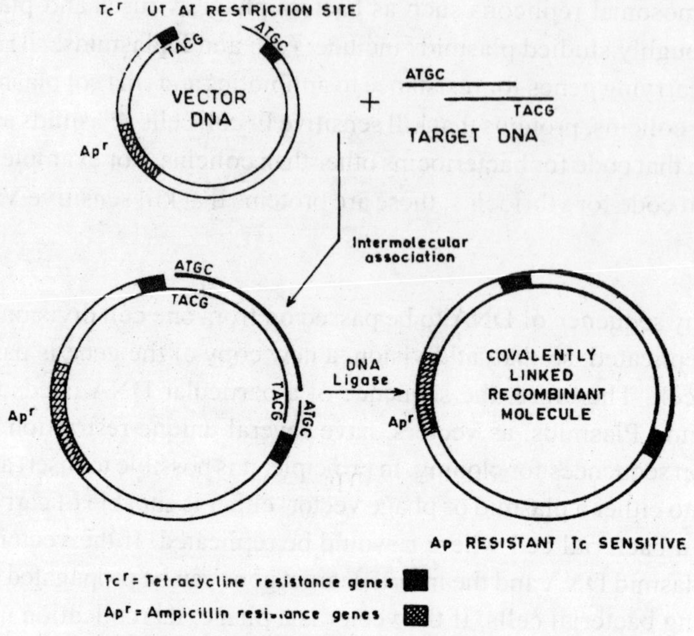


**Fig.** Ligation of vector and target DNA

Generally, plasmid vectors are chosen that carry two genes which code for antibiotic resistance. As described above, one of these genes is inactivated when any target DNA is inserted into it, but the second antibiotic gene remains intact and functional and this is useful for selecting the recombinants. (Fig.). After transformation with the mixed recombinant plasmids, the bacteria can be plated on to agar con.aining an antibiotic which will kill all untransformed cells (which did not accept the plasmid along with target DNA). The only colonies which appear after antibiotic selection will be those carrying a plasmid and those carrying plasmid containing the target DNA.

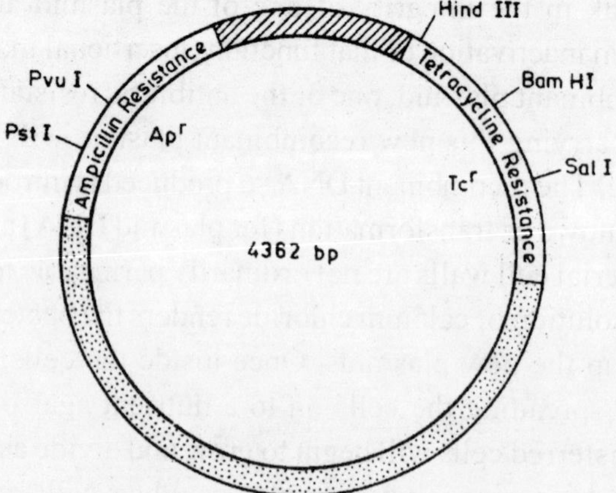


**Fig.** Schematic representation of a plasmid with genes resistant to antibiotics

As shown in Fig., insertion of DNA into the restriction site within the second antibiotic gene (say tetracycline) causes loss of resistance to tetracycline. Thus, the recombinant plasmid, unlike religated vector, cannot confer this resistance to any transformed cell. To determine which particular colony carries the recombinant plasmids, the colony is challenged with tetracycline. Recombinants are killed by the second antibiotic (tetracycline) and this indicates that these carry a vector containing an insert of foreign DNA. The recombinant plasmids will carry genes which may or may not be expressed in E. coli. If the foreign genes are expressed in E. coli, then it may be possible to select the desired recombinant by transforming specific mutants of E. coli with the test plasmids. However, certain genes (say for insulin synthesis) introduced into bacterial cells may not be expressed. Therefore, different methods are used to isolate recombinants with specific genes to be cloned. Generally, techniques such as complementation and immunogenetic are employed depending upon whether the gene to be cloned is generally present in E. coli or totally alien to the cell.

**Probing and Cloning**

Finally, a nucleic acid probe (a small piece of target gene) is needed to detect the cloned gene in the recombinant. The probe could be derived from a related clone produced from DNA of another species or it could be chemically synthesized as an oligonucleotide. The latter often relies on knowing the complete or partial amino acid sequence of the protein and product and working backwards using the genetic code to define possible DNA sequences. Whatever the source, the probe can be radioactively labelled to simplify subsequent detection of positive clones. Because of the base pairing rules, a singlestranded probe sequence will be completely complementary to a stretch of sequence on the corresponding target gene.

A second procedure for the in vitro construction of recombinant DNA molecules involves the synthesis of complementary single-stranded tails on the 3' ends of strands of DNA using the enzyme terminal transferase. This procedure has greater advantages than restriction enzyme procedures, because it can be used to join DNA molecules that terminate with any DNA sequence as well as fragments produced by cleavage with restriction enzymes that cut both strands of DNA at the same point. Once recombinant DNA molecules are produced using terminal transferase, they are cloned as described in the other procedures (by restriction enzymes). There are several other modifications of the two methods for synthesizing recombinant DNA.

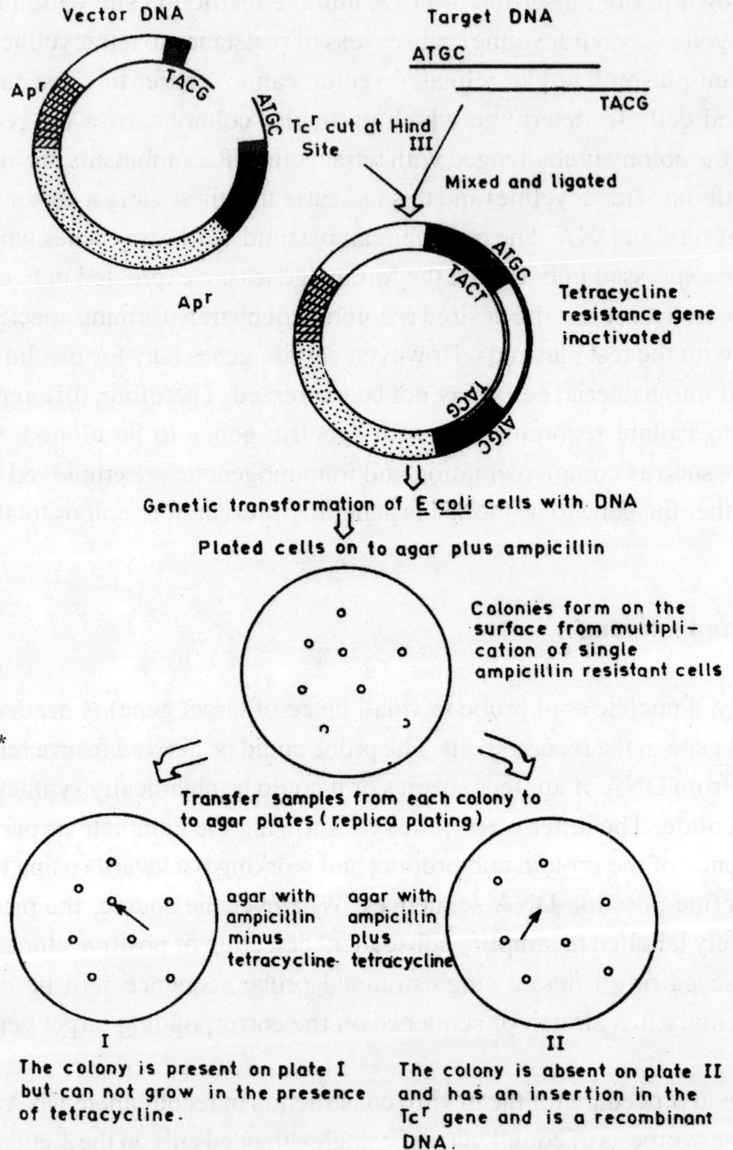


**Fig.** Selection technique for recombinant plasmid

Another approach to cloning has been to prepare clonable sized fragments of DNA by digestion of the entire genome of an organism with one or more restriction enzymes or by shearing forces (such as ultrasonication). Then these fragments are cloned at random. This technique is referred to as shot gun cloning—the isolation of a specific DNA sequence from a large population by cloning random DNA sequences and screening for the desired phenotype. However, it is quite difficult to identify a cell or clone containing the gene or sequences of interest.

A number of different strategies are used to clone a particular DNA fragment or gene depending on the nature of the target DNA and the availability of screening methods and selection schemes. There are two basic categories (i) cloning of specific fragments of DNA (ii) construction of gene banks or libraries. A gene bank or library is a large collection of independently isolated random DNA from a single organism, each associated with a plasmid vector. A library of Drosophila DNA fragments was first produced by D. Hongness and co-workers by sonic fragmentation of the DNA. As a general rule, the vectors and cosmids are used to clone large DNA fragments and for the construction of gene libraries, while the plasmid vectors are used to clone smaller DNA's (< 15 kb). The nature of the DNA to be cloned is another factor to consider when choosing a cloning strategy. For example, a gene such as human ß-globin can be cloned in two forms including (i) Genomic copy and (ii) cDNA copy. A cDNA copy made from processed ß-globin mRNA by using reverse transciptase (RNA directed DNA synthesis) will differ from the genomic copy. The cDNA copy will lack the regulatory signals located at 5' : non transcribed region of the gene as well as the introns (Introns are the DNA sequences which do not code for a particular protein and are flanked on either side by exons which code for that particular protein).

Introns are characteristic of many eukaryotic genes and are removed from the mRNA before translation begins. Consequently, a cDNA copy of a gene will be small (< 2-4 kb) relative to the maximum cloning capacities for X and the cosmids. Therefore, plasmid vectors such as pBR 322 (with genes for resistance to pencillin and tetracycline) are commonly used to clone cDNA fragments.

*Genetic Engineering of Salinity and Drought Resistant Plants*

Molecular techniques such as recombinant DNA technology may ultimately have their most significant impact on agriculture. Recent advances raise the possibility of the development of new plant germplasm through the introduction of any gene front any organism into plants. Several leading laboratories have achieved the transfer and expression of bacterial and foreign plant gene in plant cells. The use of recombinant DNA technology to isolate and transfer genes governing agriculturally important characteristics such as salinity and drought resistance is receiving a lot of attention.

Plant biologists hope to develop new sources of stress-resistant germplasm for crops. The ultimate goal of producing stress-tolerant plants is so challenging that, in practice, an approach fully integrating molecular and other methods is most likely to succeed.

The identification of salinity-tolerance genes, which can be transferred into plants using the new generations of vectors now being developed, is a critical aspect of the new molecular technology.

Many organisms use a simple rule of chemistry to live in a world deficient in available water. They have evolved sophisticated mechanisms for balancing their osmotic strength (inside) with that of their surroundings (outside): They are able to avoid dehydration by taking up or synthesizing molecules that work as osmotic balancing agents. Bacterial studies have recently shed light on the mechanisms of cellular adaptation to osmotic stress. Of approximately 150 metabolites tested so far in bacteria, only the betaine series (amino acid derivatives often found in seeds) have possessed potent biological activity in promoting growth under strongly inhibitory levels of osmotic stress. The most active molecules include glycine betaine and its precursors, proline betaine; free proline is also active but under lower levels of stress. Trimethyl-y-aminobutyrate is also active.

Plants have evolved a variety of mechanisms for adapting to osmotic stress, one of which is cellular adaptation, or osmoregulation. Several researchers have hypothesized that the same class of osmoprotective molecules that work in bacteria (glycine betaine, proline betaine, praline, etc.) and are found in scores of higher plants, also behave as osmoprotectant cells during osmotic stress and to prevent damage from cellular dehydration by balancing the osmotic strength of the cytoplasm with that of the environment. Evidence is now available that the increase in glycine, betaine or proline betaine accumulation in a wide array of higher plants is correlated with the increase in osmotic strength of the environment caused by sodium chloride, and the like.

Some encouraging results are coming from studies in which osmoprotection has been achieved at the whole-plant level by addition of these compounds directly to the nutrient solution. For example, the French researcher, D. Rudulier, recently reported preliminary experiments in which symbiotic nitrogen fixation by nodulated alfa seedlings was protected by the addition of glycine betaine or proline betaine; in this case, stressed plants (11,600 mg/L sodium chloride) lost essentially all of their nitrogen-fixation capacity, whereas plants treated with 10 mM (1,350 mg/L) betaines recovered up to 30 percent of their capacity when compared with the untreated controls. These workers have also reported that the symbiotic root nodule bacterium for alfa, Rhizobium meliloti, is protected to occur in nature. Valentine's group from Davis have attempted insertion of recombinant plasmids in rhizobia that overproduce osmoprotectants within the nodule itself.

The genes governing production of osmoprotective molecules-osm genes-are present in plants but have not yet been studied in detail from this source. It is more convenient to study these genes in bacteria.

Many recent technological breakthroughs in genetic engineering have involved plasmids, small circular DNA molecules that behave as tiny independent chromosomes. Recombinant DNA techniques have been used to insert virtually every imaginable gene into plasmids, including the stress-tolerance gene. Davis group selected a group of plasmids with an unusual property—the ability to multiply in a wide number of bacterial hosts including symbiotic root-nodule bacteria. A broad-host-range plasmid was chosen that is capable of shuttling back and forth between the bacterium E.scherichia coli and root-nodule bacteria. In essence, the osm genes conferring stress tolerance have been spliced into a broad-host-range plasmid. They have found the osm genes) in this case to be carried as a segment of DNA of approximately 10,000 base pairs.

This recombinant plasmid has at least two properties (1) the segment codes for the enzymes catalyzing the first two steps in the proline pathway, and (2) biochemical analysis of the first enzyme of this pathway (r-glutamyl kinase) isolated from cells carrying this plasmid reveals that the enzyme has lost its sensitivity to feedback inhibition by proline. This second property accounts for proline overproduction and osmotic tolerance.

A new vector system for genetic engineering of stress-tolerant plants is being developed by workers at UC Davis. Gallforming bacteria (such as Agrobacterium tumefaciens) and root-nodule bacteria (rhizobia) are potential vectors for delivering useful compounds such as the osm gene to leguminous plants. Evolutionarily speaking, these vectors belong to the same tribe of bacteria. They work differently, however, the agrobacterial system delivers foreign genes to host cells, whereas the rhizobial system does not transfer DNA to the host itself but functions as a chemical factory converting raw materials into useful products that are passed on to the host plant (such as fixed nitrogen for plant growth).

The time required for genetic engineering of drought-and salinity tolerant plants might be shortened by harnessing root nodule bacteria that have already evolved a highly effective symbiosis with leguminous plants. Indeed, the root nodules of a soyabean plant may harbor tens of billions of symbiotic bacteria (bacteriods) each working for the host plant. Harnesing a small portion of their biosynthetic capacity to produce a supply of osmoprotective compounds for protecting the sensitive nitrogen-fixation machinery of bacteriods, nodule tissue, and perhaps even the whole plant against stress is proposed. Proline is receiving the most attention as an osmoprotectant, since raw materials for its synthesis are produced in large amounts in bacteroids, details of this gene are already known, and recombinant DNA plasmids capable of transfer and replication in root nodule bacteria have been constructed.

Cellular adaptation to salinity (osmoregulation) is a fertile area for basic research in plant science. The concept of osmosensory or turgor-sensing proteins linking environmental changes in salinity to dynamic changes in membrane, biochemical, and genetic activities of the cell is one such topic of interest to molecular biologists and biochemists. The discovery of osmoprotective molecules and new classes of osm genes may lead to applications such as genetic enhancement of salinity and drought tolerance in crop plants. (Fig.).

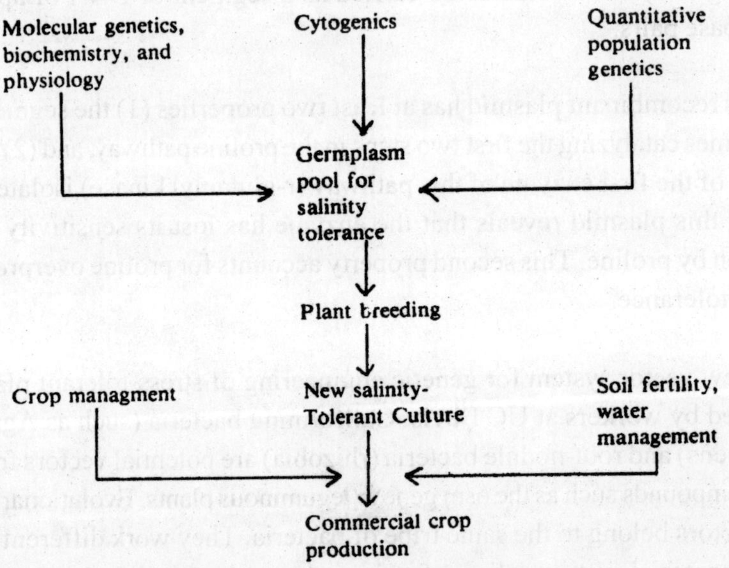


**Fig.** Integrated Approach for Development of Salinity-Tolerant Germplasm

## Prospective Outlook

Although biotechnology is an old discipline including traditional areas such as farming and brewing, it has been totally transformed with the advent of genetic engineering during the last 10-12 years.

At present, any gene can be expressed to some degree in any organism ranging from bacteria, yeast and insects to transgenic sheep. Specialized vectors have been developed for efficiently introducing DNA into cells and for stably maintaining the DNA either as an apisome or integrated into the genome. Transcription and translation signals have also been manipulated to maximize the expression of a heterologous gene in each type of host cell. The choice of host-vector system for the commercial production of mammalian proteins is no longer dictated by the availability of the system. It is now, more or less, regulated by economic and logistic factors and, in some cases, by the necessity of producing proteins with special post -translational modifications. Monoclonal antibodies (discussed in the following chapter) and DNA probes have been under investigation for several years in both the public sector and industry with a view to developing useful reagents for either monitoring clinical outcomes or elucidating clinical conditions.

The availability of a raw material can stimulate a search for fermentation processes to transform the material into useful products. For example, straight chain hydrocarbons which are by-products of the petroleum industry can be used to make single all protein. Genetic engineering now offers the opportunity to improve this conversion. The application of genetic engineering will certainly affect the food processing industry in a piecemeal fashion. Genetic engineering can be employed in the food processing industry in two ways: to design microorganisms that transform the edible biomass into food for human consumption or for animal teed; and to design organisms that aid in foodprocessing, either by acting directly on the food itself or by providing materials that can be added to food. Changing technologies in the brewing industry and increased sophistication in the molecular genetics of yeast have made it possible for researchers to achieve novel goals in yeast breeding. One strain that has already been constructed can produce a low carbohydrate beer suitable for diabetics (Fig.).

Genetic engineering offers to alter the pre-existing proteins in desirable ways; for example, by increasing the lysine content of the major wheat proteins and their nutritional value can be increased. Developments in enzyme engineering could lead to a completely new type of chemical industry utilizing moderate conditions of pH, temperature and pressure to produce from wastes or by-products, high yields of complex chemicals without generating new waste products that pollute the environment. For a variety of reasons, a number of the appropriate enzymes are not yet commercially usable. Recently microorganisms, plants and animals have been used to provide a variety of different enzyme preparations useful in the chemical and food industry (Tables). An early example of the application of recombinant DNA technology to the food industry was the production of calf chymosin by E. coli. Chymosin (rennin or rennet) is a milk-clotting enzyme used in cheese manufacture. Recombinant DNA has greatly stimulated commercial and academic interest in biologically active proteins. This has increased the demands on the protein chemist to produce a large variety of highly purified proteins. In turn, the established purification methods have been improved, for example, monoclonal antibody (discussed in detail in the next chapter) for immunopurification and high performance liquid chromatography (HPLC) ion exchange. The purification of proteins for their biological activities rather than their nutritional value is a key area in the rapidly developing biotechnology industry. Genetic manipulation of enzymes will have far

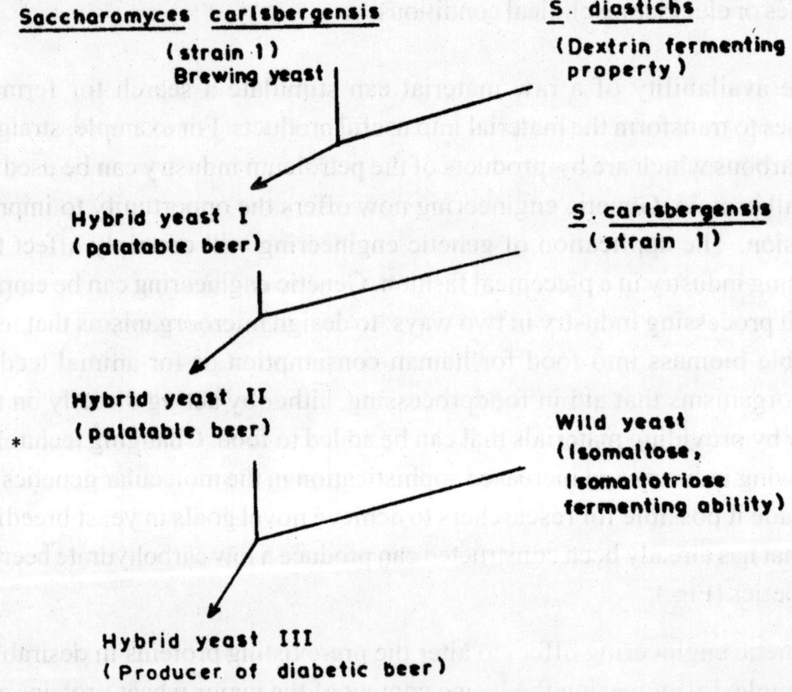


**Fig.** Yeast hybridization resulting in production of Low-carbohydrate beer

reaching applications in pharmaceutical, chemical and food industries. With advances in protein engineering, it might also prove possible to "programme" new enzymic or structural properties into proteins. Eventually, the construction of entirely novel proteins can be envisaged.

Although most genetically engineered microorganisms are being designed for contained facilities like fermenters, some are being examined for purposes such as mineral and oil recovery and pollution control. Two other aspects of microbial physiology deserve attention: the microbial production of oil muds or drill lubricants and the treatment of oil once it has been recovered. Drilling muds are suspensions of clays and other materials that serve both to lubricate the drill and to counterbalance the upward pressure of oil. Microbially produced polysaccharides have been developed for this use. EXXON holds a patent on a formulation based on the production of xanthan gum, from Xanthomonas campestris, while the Pillsbury Co. has developed a polysaccharide (glucan) from various species of Sclerotium. At least two of the small genetic engineering firms have begun research programmes to develop biologically produced polysaccharides with the desired lubricant qualities.

**Table**

Selected Enzymes of Industrial Importance and Their Utility

| *Enzyme* | *Source/Sub* | *Utilization* |
|---|---|---|
| Amylase | Animal (pancreas) | Digestive aid |
| | Plant | Brewing and baking |
| | Fungi | Brewing and baking |
| Papain | Plant | Meat tenderiser |
| Protease | Animal (Pancreas) | Dairy, food, leather |
| | Animal (Pepsin) | Beer stabilizer |
| | Animal (Renin) | Cheese |
| Pectinase | Fungi | Meat tenderiser |
| Invertase | Fungi | Clearing Wine |
| | Yeast | Confectionery |
| Proleinase | Protein | Production of Peptone |
| | (Milk, soyabean) | Pretreatment of soy sauce |

**Table**

Enzyme Engineering and Selected Enzymes of Industrial Use

| *Enzymes* | *Substrate* | *Utilization* |
|---|---|---|
| α-Amylase | Starch | Production of dextrin |
| Exol-1, 4-a-Glucosidase Proteinase | Protein | Production of Peptone |
| Papain | Protein in Beer | Removal of Turbidity |
| Renin | Casein | Production of cheese |
| Cellulase | Cellulose | Saccharification |
| Polygalacturonase | Pectin | Production of Fruit juice |
| Triacylglycerol lipase | Lipid | Hydrolysis of Lipid |
| β-Fmctofuranosidase | Sucrose | Production of Inverted Sugar |
| β-Galactosidase | Lactose | Decomposition of Lactose |
| Anthocyanase | Antocyan | Decoloration of antocyan glycosid |
| AMP deaminase | Adenylic acid | Production of Insosinic acid |
| Aminoacylase | D, L-Acyl-amino acid | Production of L-amino acid |
| Glucose isomerase | Glucose | Production of high-fructose syrup |

**Table**

Examples of Proteins Produced By Microorganisms

| *Protein* | *Application* |
|---|---|
| Interferons | Possible treatment of virus infections and cancer |
| Human Growth Hormone | Pituitary dwarfism |
| Insulin | Diabetes |
| Tissue Plasminogen Activator | Thrombosis |
| B-endorphin | Analgesic |
| Interleukin 2 | Cancer chemotherapy |
| $a_1$ antirypsin | Emphysema |
| Relaxin | Facilitate childbirth |
| Chymosin | Cheese production |
| Thaumatin | Sweetener |
| Poly aspartyl phenylalanine | Preparation of aspartame—sweetener |
| Ricin | Immunotoxin |

Recombinant DNA technology also allows established products to be expressed in new environments. For example, it is now possible to transform certain plants with genes encoding an insecticidal protein derived from the bacterium Bacillus thuringiensis.

Now-a-days, the combination of genetic engineering and mass production of a biological product can often be an essential step towards the evaluation and establishment of new products. A good example of this is seen in the tentative clinical trials of tumour necrosis factor (TNP). A large number of the therapeutic enzymes and cofactors are also under clinical evaluation ('Cables 7.6, 7.7, Fig.). In addition, several human recombinant hormones and lymphokines which are used clinically have been on the forefront for commercialization (Table).

**Table**

Recombinant Enzymes And Cofactors Useful in Clinical Laboratories

| *Protein* | *Therapeutic role* |
|---|---|
| α-I Anti-trypsin | Prevents emphysema |
| Tissue Plasminogen Activator (TPA) | Dissolves blood clots |
| Factor VIII | Aid blood clotting process |
| Factor IX | Aid blood clotting process |

**Table**

Human Recombinant Hormones

| ***Hormone*** | ***Therapeutic Role*** |
|---|---|
| Insulin | Lowering blood glucose |
| Insulin-like Growth Factor I (1 GFI) | Stimulating Growth hormone release |
| Calcitonin | Increases blood calcium level |

**Table**

Lymphokines in Clinical Use

| ***Gymphokines*** | ***Therapeutic role*** |
|---|---|
| α interferon | Treating chronic Leukaemias |
| β interferon | Treating renal cell carcinoma |
| γ interferon | Treating Karposi!s sarooma |
| interleukin I | T cell activation |
| interleukin 2 | T cell growth stimulation |
| interleukin 3 | Colony stimulation |
| interleukin 4 | B cell activation |
| interleukin 5 | B cell activation |
| Tumour necrosis Factor | Kills Tumour cells |
| Lymphotoxin | Kills Tumour cells |

Genetic Engineering clearly offers a vast range of new options of medicines, monoclonal antibodies, new vaccines, DNA probes, biosensor etc. All DNA probes are not created equal, or in fact in the same way. "Long probes (Fig.) are the more traditional entity, while `short' probes (Fig.) are the more recent development produced synthetically, often on DNA synthesizeres, currently manufactured by eight to ten companies.

Since the mid 1970's genetic engineering and one of its scientific precursors, recombinant DNA research have also been the focus of controvercy Issues such as work-place safety, University-Industry relations, changes ' patent law, gene conservation and the environmental impact of various projects have been vigorously debated. In some cases, it has been apparent the genetic engineering has been taken as a symbol of a certain kind of innovation and resisted as such. This is most obviously true at the moment in the Federal Republic of Germany, where some supporters of Green Politics pose commercial genetic engineering as a continuation of an oppressive industrial system.

Isolation of DNA fragments from target molecule

↓

Selective restriction, isolation and testing of target fragment

↓

Insertion of DNA into a vector molecule such as plasmid

↓

Propagation of host molecule

↓

Removal and isolation of DNA

↓

Enzymatic restriction of DNA

↓

Purification (gel, sedimentation)

↓

Identification and isolation of desired probe fragements

↓

Enzymatic incorporation of reporter group into probe

↓

Isolation of labelled probe

**Fig.** Production of Long Probes

Obtain target sequence

↓

Chemically synthesize modified probe

↓

Purify (HPLC, electrophonesis)

↓

Allach reporter groups

**Fig.** Production of short Probes

# FOUR

# Gene Amplification

The techniques of recombinant DNA and gene cloning, permit us to obtain an unlimited supply of identical copies of a gene sequence or DNA segment, which is cloned in a prokaryotic or a eukaryotic cell with the help of a vector. This technique, discovered around 1975, proved extremely useful in all experiments of molecular biology and genetic engineering and thus became an essential tool in all molecular biology laboratories. In 1985, yet another remarkable tool in molecular biolgoy was discovered, which is known as polymerise chain reaction (PCR) nick named now as people's choice reaction. The methodology is 'so important that the prestigeous journal Science published from USA considered PCR as the major scientific development of the year 1989, and had chosen Taq DNA polymerise, the enzyme used in PCR, as the molecule of the year 1989. (Kary Mullis, who discovered this reaction in 1985, described in his article published in April 1990 issue of Scientific American, how he got the basic idea of this reaction, while driving through the red wood mountains of California, USA). The polymerise chain reaction is such a powerful technique that it may replace completely the gene cloning with vectors in due course of time.

## Polymerise Chain Reaction (PCR)

In order to understand PCR, the readers should be familiar with the mechanism of DNA replication within the cells. The DNA replication involves polymerization of nucleotides using a template DNA strand with the help of an enzyme DNA polymerise, but this reaction invariably requires a primer strand to which further nucleotides can be added using' the DNA polymerise enzyme. If the primer strand is not available, the reaction can not proceed. In the living cells, this primer strand is not in DNA strand but is a small single stranded RNA molecule synthesized with the help of RNA polymerase enzyme. In PCR, a similar reaction takes place in an 'eppendorf tube', where the primer strand is added from outside in the form of a deoxyoligonucleotide, and DNA polymerase enzyme is added to help in polymerization. Unlimited supply of amplified DNA is obtained by repeating the

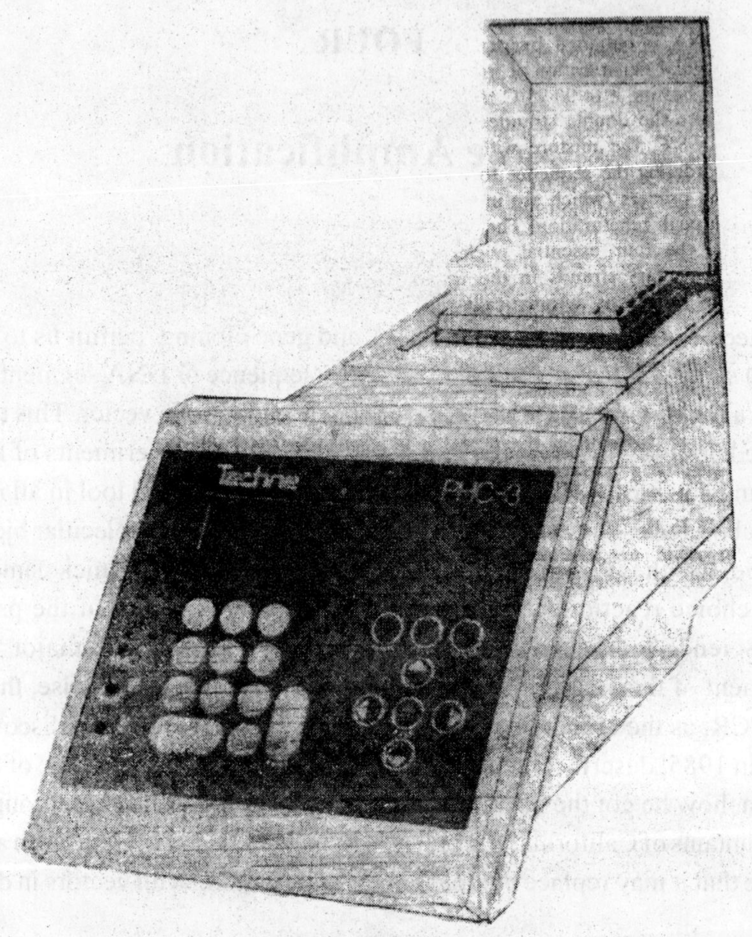


**Fig.** A thermal cycler for PCR amplification
(the instrument has a microprocessor—controlled temperature cycling needed for PCR).

reaction, which is made possible by regular denaturation of freshly synthesized double stranded DNA molecules by heating it to 90-98°C. At this high temperature, the two strands separate. Once the double stranded DNA is made single stranded by heating upto 90-98°C, the mixture with two primers, recognising the two strands and bordering the sequence to be amplified is cooled to 40-60° C. This allows the primers (which are in excess) to bind to their complementary strands through renaturation. The presence of 'Tay DNA polymerase' enzyme and all the four essential nucleoside triphosphales allow synthesis of complementary strands in the usual manner. In a thermal cycles (Fig.), this process is automatically repeated 20-30 times (as predetermined by a computer device), so that in a single afternoon, a billion copies of the sequence flanked by the left and right primers can be produced (Fig.). In order to continue the synthesis, the temperature of the mixture is alternately

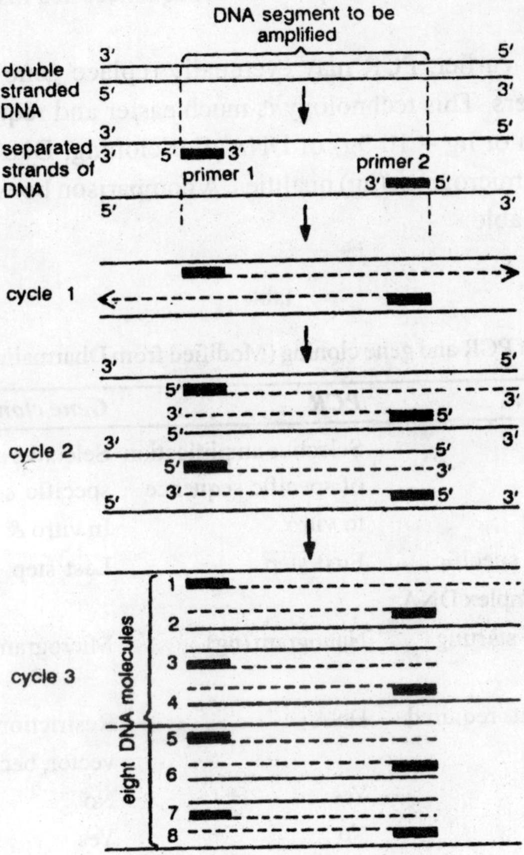


**Fig.** Basic reaction of the PCR (only three cycles of PCR are shown; in each cycle primers are shown by solid boxes, template strands are shown by continuous lines and newly synthesized strands are shown by broken lines). Modified and redrawn from Dharmalingam, K. –Biospectra, Nov-Dec. 1990. pages 3-8).

increased (for denaturation) and decreased (for renaturation) once every 1-3 minutes (as fixed by the computer device). Therefore, during temperature rise the enzymatic activity of DNA polymerase should not be destroyed, otherwise one may have to add a fresh aliquot of enzyme in each cycle of amplification. This became possible only by the discovery of thermostable enzyme Taq DNA polymerase, isolated from Thermus aquaticus growing in hot springs. This enzyme acts host at 72°C and the denaturation temperature of 90°C does not destroy its enzymatic activity. Later, other thermostable enzymes like Pflu DNA polymerase isolated from Pyrococcus furiosus and Vent polymerise isolated from Thermococcus litoralis, were discovered and were found more efficient. These enzymes allowed automation of the entire process and automatic PCR thermal cyclers are now available ( each for a price of around Rs 2. 00 lakhs), which can amplify DNA sequences at a fast speed unattended.

As pointed out earlier, PCR may eventually replace gene cloning technique discussed in Chapters. This technology is much easier and requires much smaller quantity (nanogram or ng = 10-9g) of DNA. For cloning, DNA is needed, as the starting material, in microgram (ug) qualities. A comparison between PCR and gene cloning is given in Table.

**Table**

Comparison of PCR and gene cloning (Modified from Dharmalingam, 1990).

| *Parameter* | *PCR* | *Gene cloning* |
|---|---|---|
| 1. Final result | Selective amplification of specific sequence | Selective amplification of specific sequence |
| 2. Manipulation | to vitro | In vitro & in vivo |
| 3. Selectivity of the specific segment front complex DNA | First step | Last step |
| 4. Concentration of starting material | Nanogram (ng) | Microgram (μ g) |
| 5. Biological reagents required | DNA | Restriction enzyme, lipase, vector, bacterial cell |
| 6. Automation | Yes | No |
| 7. Labour intensive | No | Yes |
| 8. Error probability | Less | More |
| 9. Applications | More | Less |
| 10. Cost | Less | More |
| 11. User's skill | Not required | Required |
| 12. Time for a typical experiment | Four hours | Two to four days |

## Different PCR Schemes

In the previous section, we discussed the basic outline of PCR technique, outlining how billions of copies of a DNA sequence can be obtained within a few hours, without the use of a vector and a host as done in gene cloning. However, there are several variations to this basic PCR technique, which allow varied applications of this technique.

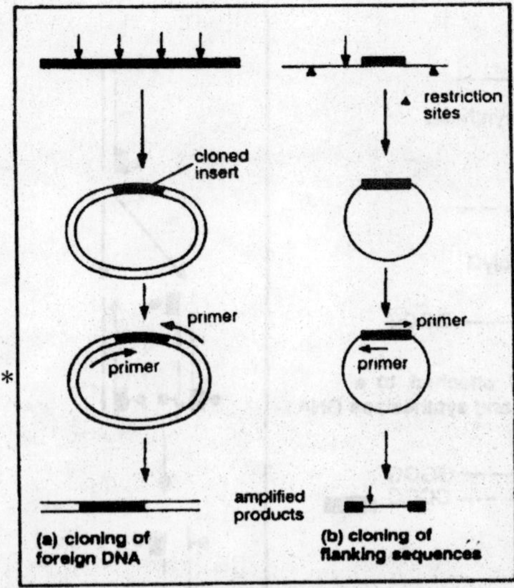


**Fig.** Inverse PCR (a) amplification of genes using flanking vector sequences as primers; (b) amplification of sequences flanking the gene, using sequences at the two ends of the gene as primers. (Modified and redrawn-from Dharamalingam, K.–Biospectra, Nov.-Dec. 1990 page 3-8)

*Inverse PCR*—In this technique, the amplification of those DNA sequences takes place, which are away from the primers and not those which are flanked by the primers. For instance if the border sequences of a DNA segment are not known and those of a vector are known, then the sequence to be amplified may be cloned in the vector and border sequences of vector may be used as primers in such a way that the polymerization proceeds in inverse direction i.e. away from the vector sequence flanked by the primers and towards the DNA sequence of inserted segment (Fig.). Similarly, if the gene sequence is known, one can use its border sequences as primers, for an inverse PCR reaction, to amplify the sequences flanking this gene e.g. the regulatory sequences (Fig.).

*Anchored PCR*— In the basic PCR technique and the inverse PCR, one has to use two primers representing the sequences lying at the ends of sequences to be amplified. But sometimes, we may have knowledge about sequence at only one of the two ends of the DNA sequence to be amplified. In such cases, anchored PCR may be used, which will utilize only one primer instead of two primers. In this technique, due to the use of one primer, only one strand will be copied first, after which a poly G tail will be attached at the end of the newly synthesized strand. This newly synthesized strand synthesis utilizing an anchor primer with which a poly C sequence is linked to complement with poly G of the template. In the next cycle, both the original primer and anchored primer will be used for gene amplification (Fig.).

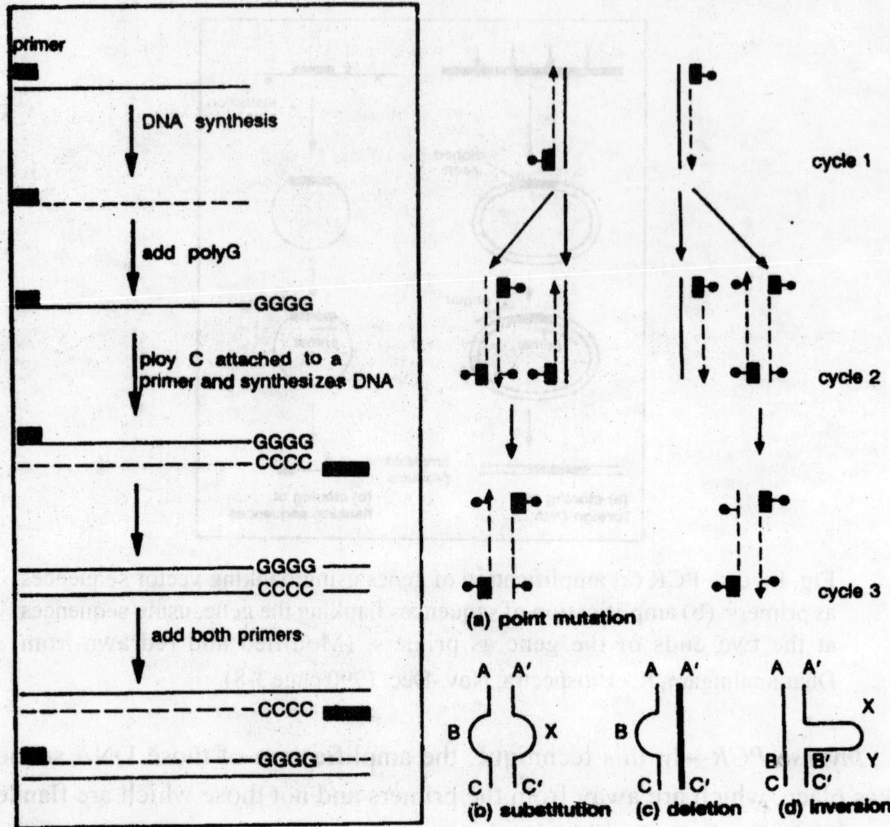


**Fig.** Anchored PCR, using a single primer in the first cycle (see text for details). (Modified and redrawn from Dharmalingam, 1990).

**Fig.** Site directed mutagenesis using PCR (minor changes are introduced in primers; these minor changes do not interfere with PCR) (Modified and redrawn from Dharmalingam, 1990).

*PCR for site directed mutagenesis*—This technique is used for introducing mutations at the desired place in a DNA sequence by altering the sequences of primers (Fig.). Since mutations are introduced only through primers, mutations are limited to the ends of the gene sequence. A variation of this technique allows mutations to be introduced at any place of interest in the gene- the method is described as overlap extension which works as fallows: In two separate PLR reactions, a particular gene is amplified into two separate segments. In each of these two reactions there is one primer at the end of the gene and the other (with desired alteration for mutation) internal to the sequence. The internal modified primers in two reactions are complementary to one another, so that the amplified products will have their ends internal to the original sequence. These internal ends of products in two reactions will overlap. The internal ends of the two PCR products, when mixed, function as primers. Extension of these primers by taq polymerase results in the formation of a complete gene, with mutations incorporated at internal sites *(Fig.)*

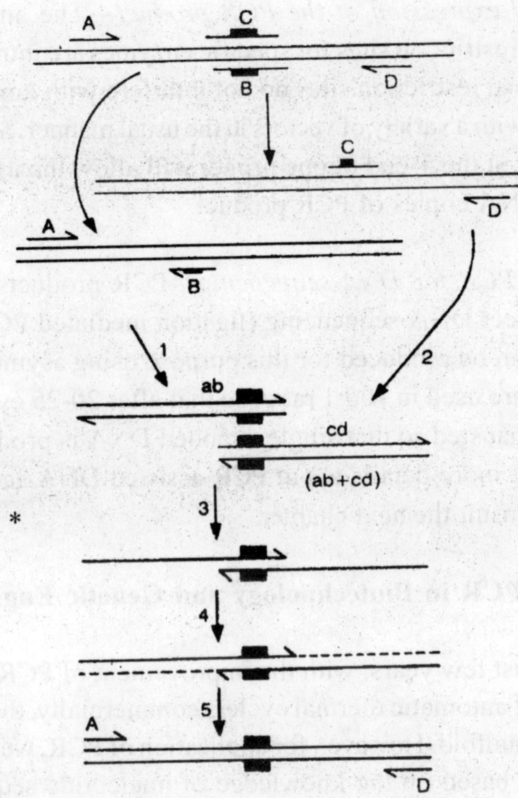


**Fig.** Introduction of mutation at an internal site using overlap extension (two pairs of primers A B: 11 and C D are used. B and C represent internal sites and overlap each other). (Modified and redrawn from Dharmalingam, K-Biospectra, Nov. -Dec., 1990 page 3-8).

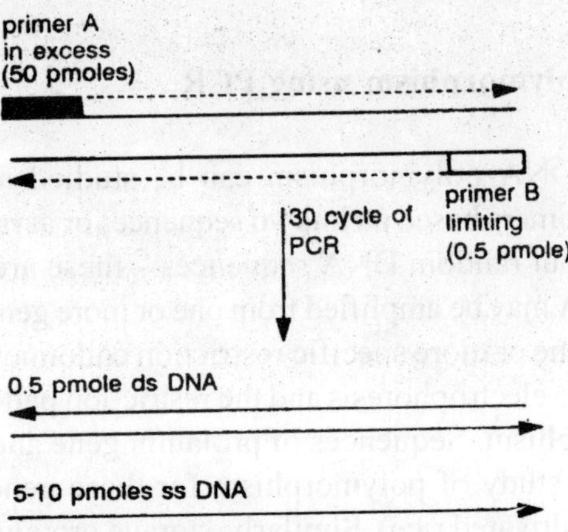


**Fig.** Asymmetric PCR for generating single stranded DNA for DNA sequencing

*Cloning and expression of the PCR product*—The amplified PCR product can be cloned, if restriction sites for specific enzymes are introduced at the 5' end of each primer. These restriction sites do not interfere with amplification and can be used for cloning with a variety of vectors in the usual manner. Similarly, incorporation of a T7 promoter at the 5' end of one primer will allow the use of RNA polymerise enzyme to get RNA copies of PCR product.

*Asymmetric PCR for DNA sequencing*—PCR products can also be used as templates for direct DNA sequencing (ligation mediated PCR =LMPCR). Single stranded DNA can be produced for this purpose using asymmetric PCR, in which the two primers are used in 100:1 ratio, so that after 20-25 cycles of amplification, one primer is exhausted so that single stranded DNA is produced in the next 5-10 cycles (Fig.). For more details about PCR assisted DNA sequencing, the readers are advised to consult the next chapter.

## Application of PCR in Biotechnology and Genetic Engineering

During the last few years, with the improvement of PCR protocols and due to the availability of automatic thermal cyclers commercially, the applications of PCR have increased manifold. However, for application of PCR, we need a pair of primers which should be based on the knowledge of nucleotide sequence of DNA to be amplified. Therefore, non-availability of this information about the DNA segment or gene to be amplified, becomes a limitation in the applications of PCR, although this difficulty has been overcome for the study of DNA polymorphism, through the use of random DNA primers. Some of the applications of PCR technology are described in this section.

### Study of DNA polymorphism using PCR

Using PCR, DNA polymorphism can be studied at loci with known DNA sequence (using primers based on known sequence) or at random sites (using primers having unknown but random DNA sequences—these are called random primers). In either case, DNA may be amplified from one or more genomic DNAs and subjected to digestion with one or more specific restriction endonucleases. The digested DNA may be subjected to electrophoresis and the restriction patterns may be photographed to reveal polymorphism. Sequences of prolamin gene and phytochrome gene have been used for the study of polymorphism for these genes in different species of Oryza (wild and cultivated rice). Similarly, storage protein genes in wheat and many genes in other animal and plant species have been used, for the study of DNA polymorphism.

The non-availability of sequences of specific genes for designing primers for DNA amplification has been overcome by a variety of approaches including the following: (i) In case of multigene families like those for rRNA, tRNA and heat shock proteins, there are conserved consensus sequences known, which are used for designing universal primers. Similarly for amplification of antibody genes also, universal primers have been used in PCR. (ii) Genebank Sequence Data have also been used for designing primers for amplification of microsatellites like poly (A), poly (G ), poly (TC) and poly (TG), etc.; this is used for a study of length polymorphism in these microsatellites, also described as 'short tandem repeats' or ' simple sequence repeats = SSR. (iii) The consensus sequences for intron-exon junctions or intron splice junctions (ISJ) have also been used for designing primers which have been used for amplification of intron or exon sequences to study DNA polymorphism. (iv) Conserved sequences at the ends of SINEs (short interspersed elements) and LINES (long interspresed elements) have also been used for a study of polymorphism in the distribution and lengths of these repeat sequences. (v) In certain viruses, DNA polymorphism could be studied by designing degenerate primers on the basis of amino acid sequence of a protein using the genetic code.

PCR and RAPD markers—The most important use of PCR for the study of DNA polymorphism, however, involves the use of random primers giving Random Amplified Polymorphic DNA (RAPD, pronounced as rapid Fig). This technique is considered simpler and offers several advantages over more commonly used technique known as RFLP (see Chapters). RAPDs have actually been used for a study of polymorphism and chromosome mapping in maize, soyabean, mouse, humans, etc.

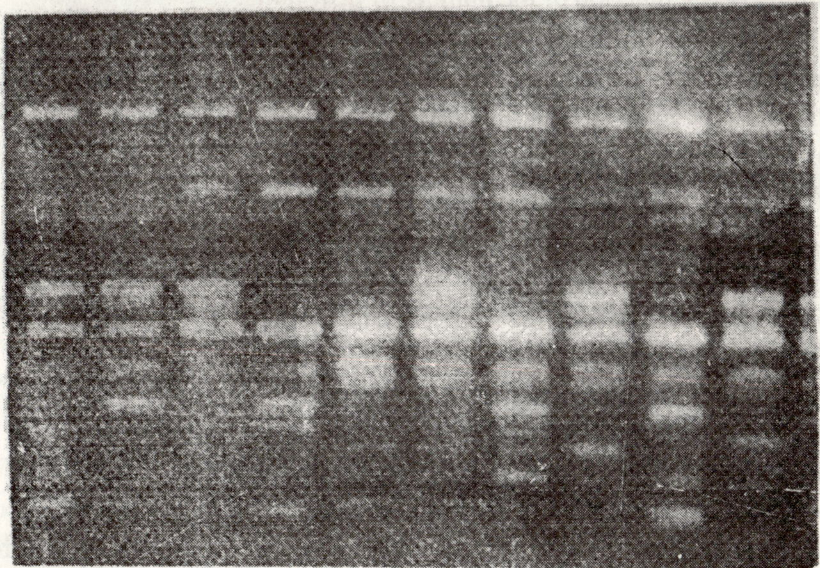

**Fig.** RAM generated through the use of random primers in PCR.

PCR and VNTR loci—Alec Jeffreys and his colleagues in U.K. suggested the use of PCR for study of DNA polymorphism at VNTR (variable number tandem repeats, also known as 'minisatellites' ) loci in humans. Primers were developed for the conserved flanking regions of VNTR loci, so that all available VNTR loci could be amplified. The PCR products differ according to the number of repeat units present in the different VNTR loci and this variation is visualized in the positions occupied by different bands after electrophoresis.

PCR and SSR loci—The VNTR loci discussed above consist of repeat units in the range of 11-60 base pairs in length. A much simpler type of repeat units exist in the form of dinucleotides, trinucleotides or tetranucleotides, e.g. (CA)n, (GT)n (AAT)n or (AGAT)n. These repeat units were described as microsatellites, but later also called short tandem repeats or simple sequence repeats (SSR). Some of these SSR loci, such as (CA)n or (GT)n occur in the human genome as many as 50,000 times, within varying from 10 to 60.

The DNA sequences flanking SSRs are conserved, allowing the selection of PCR primers that will amplify the intervening SSR in all genotypes of the target species. PCR reaction includes a small amount of $^{32}$P labelled nucleotide (only one of the four nucleotides is labelled) or it may include primers, one or both of them being end labelled. This allows visualization of PCR products using autoradiography after electrophoresis on a standard sequencing gel (Fig.). Variation in the length of PLR products is a function of the number of SSR units.

The principle underlying the detection of SSR length polymorphism using two parents and their F1 hybrid is illustrated in Figure. Markers resulting from SSR length polymorphisms are placed on genetic maps in relation to RFLP's, RAPD's, VNTR's, other SSR's and even the morphological markers.

A comparison of very limited sequence data from plants (including algae) with those of vertebrates (including humans) indicated that SSR's in plants are found with the same frequency as in vertebrates. Recently DNA sequence data for five tree species and *Zea mays* have been screened for the presence of (AC)n and (AG)n and the frequency in these six species was found to be close to one microsatellitc (SSR) per 5 x 103 to 3 x 105bp. Other plant and animals species have also been examined for the presence of a number of SSR's and a wide occurrence of these markers in plants and animals was recorded suggesting their utility in preparing saturated molecular maps.

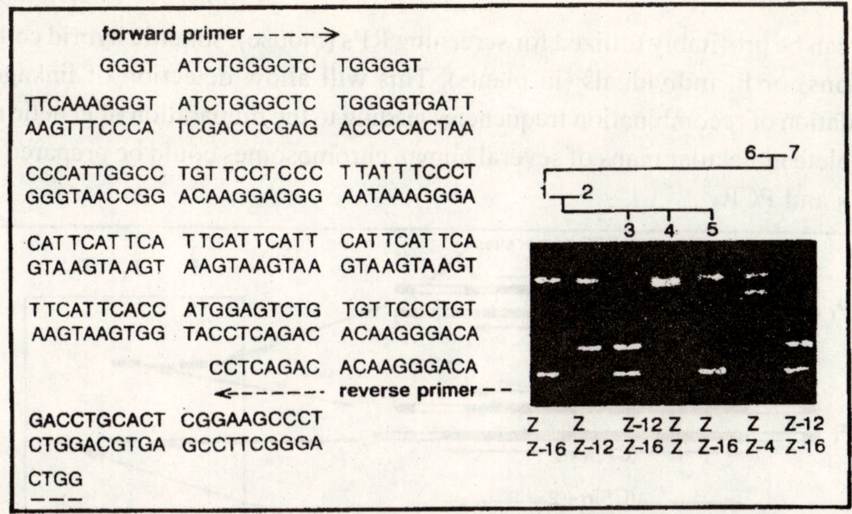


**Fig.** Generation and visualization of simple sequence repeat (SSR) length polymorphism.

## Molecular Mapping

Genetic and physical chromosome maps have also been prepared, both in plants and animals, using PLR. Once RAPD's are detected, recombinant inbreds (RI's) or $F_2$ / backcross segregating populations, or doubled haploids (DH) derived from haploids (in plants) can be used to detect linkage and recombination frequencies. The molecular markers which cosegregate in RI's or $F_2$'s or DH's suggest linkage and the frequency of cosegregation among RIs or $F_2$ / backcross plants will be a measure of the strength of linkage. The molecular markers identified through PLR, can also be assigned to specific chromosomes using aneuploids or alien addition lines (e.g. wheat-barley additions =21 pairs of wheat chromosomes + 1 pair of a specific barley chromosome). A number of molecular markers have been mapped using PCR in maize, soyabean, barley, mouse, humans, etc.

Sequence tagged microsatellite sites (STMS) or SSR loci have also been examined and mapped using PCR. The gene sequence databases are screened for a microsatellite and the flanking sequences are used for designing primers for PCR. Several STMS or SSR loci using this technique have been mapped in mouse, where as many as 200 microsatellites have been detected in the genome database.

Besides SSR's or STMS's, other markers including sequence tagged sites or STS's (a short DNA sequence in a molecular genomic DNA marker) and expression sequence tags or EST's (a short DNA sequence from a cDNA clone representing an expressed gene) have also been utilized for genetic mapping. In both these cases,

PCR can be profitably utilized for screening RI's (mouse), somatic hybrid cell lines (humans) or $F_2$ individuals (in plants). This will allow detection of linkage and calculation of recombination frequencies leading to the preparation of genetic maps. Complete molecular maps of several human chromosomes could be prepared using STS's and PCR.

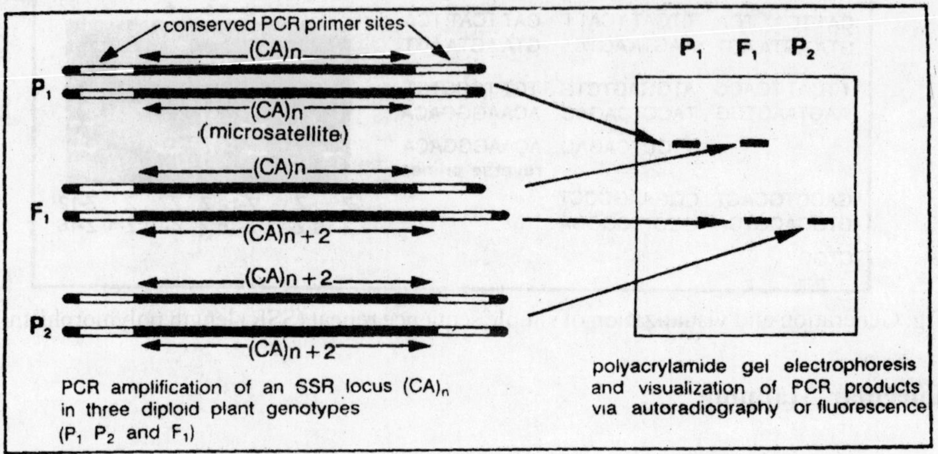


**Fig.** The principle involved in the detection of SSR length polymorphism, ($P_1$ and $P_2$ are parents and $F_1$ is hybrid between them)

## Gene Tagging

PCR has also been utilized for developing molecular markers closely linked to specific genes of economic importance. For instance in tomato, 144 random primers were used to produce 625 PCR products from a set of near isogenic lines (NIL'S, which differ only for the presence and absence of a specific gene) for Pseudomonns resistance. At least three of these PCR products were present in one and not in the other of a pair of NIR's showing linkage with Pseudomonas resistance gene (Pto). These PCR products could be mapped close to *Pto.* Such tagging of genes through PCR will be used in future for plant breeding (selection of desirable plants in segregating populations) and also for isolation of genes.

## Confirming the Presence of Transferred Gene

When a gene is transferred with a vector to cultured cells or organisms, primers can be designed to conduct PCR for amplification of the gene sequence so transferred. This technique allows the confirmation of transfer and maintenance of the gene of interest. In gene therapy experiments, the transfer and presence of a marker gene for neomycin resistance (NeoR) could be detected in the blood of patients, even after 60 days, although the frequency of marked lymphocytes gradually decrease necessitating a second transfusion of such marked cells. Similarly, ADA (adenosine deaminase deficiency) gene was transferred for therapeutic purposes and PCR was used to detect the presence of this gene upto 6 5 months, so that the patient (4 year old girl) was then placed on a programme of maintenance infusions at 3-5 months intervals. Thus, PCR proves to be of immense help in monitoring a gene in genetic engineering or gene therapy experiments.

## Human Genetics

PCR has also found extensive use in human genetics. Some of these uses include the following:

Prenatal diagnosis using PCR—Prenatal diagnosis of sickle cell anemia with enhanced sensitivity was perhaps the first application of PCR (Saiki, 1985; Science 230:1350-1354). Using PCR, this is done in less than one day in contrast to several weeks needed, when Southern blots are used for hybridization with a probe. The test has also been used for diagnosis of phenylketonuria (PKU), 8- thalassemia, hemophilia, etc. The PCR product in all these cases is examined using a labelled probe to suggest whether or not mutant sequence causing the disease is found or not. Sometimes, RFLP pattern of PCR products in healthy and defective feti (singular= fetus) differ, thus enabling prenatal diagnosis. In still other cases, PCR product may be sequenced to reveal the difference (for sequencing with PCR, consult-next chapter).

Recombination data using PCR—PCR can be used for DNA amplification, utilizing individual sperms, which are the products of meiosis and recombination. Therefore, by studing PCR products from a large number of individual sperms one can calculate the proportion of sperms that are recombinant for linked markers. This allows construction of genetic maps at a resolution which is not possible with pedigree analysis. Microdissected chromosomes bands are also used for PCR to allow physical location of genetic lesions.

Sexing of embryos using PCR—Since DNA sequences in single cells can be studied using PCR, sex of human or livestock embryos, fertilized in vitro, can be determined before implantation. For this purpose, PCR primers and probes specific for sex chromosomes (e.g. ZFY or ZFX sequences) can be used. Y-specific primers and probes in humans are available for this purpose. This technique can also be used to detect sex-linked disorders in the fertilized embryos.

## DNA Fingerprinting

DNA fingerprinting is used in forensic medicine for identification of criminals and disputed parentage. Some well characterized sequences of microsatellites are being used for designing primers, so that DNA fingerprinting may be achieved through PCR. PCR allows amplification of DNA from individual hairs, stains of blood or seminal fluid having partially degraded DNA, for characterization of individuals, which was not possible earlier.

# FIVE

# Gene Cloning

Vectors of late, techniques for manipulating prokaryotic as well as eukaryotic DNA have witnessed a remarkable development. This has allowed breakage of a DNA molecule at two desired places to isolate a specific DNA segment and then insert it in another DNA molecule at a desired position. The product, thus, obtained is called recombinant DNA and the technique often called genetic engineering. Using this technique, we can isolate and clone single copy of a gene or a DNA segment into an indefinite number of copies, all identical. This became possible because vectors like plasmids and phages reproduce in a host (e.g. E. coli) in their usual manner even after insertion of foreign DNA, so that the inserted DNA will also replicate faithfully with the parent DNA. This technique is called gene cloning. With this technique, genes can be isolated, cloned and characterized, so that the technique has led to significant progress in all areas of molecular biology.

A variety of vectors have been developed which not only allow multiplication, but may also be manipulated in such a way that the inserted gene may express in the host. Due to the importance of a variety of these cloning and expression vectors in genetic engineering experiments, they are discussed in some detail in this chapter. The techniques used for inserting foreign DNA in these vectors and the development of chimeric DNA molecules (for developing molecular probes, gene libraies, etc.) will be discussed in the next chapter. The commercial use of these techniques will be discussed later.

### Cloning Vectors

One of the most important uses of recombinant DNA technology is the cloning of (i) random DNA or cDNA segments, often used as probes (ii) specific genes, which may be either isolated from the genome or synthesized organochemically or in the form of eDNA from mRNA. This cloning of DNA possible only with the help of another DNA molecule, which is capable of replicating in a host. This other DNA molecule is often used in the form of a vector, which could be a plasmid, a bacteriophage, a derive cosmid or phagemid (see later in this section), a transposon or even virus. Techniques should also be available, which will allow selection chimeric genomes obtained after insertion of foreign DNA from a mixture of chimeric and the original vector (see later). Another critical desired feature of any cloning vector is that it should possess a site at which foreign DNA can be inserted without disrupting any essential function. Therefore, in each case, an enzyme will also have to be selected which will cause a single break. Sometimes vectors are modified by inserting DNA segment to create unique sites for one or more enzymes to facilitate; its use in gene cloning. This inserted DNA with restriction sites for several enzymes is sometimes called a polylinker.

**Plasmids as vectors**

Plasmids are defined as autonomous elements, whose genomes exist in the cell as extrachromosomal units. They are self replicating circular (cx rarely linear) duplex DNA molecules, which are maintained in a characteristic number of copies in a bacterial cell, yeast cell or even in organcelles found in eukaryotic cells. These plasmids can be single copy plasmids that are maintained as one plasmid DNA per cell or multicopy plasmids, which are maintained as 10-20 genomes per cell. There are also plasmids which are under relaxed replication control, thus permitting their cumulation in very large numbers (upto 1000 copies per cell). These are the plasmids which are used as cloning vectors, due to their increased yield potential.

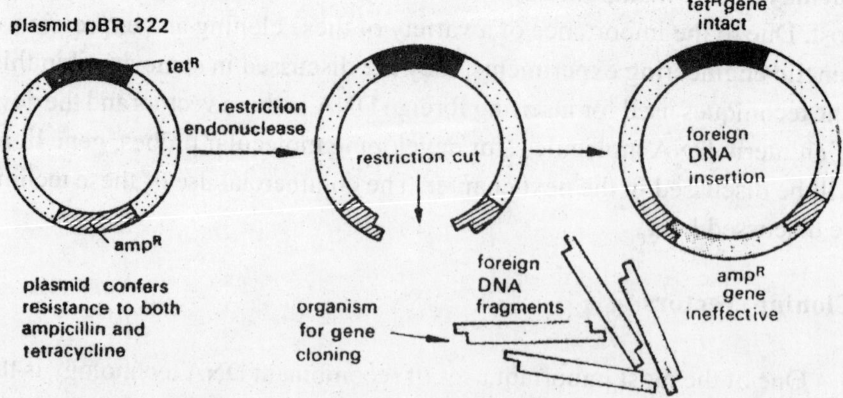


**Fig.** The concept of cloning a foreign DNA segment in plasmid vector.

Circular plasmid DNA which is used as a vector, can be cleaved at one site with the help of a restriction enzyme to give a linear DNA molecule. A foreign DNA segment can now be inserted by joining the ends of broken circular DNA to the two ends of foreign DNA, thus regenerating a bigger circular DNA molecule that can now be separated by gel electrophoresis on the basis of its size (Fig.). Selection of chimeric DNA is also facilitated by the resistance genes, which the plasmid may carry against one or more antibiotics. If a plasmid has two such genes conferring resistance against two antibiotics and if the foreign DNA insertion site lies within one of these two genes, then the chimeric vector loses resistance against one antibiotic, the gene for which foreign DNA has inserted within its structure. In such a situation, the parent vector in bacterial cells can be selected by resistance against two antibiotics and the chimeric DNA can be selected by retention of resistance against only one of the two antibiotics pBR322 and pB12327 vectors. The naturally occurring plasmids may not possess all the above and other essential properties of a suitable cloning vector. Therefore, one may have to restructure them by inserting genes of relaxed replication and genes for antibiotic resistance. This has actually been done and

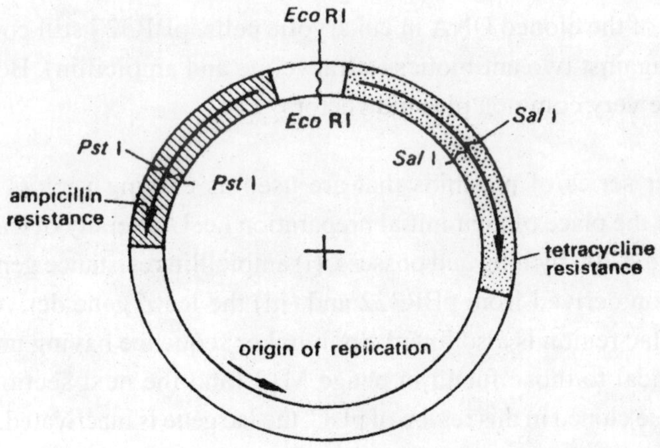


Plasmid pBR322 and its major characteristic features

suitable plasmid vectors have been obtained. One of the standard cloning vectors widely used in gene cloning experiments is pBR322 (derived from E. coli plasmid WEI), which is 4,362 by DNA and was derived by several alterations in earlier cloning vectors (pBR322 was named after Bolivar and Rodriguez, who prepared this vector). It has genes for resistance against two antibiotics (tetracycline and ampicillin), an origin of replication and a variety of restriction sites for cloning of restriction fragments obtained through cleavage with a specific restriction

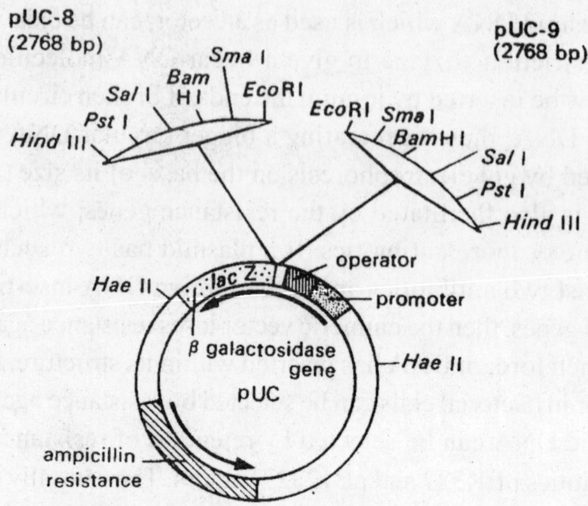


Plasmid pUC and its major characteristic features

enzyme (Fig.). Another vector pBR327 was derived from pBR322, by deletion of nucleotides between 1,427 to 2,516. These nucleotides are deleted to reduce the size of the vector and to eliminate sequences that were known to interfere with the expression of the cloned DNA in eukaryotic cells. pBR327 still contains genes for resistance against two antibiotics (tetracycline and ampicillin). Both pBR322 and pBR327 are very common plasmid vectors.

Another series of plasmids that are used as cloning vectors belong to pUC series (after the place of their initial preparation i.e. University of California). These plasmids are 2,700 by long and possess, (i) ampicillin resistance gene, (ii) the origin of replication derived from pBR322 and (iii) the lac Z gene derived from E. coli. Within the lac region is also found a polylinker sequence having unique restriction sites (identical to those found in phage M13 (and the next section). When DNA fragments are cloned in this region of pUC, the lac gene is inactivated. These plasmids when transformed into an appropriate E. coli strain, which is lac'(e.g. JM103, JM109), and grown in the presence of IPTG (isopropyl thiogalactoside, which behaves like lactose, and induces the syhthesis of 6 galactosidase enzyme) and X-gal (substrate for the enzyme), will give rise to white or clear colonies. On the other hand, pUC having no inserts and transformed into bacteria will have an active lac Z gene and therefore will produce blue colonies, thus permitting identification of colonies having pUC vector with cloned DNA segments. The cloning vectors belonging to pUC family are available in pairs with reversed orders of restriction sites relative to lac Z promoter. pUC8 and pUC9 make one such pair (Fig.). Other similar pairs include pUC12 and pUC13 or pUC18 and pUC19.

As discussed above, in pBR322 and pBR327, the DNA is inserted at a site located in one of the two genes for resistance against antibiotics, so that it will inactivate one of the two resistance genes. The insert bearing plasmid can be selected by their ability to grow in a medium containing only one of the two antibiotics and by their failure to grow in a medium containing both the antibiotics. The plasmids carrying no insert on the other hand, will be able to grow in media containing one or both the antibiotics. In this manner, the presence of lac Z gene in pUC' and resistance genes against ampicillin and tetracycline in pBR322 and pBR327 allow selection of E. coli colonies transformed with plasmids carrying the desired foreign cloned DNA segment.

Yeast plasmid vectors—Special vectors are also known for introducing DNA segments in yeast cells, a eukaryotic system that has been used extensively for developing genetically engineered yeast cells. Although E. coli plasmids or phages can be used for transfer of genes to yeast cells, the freqhency of transfromation is rather low. For instance, yeast cells that are LEU (lacking the ability to synthesize leucine) could be transformed using chimeric plasmid carrying LEU + gene from yeast. This involves integrative transformation (crossing over) having low frequency. To overcome this difficulty of low frequency transformation, efforts were made to prepare a variety of vectors for yeast.

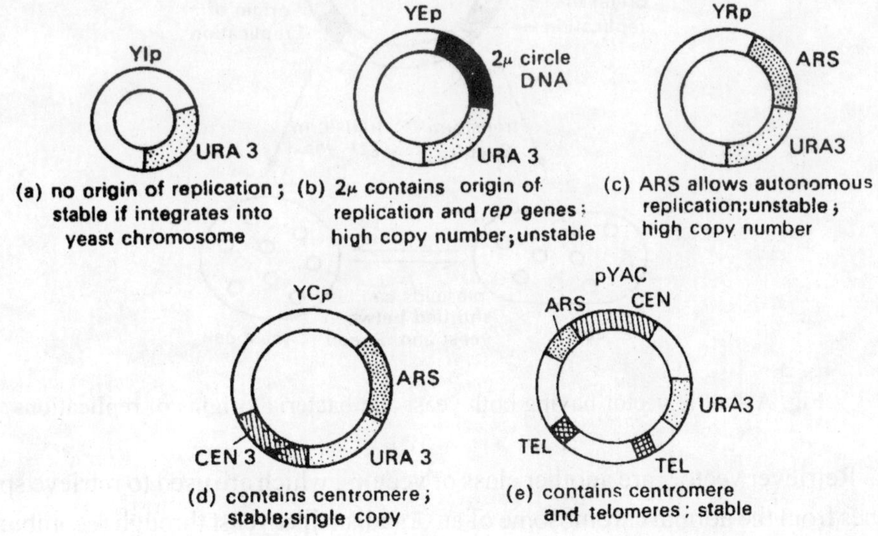


**Fig.** Five different yeast plasmid vectors

The possible examples of the yeast plasmid vectors are : (i) Ylp or yeast integrative plasmids, which allow transformation by crossing over and have no replication origin; (ii) YEp or yeast episomal plasmids, which carry 2μ DNA sequence including the origin of replication and rep genes; (iii) YRp or yeast replicating plasmids, which carry autonomously replicating sequence; this sequence is very common with many yeast genes so that stable transformation can be achieved by crossing over; (iv) YCp or yeast centromere (CEN) containing plasmids, which function as true chromosomes and segregate caring mitosis and meiosis; (v) pVAC or yeast artificial yeast chromosome to vector which carries both centromere and telomere sequences and, therefore, can be used to obtain artificial chromosomes.

The above vectors have been extensively utilized for a study of yeast genome and are also used as shuttle vectors, which allow genie sequences to be routinely transferred back and forth between yeast and E. coli cells, provided they contain origins of replication active in both yeast and E. coli. These vectors also have markers enabling the selection of E. coli cells or colonies transformed with these vectors. The vectors can therefore be amplified in E. coli and then used to transform yeast cells (Fig.).

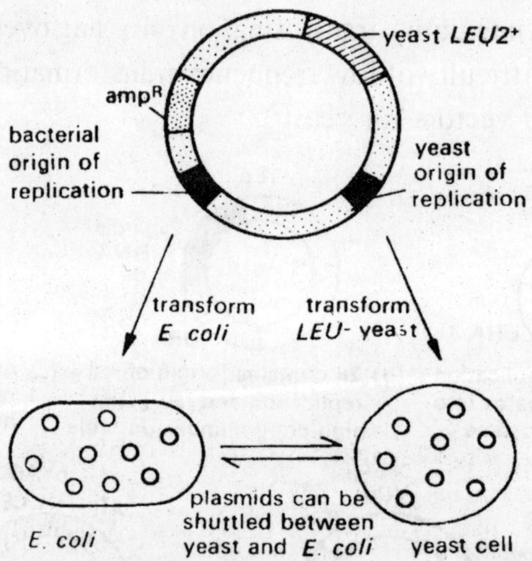


**Fig.** A shuttle vector having both yeast and bacterial origins of replications

Retriever vectors are another class of vectors, which are used to retrieve specific genes from the normal chromosome of an organism like yeast through recombination. They will multiply both in E. coli and yeast. In such a vector, the gene in question is removed by enzymes to produce gapped vector having flanking sequences. Through recombination, they acquire lost sequences from the host. Retriever vectors are very useful in isolation of genes from yeast for further molecular studies including sequencing.

Ti and Ri plasmids as vectors for higher plants. Among higher plants, Ti plasmid of Agrobacterium tumefaciens or Ri plasmid of Agrobacterium is the best known vector. TDNA, from Ti or Ri plasmid of Agrobacterium, is considered to be a very potential vector for cloning experiments with higher plants. This will involve the following steps: (i) foreign DNA has to be first cloned into T-DNA of Ti or Ri plasmid; (ii) modified hybrid T-DNA can be transferred to the genome of plant cells by Agrobacterium infection. Recombinant Ti plasmids can also be obtained by cloning of a large fragment of Ti plasmid in pBR322. These recombinant Ti plasmids can then be used for transformation of higher plants. This system can be widely used, since Agrobacterium infects nearly all dicotyledonous plants. Such cloning in plants has proved to be of immense use to modify agricultural plants to increase their productivity. Monocotyledonous plants to which our cereal crops belong are, however, difficult materials to be used for improvement through such techniques; some success, however, has been achieved in cereals also in recent years.

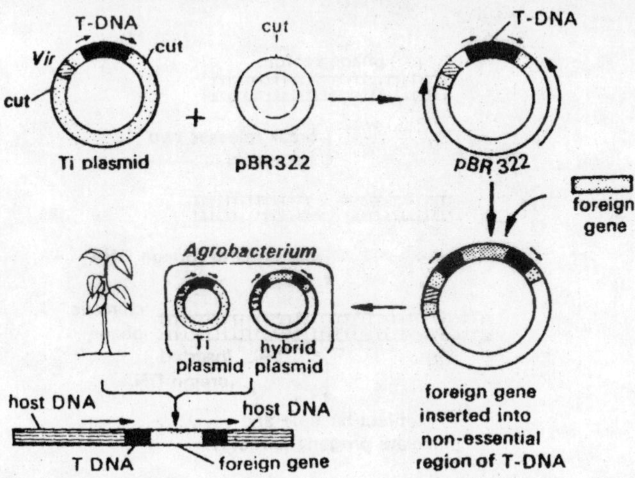


**Fig.** Derivation of recombinant Ti plasmid and its use in transformation of higher plants

## Bacteriophages as vectors

Bacteripohages provide another source of cloning vectors. Since usually a phage has a linear DNA molecule, a single break will generate two fragments, which are later joined together with foreign DNA to generate a chimeric phage particle (Fig.). The chimeric phage can be isolated after allowing it to infect bacteria and collecting progeny particles after a lytic cycle. The use of phage particles as vector imposes a limitation on the size of foreign DNA, which can be cloned, because the capacity of phage head is only limited, and if the size of foreign DNA is too long, size of phage DNA may not be accommodated in phage head. Inorder to overcome this problem, those segments of DNA, which do not contain essential genes, may be removed. Such a technique has been followed in phage lambda (i) to create a smaller vector, genomic having single restriction site for the enzyme Eco RI. Since the reduced wire also fails to be adequately packed in phage head (there is also a requirement of a minimum size of DNA), this automatically provides a selection method, in which only the chimeric particles will be obtained in the phage progeny, and the vector particles lacking cloned segment will be eliminated due to its reduced size.

Lambda phage vectors (Agt10, igtll, EMBL3, EMBL4, (baron). Plasmid vectors described in the previous section are often used for cloning DNA segments of small size (upto 10 kilobases). However, while preparing a genomic library in a eukaryote, the cloned fragements should be large enough to contain a whole gene. This will also allow cloning of the whole genome into a number, which will not be unreasonably large and therefore can be screened without serious difficulty.

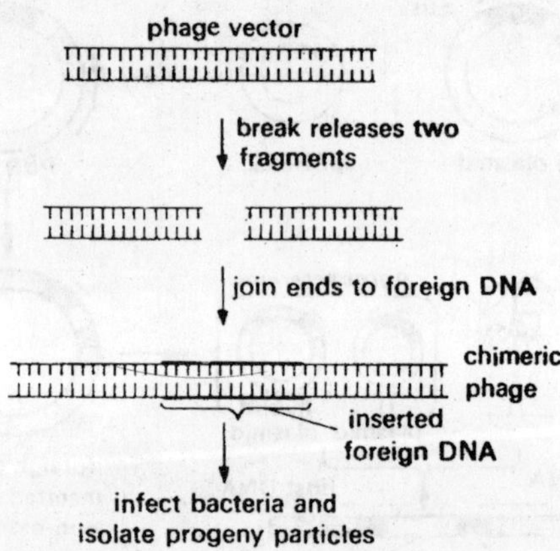


**Fig.** Cloning of foreign DNA in a nonessential region of phage vector

The above properties and other requirements of cloning whole genome in eukaryotes are fulfilled by the phage lambda and cosmid vectors, the former permitting cloning of segments upto 20-25kb long (kb =kilobases) and latter accommodating segments upto 45kb long. Phage lambda (λ), however, is easier arıd more efficient for making genomic and cDNA libraries.

(a) lgt 10 and λgt11. λgt 10 and λgt11 are modified lambda phages designed to clone cDNA fragments. The major difference between these two vectors is that .l gt11 is an expression vector, where inserted DNA is expressed as ß galactosidase fusion protein.

lgt10 is a 43 kb double stranded DNA for cloning fragments that are only 7 kb in length. The insertion of DNA inactivates cl+ (repressor) gene generating a cl bacteriophage. Nonrecombinant lgt10 is cl+ and forms cloudy plaques on appropriate E. coli host, while recombinant cl .1gt10 forms clear plagues permitting screening of recombinant plaques. Further, in an E. coli strain carrying hf IA150 mutation (high frequency, lysogeny mutation) only cl phage will form plaques, because cl+ will form lysogens (integrate with bacterial genome) and will not undergo lysis to form any plaques. Recombinant Agr10 plaques can thus be easily selected.

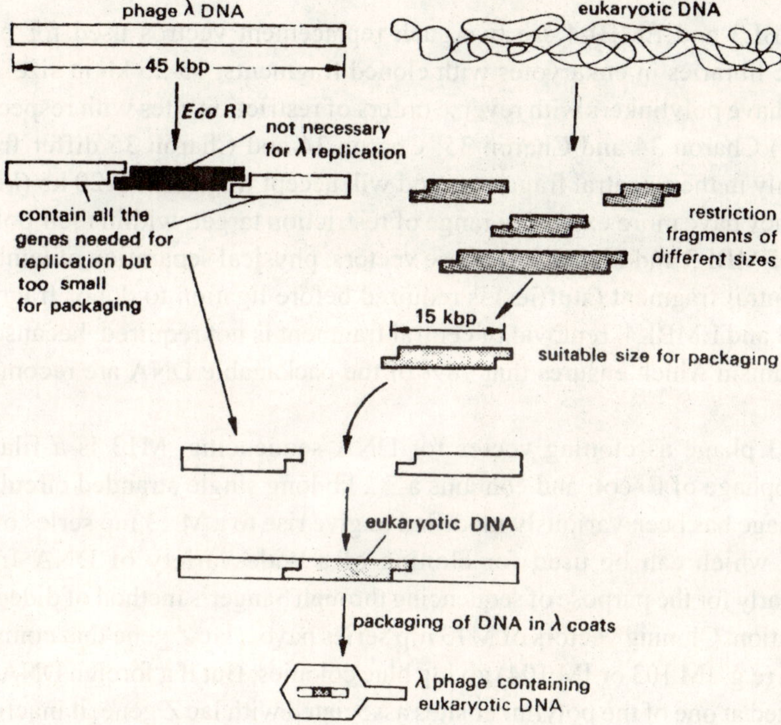


**Fig.** Use of lambda (A) phage as vector for cloninb a eukaiyotic DNA fragment.

*lgt*11 is a 43.7 kb double stranded A phage for cloning DNA segments, which are less than 6 kb in length (usually for cDNA). Foreign DNA can be expressed as 6 galactosidase fusion proteins. Recombinant lgt 11 can be screened using either nucleic acid or antibody probes lgt10 can not be screened using antibodies). The recombinant .lpl l becomes gal-,while non-recombinant lgt11 remains gal+, so that an appropriate E. coli host, with recombinant phage (,gal-) will form white or clear colonies and that with non-recombinant phage (gal+) will form blue colonies permitting screening in the presence of IPTG (inducer) and Xgal (substrate). (b) EMBL3 and EMBL4 (replacement vectors). EMBL3 and EMBL4 are two vectors that are designed so that a central non- essential part of 44 kb long phage can be replaced by a foreign DNA. Cleavage of the phage with an appropriate enzyme generates three fragments (left arm, right arm and a central fragment called staffer). The central fragment representing 40% of the phage genome is nonessential for propagation of the phage and can be replaced by foreign DNA that may be as long as 20-23 kb. Nonrecombinant DNA resulting from fusion of left arm and right arm will give a DNA that is too small for packaging, so that there is automatic selection against non-recombinant phages.

EMBL3 and EMBL4 are two such replacement vectros used for preparing genomic libraries in eukaryotes with cloned fragments, 15-25 kb in size. The two vectors have polylinkers with reverse crders of restriction sites with respect to each other (c) Charon 34 and Charon 35: Charon 34 and Charon 35 differ from each other only in their central fragments and will accept fragments 9-20 kb (kilobases) long. They have more extensive range of restriction target, within their polylinkers than do EMBL3 and EMBL4. In these vectors, physical separation of lambda arms from central fragment (stuffier) is required before ligation to donor fragments. In EMBL3 and EMBL4, removal of central fragment is not required, because there is a mechanism which ensures that 98% of the packagable DNA are recombinants.

M13 phage as cloning vector for DNA sequencing. M13 is a filamentous bacteriophage of E. coli and contains a 7.2 kb long single stranded circular DNA. M13 phage has been variously modified to give rise to a M 13 mp series of cloning vectors which can be used for cloning of a wide variety of DNA fragments particularly for the purpose of sequencing through Sanger's method of dideoxy chain termination. Cloning vectors of M13 nip series have ,, lac Z gene that complements gal host (e.g. JM 103 or JM 104) giving blue colonies. But if a foreign DNA segment is inserted at one of the polylinker sites associated with lac Z gene, it inactivates lac Z gene and no complementation is possible. Thus, on transformation, only white or clear plaques are obtained thus permitting selection of recombinant M13mp plaques. Once the foreign DNA is cloned in M13 vector, commerically available oligonucleotide primers are used for copying the insert in presence of dideoxynucleotides, so that fragments of different sizes with known termini are produced, permitting the determination of the sequence. Reversed order of the restriction sites in polylinker is present in a pair of vectors like M13mp8 and M13mp9 permitting sequencing from both the ends of double stranded DNA molecule.

The extension of the commerical primer or cloned fragement in M13 vector in the presence of radioactively labelled nucleotides will also alloy generation of radioactive single stranded probes.

Cosmids as vectors—Cosmids are plasmid particles, into which certian specific DNA sequences, namely those for cos sites are inserted. Since these cos sites enable the DNA to get packed in lambda particle, cosmid allow the packaging of DNA in phage particle in vitro, thus permitting their purification. Like plasmids, these cosmids perpetuate in bacteria an do not carry the genes for lytic development. The advantage of the use of cosmids for cloning is that its efficiency is high enough to produce complete genomic library of 106 - 107 clones from a mere 1 μg of insert DNA. The disadvantage, however, is its inability to accept more than 40-5 kbp of DNA. Bacteriophage P1 system and F-factor based vectors can allow cloning of DNA segments, as large as 100 to 1000 kbp (or 1 Mbp = $10^6$bp ) length. These are therefore, preferred.

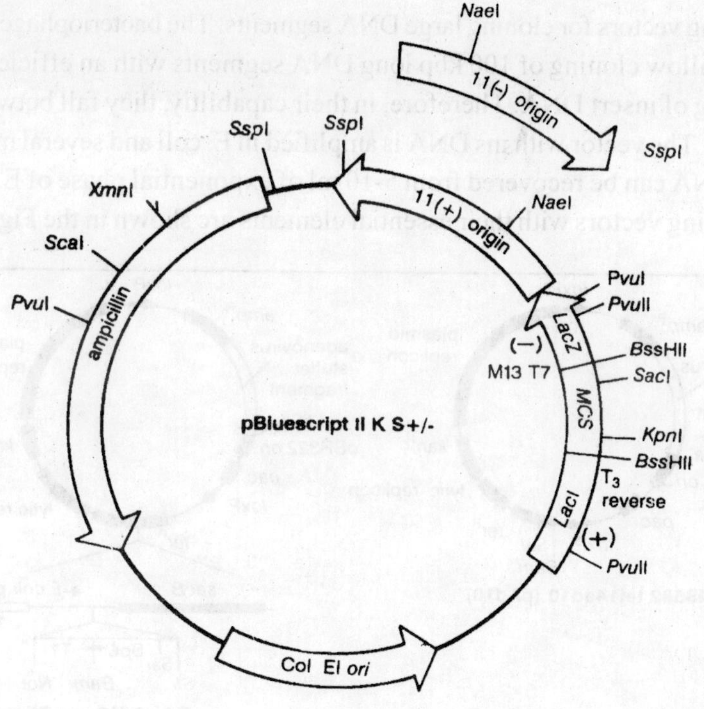


**Fig.** Phagemid pBluescript IIKS +/- and its major characteristics

Phagemids as vectors—Phagemids are prepared artificially by combining features of phages with plasmids, as the name suggests. One such phagemid, which is commonly used in molecular biology laboratories, is pBluescript II KS, which is derived from pUC19 and is 2961 by (bp = ba pairs) long. The KS designation indicates the orientation of polylinker, such that the transcription of lac Z gene proceeds from the restriction site for Kpnl to that for Sncl. The detailed structure of pBluescript II

1 (+/–) is given in Figure. It may be noted that it has the following features: (i) a multiple cloning site (MCS) flanked by $T_3$ and $T_7$ promoted to be read in opposite directions on the two strands, (ii) an inducible promoted (Lac 1), upstream of lac Z region, which complements wish coli (lac $Z^-$) and provides the facility for selection of chimaeric vector DNA (recombinant vector) using the criterion of white colonies (as against blue colonies obtained if no foreign DNA is inserted); (iii) *fl* (+) and *fl*(–) origins of replication dervied from a filamentous phage for recovery of sense (+) and antisense (–) strands of lac Z gene, when host coinfected with a helper phage; (iv) an origin of replication (CoIE l ori derived from plasmid, and used in the absence of helper phage; (iv) gene for ampicillin resistance for antibiotic selection of chimeric phagemid vector.

PI cloning vectors for cloning large DNA segments. The bacteriophage Pl cloning vectors can allow cloning of 100 kbp long DNA segments with an efficiency of $10^5$ clones per μg of insert DNA. Therefore, in their capabiltiy, they fall between YACs and cosmids. The vector with ins DNA is amplified in E. coli and several micrograms of cloned DNA can be recovered from 5-10 ml of exponential phase of E. coli cells. Two P1 cloning vectors with their essential elements arc shown in the Figure below.

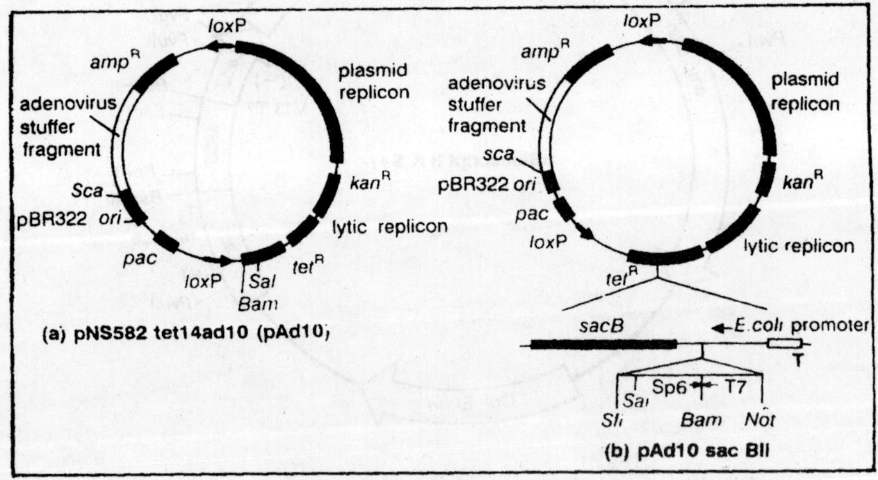


Fig. P1 cloning vectors with their major characteristics

F-factor based vectors (which behave like a phage)—F-factor based vectors have recently been developed for cloning large DNA segments (125 kbp) in E. coli. The cloning of large DNA segments is achieved by a method called chromosomal building, in which through repeated recombination, size of cloned segment can be increased. These bacterial vectors will complement the YAC (yeast artificial chromosomes; sec later in this chapter for details) vectors for cloning segr ents larger than 100 kbp in length and offers some advantages over YAC system. These vectors have already been used for cloning large DNA segments from the bithorax gene of Drosophila.

**Plant and animal viruses as vectors**

A number of plant and animal viruses have also been used as vectors both for introducing foreign genes into cells and for gene amplification and expression in host cells. In the latter case, we may be interested only in getting increased quantity of the gene product and may not be intereseted in the integration of the foreign gene in the host genome. Some of the viruses that are commonly used as vectors will be described in this section.

Plant viruses (cauliflower mosaic virus or caMV and geminiviruses). Cauliflower mosaic virus (CaMV), tobacco mosaic virus (TMV) and geminiviruses are three groups of viruses that have been used as vectors for cloning of DNA segments. CaMV infects particularly the members of Cruciferac and has a double stranded DNA molecule, 8kbp in size whose sequence is now known. Following infection, the virus spreads systematically throughout the plant in a very high copy number reaching upto 105 virus particles per cell. These features make CaMV a suitable vector for transformation of higher plants although there are instances of the use of other viruses also as vectors leading to the production of' transgenic plants.

Geminiviruses comprise a group of single stranded DNA plant viruses causing important diseases in cassava, maize and other cereals. They replicate via double stranded DNA forms and have been subdivided into two groups, one infecting Monocots and transmitted by leafhoppers and the other infecting dicots and transmitted by whitefly (Table). The dicot geminiviruses have two DNA segments in their genome, DNA A and DNA B, both essential for infection by mechanical inoculation. DNA A carries all genes responsible for replication.

It has been shown that deletions in coat protein gene do not destroy infectivity of the virus, which means (hat long deletions in this region can be used to insert foreign DNA to obtain gene constructs that will be extremely useful as amplification and expression vectors in plants. Further, since the coat proteins are the most abundant proteins synthesized by the virus, promoters of coat protein genes can be used for high level expression of cloned genes. CLV and TGMV have already been leafed for their ability to allow multiplication and expression of a cloned gene.

**Table**

Sonic important geminiviruses

| *Geminiviruses infecting monocots* | *Geminiviruses infecting dicots* |
|---|---|
| 1. Maize streak virus (MSV) | 1. Cassia latent virus (CLV) |
| 2. Wheat dwarf virus (WDV) | 2. Tomato golden mosaic virus (TGMV) |
| 3. Digtaria streak virus (DSV) | 3. Bean golden mosaic virus (BGMV) |
| 4. Chloris striate mosaic virus (CMSV) | |

Animal viruses—A number of animal viruses are also used as vectors, either for the delivery of nucleic acid into cultured cells followed by its integration with host genome or for the amplification and high level expression of foreign gene, using the promoters from virus genes. These cloned genes can have a variety of uses including gene therapy in mammalian cells and synthesis of important proteins by cloned genes in cell cultures. Some of these viruses will be discussed later in this chapter, because they are used as expression vectors.

## Transposons as vectors

Transposons of higher plants (Ds, Ac or Mu 1 of maize). Ac and Us are popular transposons in corn, and were earlier known to represent activator-dissociation (Ac-Ds) system. Each represents a transposon with short terminal repeats enclosing a long DNA segment, which measures more than 4500 by in Ac and about 400 by in Ds. Each possesses genes including the gene for transposes enzyme responsible for transposition. Part of this region can be deleted and the transposon can be used for cloning of foreign DNA segment in much the same way as in other cases.

Transposons of Drosophila (P elements) P element of Drosophila consists mainly of 31 by (base pairs) terminal inverted repeats enclosing a 3 kilobase protein coding region, coding for transposase and a repressor of transposition. Many internally deleted P elements are also found, where the transposase derived from other complete P elements help their mobilization. In these deleted P elements, gene for transfer can be inserted to yield recombinant P element, which can be microinjected into the fertilized egg, along with normal P element (helper P to supply transposases). The inserted gene is transposed onto an embryonic chromosome.

## Artificial chromosome (YAC and MAC) vectors for cloning large DNA segments

We know that in plasmids, sequences upto IU-15 kbp in lambda (a) phage sequences upto 22kbp and in cosmids sequences upto 40kbp can be cloned. Yeast artificial chromosome (YAC) vectors, mentioned earlier allow cloning of sequences that are several hundred kilobase pairs (upto 1000 kbp— 1 Mbp) long (Fig.). These long molecules of DNA which may represent whole chromosomes can be cloned in yeast by ligating them to vector sequences that allow their propagation as linear artificial chromosomes. These long DNA molecules can be generated and allow construction of comprehensive libraries (with large DNA segments) in microbial hosts. With the isolation of mammalian telomere and centromere, mammalian artificial chromosomes (MACS) will also be produced in future.

YACs have two disadvantages. (i) Cloning efficiency is low (1000 clones/µg DNA as against $10^6 - 10^7$ clones/µg DNA for cosmids), thus making it impractical to generate complete genomic libraries through the use of YACs. (ii) It is not possible to recover large amount of pure insert DNA from individual clones; selective amplification of YAC's DNA has recently allowed this problem to be overcome. Both these problems have been overcome in bacteriophage Pi system, described earlier.

## Expression Vectors For High Level of Expression of Cloned Genes

In a number of cloning vectors described earlier in this chapter, the main utility has been to clone a gene or a DNA sequence. In other words, they are used for obtaining millions of copies of the cloned DNA stagment for further use in either the genetic engineering experiments or for further basic studies. With this objective in mind, the cloned genes in these vectors need not express themselves either at the transcription level (production of mRNA) or at the translation level (production of protein). But, when the cloned gene is used for transformation to generate transgenic plants or animals or for production of microbes to be used in industry, then the cloned gene must be expressed. Sometimes, a high level of expression of gene is desirable if the product of cloned gene is to be recovered as a commercial product. This objective can be achieved through the use of promoters and expression cassettes (an expression cassette is a combination of DNA sequences, which allow faithful expression of the cloned gene).

### Promoters

Promoters from non-plant sources express poorly in plants (except promoters of T-DNA) and promoters of non-animal sources express poorly in animals. Therefore, to obtain correct expression, genes transferred to plants should be linked to plant-specific promoters and those transferred to animals should be linked to animal-specific promoters. These promoters can also be interchanged so as to confer on the transferred gene a specific pattern of expression (tissue specific or stimulus specific or devlopmental stage specific expression). To illustrate this, some examples of promoters used in plants and insects will be discussed

Nopaline synthase (nos) promoter from T-DNA—This promoter is 200 by long and contains several DNA sequence motifs which direct the expression of linked gene. The upstream element from -97 to - 130, when duplicated, increases the expression three fold, suggesting that this may be an enhancer. The nos promoter is most active in basal regions of plants, its activity slowing down in vegetative parts at the onset of flowering, even though in the flower its activity is increased.

Dual promoter of mannopine synthase (mas) genes 1 and 2—These promoters are present in a T-DNA fragment 407 by long. The expression seems to be devlopmentally regulated and the linked genes seem to be most active in the basal region of the plant. Expression is also induced on wounding. Since they are closely linked, if a marker gene is linked to one of these two promoters and a foreign gene is linked to the other promoter, selection for high expression of marker gene will lead to high level of expression of the foreign gene.

35S RNA promoter of CaMV—This is the most extensively used promoter in a wide variety of plants. The promoter is 343 by long and contains a strong transcriptional enhancer, which on duplication results in 10 fold increase in expression, even at a distance of 2 kbp. High expression is mainly observed in leaf tissue.

Polyhedrin promoter from baculovirus—Baculovirus can synthesize polyhedrin protein in infected cells (baculovirus is an insect virus), so that polyhedrin makes upto 50% of the total protein in the cell, giving upto 1g of protein per liter of insect cell culture. In view of this, the promoter has been used to construct expression vectors to allow high level of expression of any gene under the influence of this promoter. This is being used for developing biopesticides or even for production of specific chemicals in industry.

## Expression Cassettes

The literal meaning of cassette is 'a device containing film such as magnetic tape for insertion into an equipment like tape recorder, VCR, etc.' Therefore by expression cassette, we mean gene constructs, which allow the insertion of foreign genes, either as transcriptional or translational fusions, behind specific promoters.

pRT plasmids are typical in this connection and have been derived from pUC18/19. They consist of a series of vectors (e.g. pRT100, pRT101, pRT102, pRT103, pRT104), which differ in polylinker sequences, each flanked by 35S promoter of CaMV on one end and a sequence for poly A addition on the other end (Fig.).

A variety of marker genes (cat, nptII, gus, etc. ) have been inserted into these cassettes and their expression studied, both in protoplasts (transient expression) and in stable transgenic tissues.

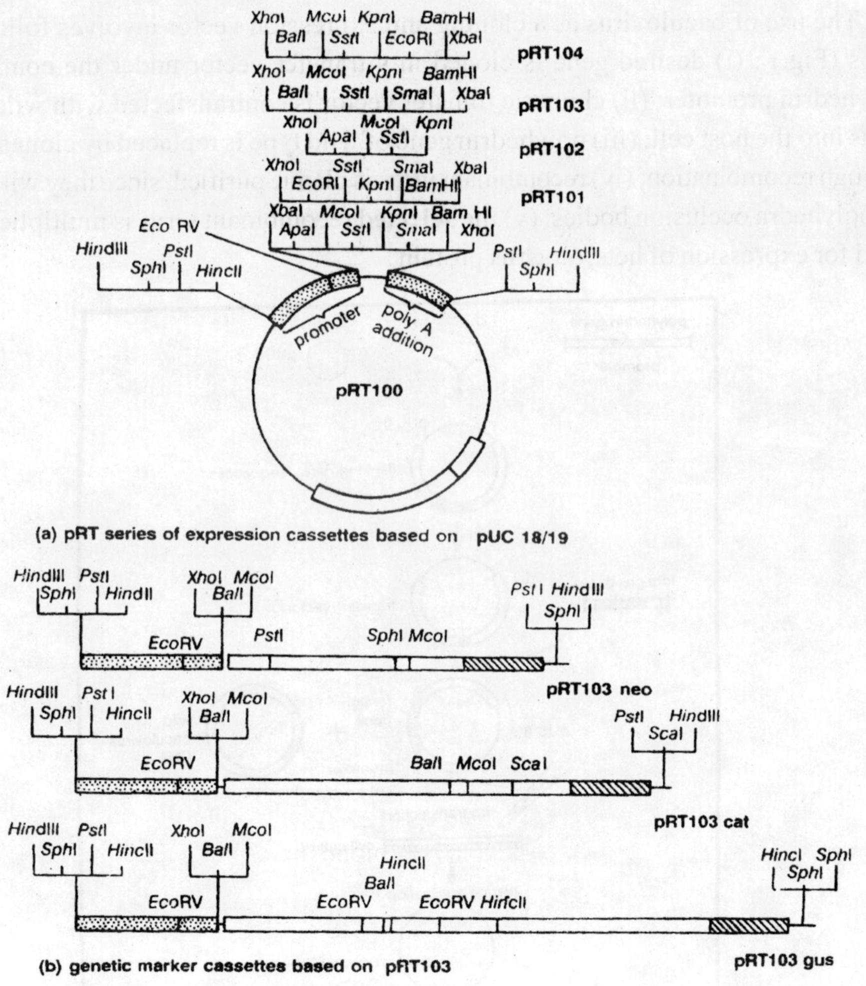


**Fig.** Expression cassettes for plant cells: (a) pRT series with different polylinkers flanked by promoter and poly A addition sequences from 35S RNA of CaMV; (b) Cassettes with three genetic markers (*nptII, cat and gus)* based on pRT 103.

**Baculovirus, an expression vector system for insect cells**

Development of baculovirus expression vector system is based on its life cycle which involves the following steps: (i) in early phase (12 hours), virus particles are budded from infected cells and spread the virus; (ii) in the late phase (2-5 days), occluded virus units accumulate in the host nucleus and get embedded in protein polyhedra; (iii) insect is decomposed releasing polyhedra; (iv) polyhedra are ingested by another insect, where polyhedra are dissolved releasing virus for multiplication.

The use of baculovirus as a cloning and expression vector involves following steps (Fig.) : (i) desired gene is cloned in a transfer vector under the control of polyhedrin promoter; (ii) chimeric transfer vector is contransfected with wild type virus into the host cell; (iii) polyhedrin gene of wild type is replaced by cloned gene through recombination; (iv) recombinant virus is plaque purified, since they will have no polyhedra occlusion bodies; (v) the selected recombinant virus is multiplied and used for expression of heterologous protein.

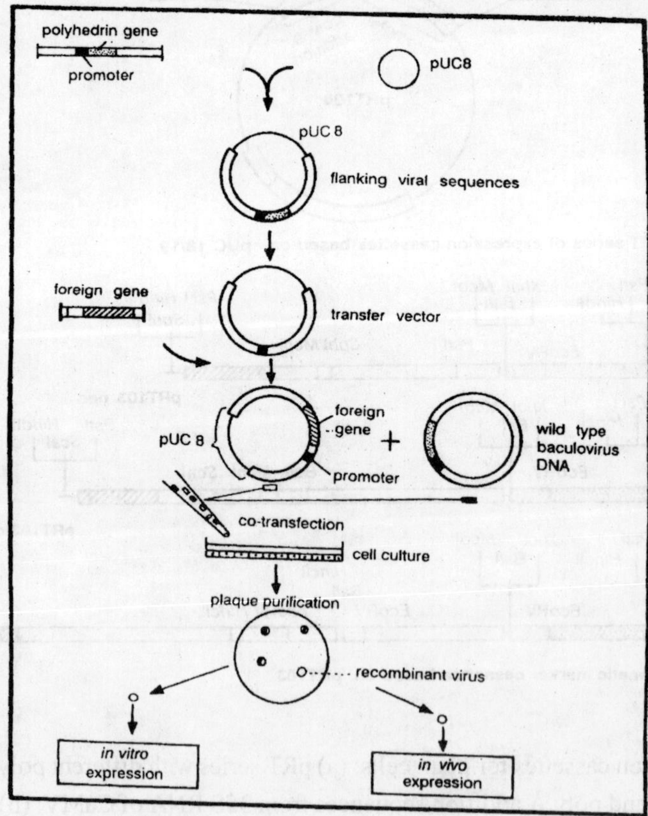


**Fig.** Construction of baculovirus expression vectors

**Virus expression vectors for mammalian cells**

A number of viruses have also been used as transfer or expression vectors for mammalian cells. These include the following: (i) The first viruses used for mamalian cells included DNA viruses like papovaviruses (SV40 and polyoma) and adenoviruses, but since their capacity for foreign genes is small, they can not be used for the expression of many genes which are failrly large in size; (ii) renovintses (murine and avian retroviruses) are very useful vectors, since they can infect broad variety of cell types and can be used for cloning !arge genes; however, being single stranded RNA (Table), they can not replicate without being integrated so that we will need replicating cells for expression or alternatively high-titer (high concentration) for fully differentiated cells (e.g. neurons, hepatocytes, etc.); (iii) adeno-associated virus (AAV) overcomes the limitations associated with retroviruses even though, like retroviruses, it is an integrating virus, with 5kb single stranded DNA; it is nonpathogenic, and can be concentrated to high titer (109 particles/µl); (iv) herpes virus, is another important non-integrating virus of large size (150 kb), which can be used for expression of large intact genes.

Several of the above viruses have already been used and showed expression of cloned genes, but they need to be further characterized for their long term usage. Some of them are also being used for gene therapy.

**Table**

Transforming viruses used as vectors in mammalian cells.

| *Viral class* | *Type of virus* | *Genome size* |
|---|---|---|
| 1. Adano | DNA duplex | 37 kb |
| 2. Papova | DNA duplex | 5-6 kb |
| 3. Herpes | DNA duplex | 160 kb |
| 4. Adeno associated | DNA (single strand) | 5 kb |
| 5. Retro | RNA (single strand) | 6-9 kb |

**Binary and Shuttle Vectors**

For the purpose of gene transfer in higher plants, a number of binary vectors were developed. These were based on the pPC'V (plant cloning vector) series of plasmids (Fig.). These vectors contain a conditional mini-RK2 replicon, which is maintained and mobilized by traps-acting functions derived from the plasmid RK2 replicon (origins-ori, and $ori_T$). The plasmid RK2 was introduced into both E. coli as well as Agrobacterium (GV 3101, which contains a Ti plasmid lacking T DNA) to facilitate replication of binary vectors in both these hosts, so that the vector can be maintained and shuttled between both—hence the name shuttle vector also. Similar binary or shuttle vectors for maintenance and transfer between E. coli and yeast cells have also been developed.

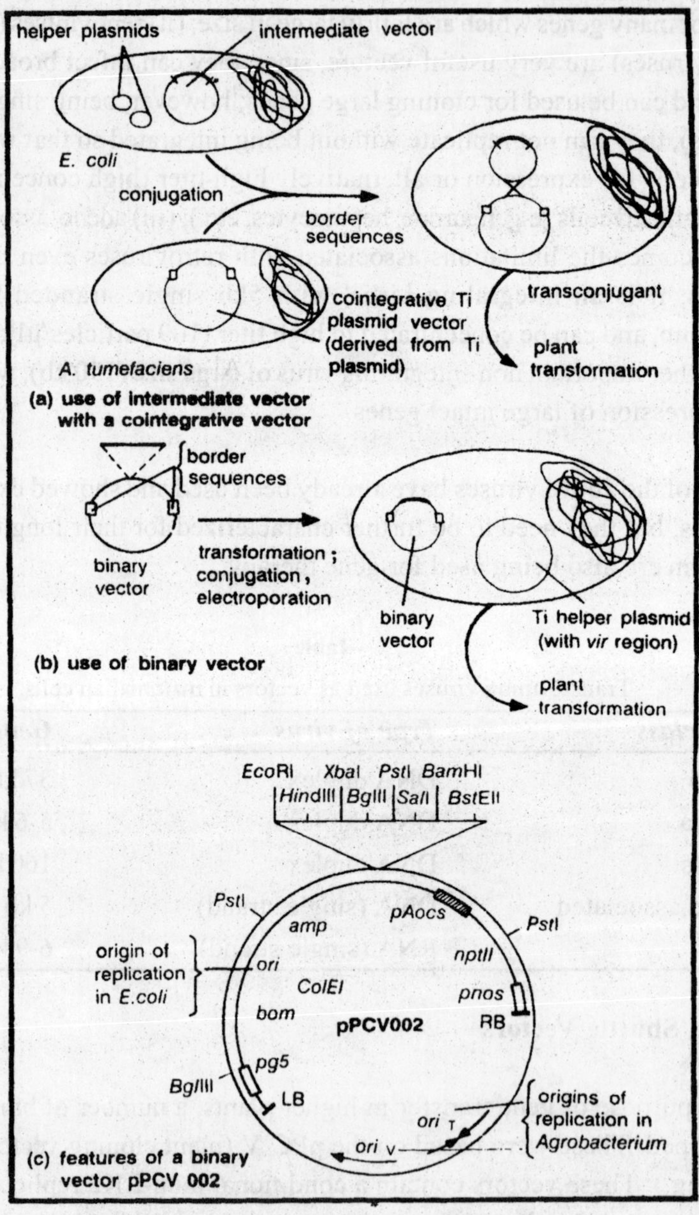


Fig. Agrobacterium based plant transformation vectors

## Restriction Mapping

We have studied above about a variety of cloning and expression vectors, which are used for a variety of experiments including (i) cloning of DNA segments or genes through construction of chimeric DNA molecules; (ii) use of cloned DNA or RNA segments as molecular probes and (iii) preparation of molecular maps and gene libraries, the latter to be used for isolation of genes. These different uses of vectors will be briefly described in this chapter. A brief account on restriction enzymes is also included in this chapter, since these enzymes are used regularly as a routine in all DNA cloning and recombinant DNA experiments.

### Restriction Enzymes for Cloning

For cloning of DNA, often we need to cut the DNA at specific sites, which are recognized and cleaved by specific enzymes (endonucleases, which cleave DNA at internal sites), described as restriction enzymes. Also by locating the positions of cleavage, sites of a number of restriction enzymes, restriction maps can be prepared (see later in this chapter). These restriction enzymes recognize short sequences of double stranded DNA as targets for cleavage. Different enzymes recognize different, but specific sequences, each ranging in length from 4 to 8 base pairs. The enzymes are named by a three letter (or four letter) abbreviation (letters are italicized) that identifies their origins.' Roman numerals (I, II, 111, etc.) are added to distinguish several enzymes with same origin. For instance EcoRI is derived from E. coli and Hpal is derived from Hemophilus parninfluenzae.

Besides cleavage, modification in the form of methylation is also brought about by some enzymes called modification enzymes (sometimes also called methylases). This methylation distinguishes genes in different states of functioning. There are also enzymes which perform the function of both restriction (cleavage) and modification (methylation). Based on these attributes, restriction enzymes have been grouped into two classes: (i) type II restriction enzyme systems (e.g. EcoRI) which have separate enzymes for modification and restriction; and (ii) type I (EcoK and EcoB) and type III (EcoP1, EcoP15) enzyme systems, in which same enzyme possesses both activities (bifunctional), although the restriction and modification sites differ in position.

Of the above two classes of restriction enzymes, type II enzymes are most important for cloning purposes. The target sites for these enzymes are 4-8 by with a symmetry (palindromes). Most of type II restriction enzymes cleave DNA at unmethylated target site. Some enzymes introduce staggered cuts, others generate blunt ends. Enzymes with 4 by target sites are used when frequent cuts are desired and those with 8 by (eg. Not I) are used when rare cuts are desired to get long DNA segments. Otherwise, majority of enzymes used have 6 by target sites. Some of them can cleave both methylated as well as unmethylated targets, but majority of them cleave only unmethylated targets. Details of some of these enzymes are listed in Table.

**Table**

Recognition sequences of several restriction endonucleases (arrow indicates the position of cal).

| *Microbial origin* | *Enzyme* | *Recognition site (two fold symmetry)* |
|---|---|---|
| 1. Escherichia coli | *Eco*RI | ↓<br>S'GAA TTC, 3'<br>3' CIT AAG 5'<br>↑ |
| 2. flemophilus influenzae | *Hind*II | ↓<br>5' GTPy PuAC 3'<br>3' CAPu PyTG 5'<br>↑ |
| | *Hind*II | ↓<br>S' AAG CTT 3'<br>3' TTC GAA 5'<br>↑ |
| 3. Homophilus parainfluenzae | *Hpa*I | ↓<br>S' GTT AAC 3'<br>3' CAA TTG 5'<br>↑ |
| | *Hpa*II | ↓<br>5' CC GG 3'<br>3' GG CC 5'<br>↑ |
| 4. Hemnphilus aegyptians | *Hae*III | ↓<br>5' GG CC 3'<br>3' CC GG 5'<br>↑ |
| 5. Moraxella spp. | *Msp*I | ↓<br>5' CC GG 3'<br>3' GG CC 5'<br>↑ |
| 6. Proteus vulgaris | *Pvu*I | ↓<br>5' CGA TCG 3'<br>3' GCT AGC 5'<br>↑ |
| 7. Bacillus amylnliquefaciens | *Bam*H1 | ↓<br>5' GGA TCC 3'<br>3' CCT AGG 5'<br>↑ |
| 8. Streptnmyces albus | *Sal*I | ↓<br>5' GTC GAC 3'<br>3' CAG CTG 5'<br>↑ |
| 9. Prnvidencia stuarhii | *Pst*I | ↓<br>5' CTG CAG 3'<br>3' GAC GTC 5'<br>↑ |

**Technique of Restriction Mapping**

The recombination frequencies, which are the function of distances between genes, are not completely independent of (i) the nature of mutants used, (ii) the position of these mutants on chromosomes, (iii) the genetic background, (iv) the environmental conditions and a variety of other factors. Consequently, the distances between genes on a genetic map may not correspond to the distances between them on the DNA molecule of which they are a part at the molecular level. Gaps may also be present on a genetic map due to nonavailability of mutants in that region.

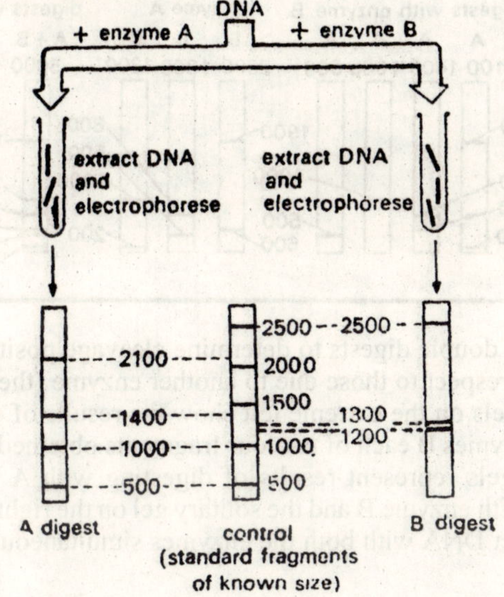


**Fig.** Cleavage of DNA by two restriction endonucleases A and B into fragments which can be separated by agarose gel electrophoresis

At the molecular level, the fine structure of a gene can be studied through determination of nucleotide sequence of the concerned DNA segment. Sometimes this is done through isolation of DNA segment corresponding to a gene followed by sequencing of the DNA segment, which needs time and energy. Instead, we can prepare a map of the DNA by its cleavage at specific sites with the help of restriction endonucleases, which recognize very short specific DNA sequences and cleave the DNA at these specific sites. These sites of cleavage can be identified and mapped to give rise to a restriction map. In a restriction map, the nucleotide sequence of DNA between sites close to one another (300bp or less) can be determined. These regions can then be connected into a sequence of the entire gene, whose nucleotide sequence then can be compared with amino acid sequence of protein coded by it.

This can be further extended to adjoining genes to saturate the physical map, an objective which is difficult, but is being attempted for human genome.

On a restriction map, is found a linear sequence of sites, each for a specific enzyme and the distances between them are measured in terms of number of base pairs of DNA. This technique can be used both in prokaryotes and eukaryotes, although, all regions of chromosomes can not be easily mapped in eukaryotes. Different steps involved in restriction mapping include the following.

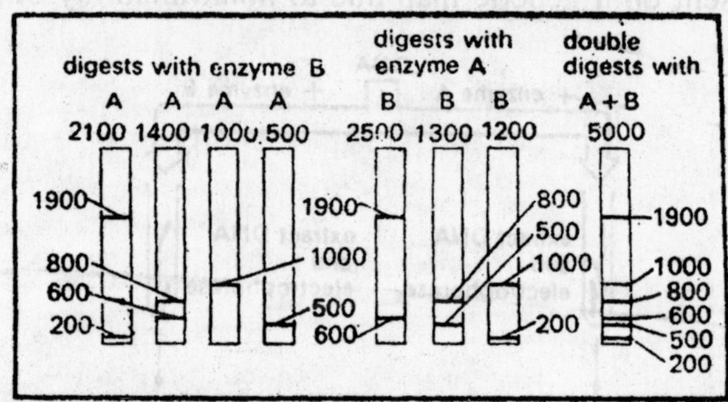


**Fig.** Technique of double digests to determine cleavage positions of DNA due to one enzyme with respect to those due to another enzyme (the two enzymes are A and B); the four gels on the extreme left show the results of electrophoresis after digesting with enzymes B each of the four fragments obtained after digestion with A; three central gels represent results of digesting with A the three fragments earlier obtained with enzyme B and the solitary gel on the right shows the results of digesting the intact DNA with both the enzymes simultaneously).

## Restriction cleavage and gel electrophoresis

If we take a particular DNA molecule or DNA sample, digest it with a specific restriction enzyme and then subject the sample to gel electrophoresis (a technique in which DNA digest is loaded on gel slab and the DNA fragments move under the influence of an electric current), we will notice a series of bands on the gel slab or cylinder. The bands can be observed under ultraviolet light after staining with ethidium bromide. The position of different bands will depend on DNA fragment size, such that smaller the fragment, more rapidly will it move, and longer the fragment, more slowly will it move. It will mean that the fragment away form the loading site will be smaller and those close to the loading site will represent longer DNA fragments. The gel can be calibrated by using a mixture of DNA fragments of known lengths so that the position of the bands on this standard gel can be compared with the bands in the experimental DNA digest and the fragment length in each band of DNA digest can thus be determined sometimes using a computer device. In Figure, we have shown the results of digestion of 5(X)Ubp long hypothetical DNA molecule digested separately by two enzymes A and B. As shown in the figure, the enryme A cleaves the DNA into four fragments of lengtsh 2100, 1400, 1000 and 500bp, while the enzyme B cleaves it into three fragments of lengths 2500, 1300, 1200bp. These data with some additional experiments can be used to generate a restriction map.

## Construction of a Restriction Map

The data on DNA digestion by more than one endonucleases as discussed above can be utilized to arrange the sites of cleavage in a defined order. This can be done by several methods and one of them is illustrated in Figure. This method involves successive digests with two individual enzymes, where we extract each fragment produced in the individual digests with either enzyme A or enzyme B and then cleave it with the other enzyme. The original DNA sample is also digested by a mixture of both the enzymes to confirm the results of individual successive digests.

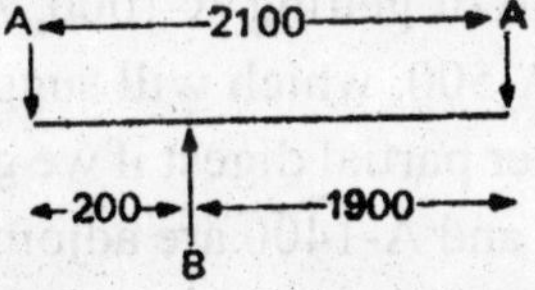


**Fig.** The sites of cleavage by enzymes A and B in one of the four fragments, 2100 bp long obtained due to cleavage by enzyme A; the fragment has a site for enzyme B at 200 bp length from one end.

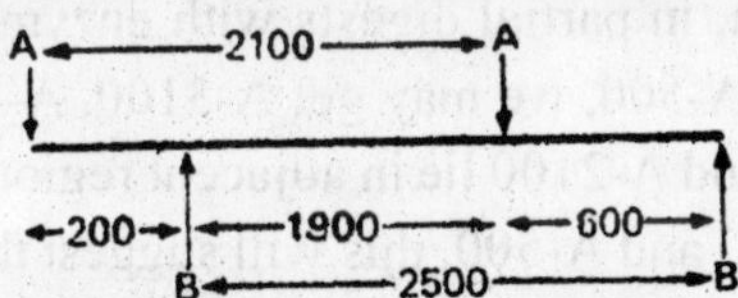


**Fig.** Reconstruction of restriction map showing restriction sites of enzymes A and B in two overlapping fragments (2100 bp fragment obtained due to enzyme A and 2500 pb fragment obtained due to B).

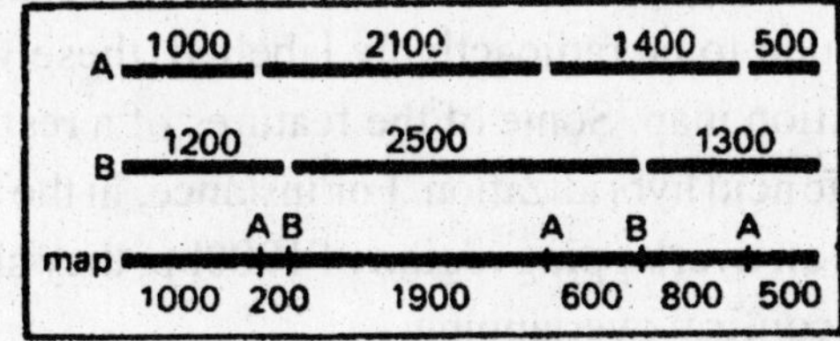


**Fig.** A restriction map, in the form of linear arrangement of cleavage sites; the upper two figures show sites due to individual enzymes and the bottom figure shows arrangement of cleavage sites of both enzymes (see text for details); sites of more enzymes can be added to this map by same technique.

In *Figure*, we have shown results of such an exercise. When the fragment A-2100 is obtained and digested with enzyme B, it is cut into two fragments of 1900bp and 200bp. Fragment B-2500 (obtained from individual digest with enzyme B) is similarly cut by enzyme A into two fragments of 1900bp and 600bp, suggesting overlap of A-2100 and B-2500 *(Fig.)*, in the region of fragment of 1900bp which is obtained by one cut with enzyme A and the other by enzyme B. The DNA fragment of 1900bp is also available, when DNA is digested with a mixture of both enzymes A and B (double digest). It may be seen in *Figure*, that in reciprocal digests (A followed by B and B followed by A) and double digests, same fragments are available. Using this information, we can find the overlapping regions in A and B digests and find out the sites of cleavage by A and B. This will then allow us to prepare the restriction map as shown in *Figure*.

**Use of partial digests, end labelling and hybridization in restriction mapping**

The above technique of individual and double digests can be supplemented with other techniques for actual construction of maps. For instance, by permitting incomplete digestion, fragments longer than those obtained by complete digestion may be obtained. These will be called partial digests. In the example used in the last section, in partial digests with enzyme A, instead of getting A-1000, A-2100, A-1400, A-500, we may get A-3100, A-1400 and A-500, which will suggest that A-1000 and A-2100 lie in adjacent regions. In another partial digest if we get A-1000, A-3500 and A-500, this will suggest that A-2100 and A-1400 are adjoining, which will mean that A-1000 and A-1400 are found on the two sides of A-200. This technique will thus allow the arrangement of fragments in a linear order.

Another useful technique involves labelling of the ends (end labelling) of DNA molecule before digestion, so that the fragments containing these ends can be identified due to labelling, even after digestion. In the above example, using end labelling, if A-1000 and A500 are found to be radioactively labelled, these will be present at the two ends of the restriction map. Some of the features of a restriction map can also be confirmed by nucleic acid hybridization. For instance, in the above example, if A-2100 and B-2500 have an overlapping region of 1900bp, they should hybridize with each other, which will confirm overlapping.

The above technique will help in accurate completion of a restriction map, but will require that we have a complete set of restriction fragments which make the entire DNA region being mapped. Once the restriction map (which is a physical map) is ready, it can be compared with the genetic map. Although large changes already located on the genetic map can be easily located on the restriction map, but point mutations can not be always and easily located on the restriction map, since the restriction sites often do not change due to mutation. In such a situation, one may like to determine the nucleotide sequences on individual fragments and compare them in normal and mutant individuals. Since sequencing of nucleotides is more laborious and can not be easily undertaken for the whole genome, another technique of molecular genetic markers in the form of restriction fragment length polymorphisms (RFLPs) has been utilized during the 1980's to generate genetic linkage maps in human, mouse, rat, pig, sheep, fruitfly (Drosophila), nematode (Coenonhahditis elegans), yeast, maize, wheat, rice, barley, tomato, potato, lettuce, rice, pepper, pea, etc.

## Construction of Chimeric DNA

Chimeric DNA molecules are produced by inserting a foreign DNA segment into the DNA molecule of a vector. These chimeric DNA molecules are the means of cloning genes or other small DNA sequences. The construction of chimeric DNA, as we know now, involves joining of the broken ends of the vector with the two ends of the sequence to be cloned. There are at least three methods available for the construction of this chimeric DNA.

## Palindromes and Staggered Cleavage

This is the most common method, where due to the presence of a palindromic sequence, a restriction enzyme causes staggered cuts producing short complementary single stranded stickly ends (a palindromic sequence is one, which has complementary sequences at the two ends of a single strand, e.g. ATC....,GAT). This can be illustrated using the example of enzyme EcoRI, which cuts the sequences at specific positions (Fig.). When another DNA molecule is similarly cut by the same enzyme, similar sticky ends having same sequences in the single stranded ends will be produced, so that when it is mixed with the previous molecule similarly treated, the two will anneal producing a chimeric DNA or chimeric plasmid if plasmid DNA is involved (Fig.). Enzyme DNA ligase helps in joining the bonds at the cut ends of two molecules.

The above technique has the advantage of regenerating two restriction sites (eg., EcoRI site) in the chimeric DNA, so that the foreign DNA segment can be retrieved rather easily from the cloned copies of chimeric DNA by cleavage again with the same enzyme. There are also following disadvantages with this technique. (i) The two cleaved ends of a vector or of a foreign DNA may join end to end before getting inserted. Therefore, while isolating chimeric DNA, one will have to select chimeric molecules each having only a single insert. This can be achieved by separating molecules of different sizes by gel electrophoresis. (ii) The recognition site, particularly in the sequence to be cloned may not tie at a convenient position, so that sometimes only a part of the desired segment will be inserted.

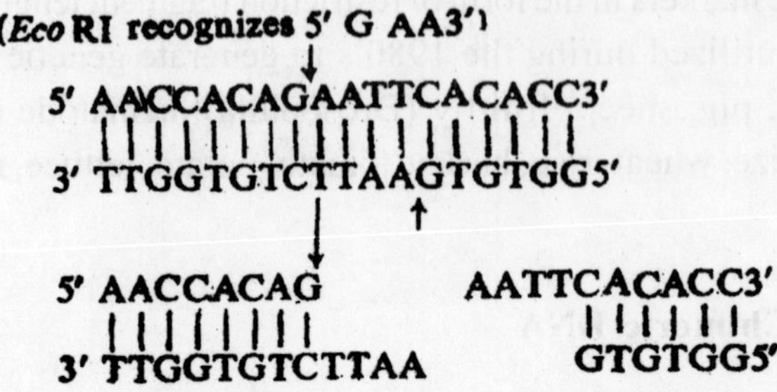


Fig. 3.6. Staggered cleavage of a double stranded DNA fragment, producing single stranded sticky ends

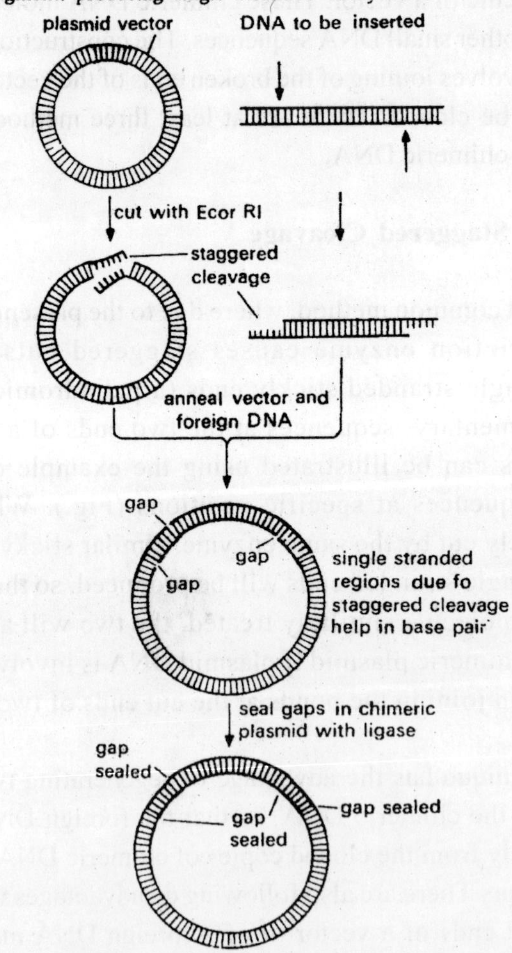


Fig. Cloning of a DNA segment in a plasmid vector (e.g. pBR322) by staggered cleavages, both in vector DNA and foreign DNA to be cloned.

**Adding poly dA at the 3' ends of the vector and poly dT at the 3' ends of the DNA clone**

In this method, DNA can be cut at the desired position both in the vector and in the clone, without staggered cleavage. Using precursor dATP, poly dA is added at both the 3' cut ends of the vector with the help of enzyme terminal transferase. Similarly, using precursor dTTI', poly (IT is added at both the 3' cut ends of the DNA sequence to be cloned. The vector and clone can then be joined by annealing the poly dA with poly dT tails and then ligating them using DNA ligase enzyme (Fly. 3-8). In this techinque, there is no chance of reannealing between the two cut ends of the same DNA molecule or of the two similar DNA molecules, so that one of the disadvantages of the first method has been overcome in this case. However, in this technique, it will not be easy to retrieve the cloned DNA, because the recognition site of the enzyme due to insertion of poly dA and poly dT has been lost in this cast. But if poly dG and poiy dC are used instead (which serve the same purpose), cloned sequence can be retrieved, since poly dG, poly dC regenerate the recognition site for the enzyme PstI.

**Blunt end ligation by T4 DNA ligase**

In this technique, a restriction enzyme is used to cut a duplex DNA at the same place in both the DNA strands. The broken ends are then used for joining with the two ends of another DNA molecule irrespective of the sequences present at the broken ends of the two DNA molecules. The T4 DNA ligase is used for this joining reaction. The disadvantage of this technique is that any two broken ends may join including those belonging to the same DNA molecule. This leads to the production of a variety of products and one will have to select the desired product from a mixture of products.

This blunt end ligation is used for developing a method in which the DNA to be cloned can be easily retrieved whenever required. This method makes use of short DNA duplexes (linkers), that contain EcoRI palindrome or some equivalent palindrome, which being small in size, can be synthesized chemically. These linkers can be linked to the blunt ends of vector DNA or of an insert by blunt end ligation (Fig.). This will allow the creation of an EcoRI site in the linker region of the vector, so that when a DNA segment is now cloned, the inserted DNA segment can be retrieved by cleavage with EcoRI. However, in this method, there is no restriction on the original choice of sites to generate the cut ends. Therefore, with this method it is now possible to insert a foreign DNA segment at a particular site in the linker region of the vector and then retrieve this foreign DNA segment whenever necessary.

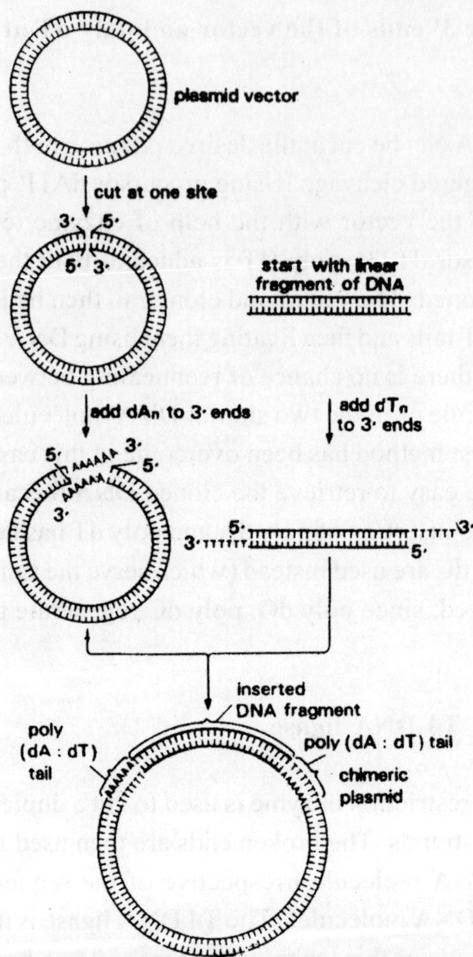


**Fig.** Cloning of DNA segment in a plasmid vector by poly dA : poly dT tailing technique

## Cloning in Bacteria and Eukaryotes

**Cloning in bacteria:** Bacteria are often used as hosts for multiplication of chimeric vectors or recombinant DNA molecules. Therefore, a bacterial strain should be available, which should be the host for the vector and should be modified or seleeted for high transformation efficiency. Bacteria can be classified into. (i) gram negative bacteria e.g. Escherichin coli. Pseudomonas, Rhizobium and (ii) gram positive bacteria e.g. Bacillus subtilis and Streptomycetes sp. Ordinarily same plasmid can not maintain itself in stable form in both classes of bacteria and, therefore, plasmid vector, for two classes usually differ. While gram negative bacteria can be used as hosts for plasmids, phages and cosmids vectors, gram positive bacteria can be used as hosts for only plasmids. Chimeric vector is inserted into bacterial cells and transformed bacterial colonies are selected and used for multiplication in suspension cultures. The chimeric DNA is retrieved, whenever required.

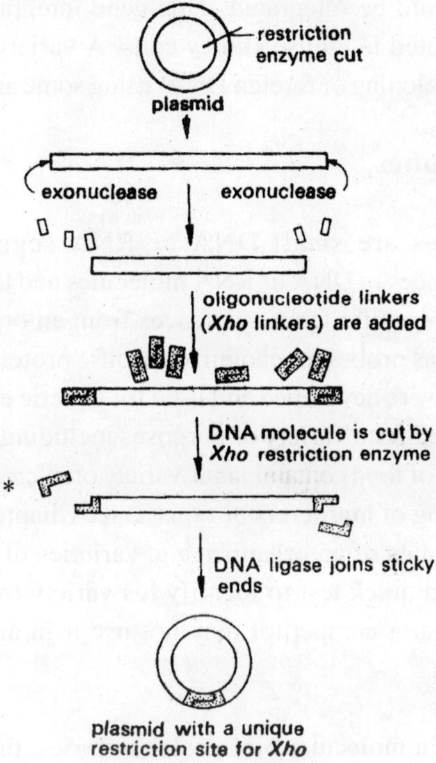


**Fig.** Addition of a "linker" (carrying a restriction site) to a vector molecule

## Cloning in eukaryotes

Since chromosomes found in the nucleus of eukaryotes are separated from the rest of cell through nuclear membrane, and since many of the genes are split genes with exons and introns, genetic engineering with eukaryotes requires new methods and tools. When eukaryotic genes are cloned in prokaryotes, the split genes can not be correctly expressed, because prokaryotes do not have the equipment necessary for splicing out the RNA transcribed from the introns of a gene. In view of this, eukaryotic cells may sometimes be needed for cloning and particularly for expression of cloned eukaryotic genes.

Among eukaryotes, DNA cloning has been done in yeast, mouse and to some extent even in some higher plant species. In yeast, a plasmid called 2μ DNA (63 bp) is found, which is an appropriate cloning vehicle. An efficient transformation method is also available, which involves protoplast production followed by PEG directed introduction of DNA into protoplasts. In animal cells like mouse cells, special animal viruses were used as cloning vehicles. Simian virus 40 (SV40) is one such virus, in which globin gene could be integrated. This gene integrated in SV40) could be transcribed and translated in mouse kidney cells. A variety of human cell lines in culture is also used for cloning of foreign DNA, using some animal viruses as vectors.

## III. Molecular Probes

Molecular probes are small DNA or RNA segments that recognize complementary sequences in DNA or RNA molecules and thus allow identification and isolation of these specific DNA sequences from an organism. Antibodies are also occasionaly used as probes to recognize specific protein sequences. Although, initially, these probes were developed and used for genetic engineering research but are now frequently used for a variety of purposes including diagnosis of infectious diseases, identification of food contaminants, variety of microbiological tests, forensic tests (e.g. fingerprinting of murderers or rapists: see Chapter), etc. Probes can also be used to identify strains of an organism e.g. varieties of a crop species (a plant breeder likes to have a quick test to identify his variety to maintain his patent or breeder's' right, so that a competitor may not use it in another name leading to infringement of rights).

For basic studies in molecular biology laboratories, the molecular probes are frequently used for identification and isolation of genes or related sequences. In theory, any nucleic acid (or rarely protein, e.g. antibody) can be used as probe, provided it can be labelled to permit identification and quantitation of the hybrid molecules formed between the probe and the sequence to be identified. In practice, double and single stranded DNAs, mRNA and other -RNA's synthesized in vitro are all used as probes. DNA/RNA probe assays are faster and sensitive; so that many conventional diagnostic tests for viruses and bacteria involving culturing of the organism, are being fast replaced by antibody and DNA probe assays. While culture tests can take days or even months, molecular probe assays can be performed within few hours or minutes. Therefore, the use of DNA probes has become today's most sophisticated and sensitive technology for a variety of uses involving biological systems both in basic and applied studies including their commerical use.

### Preparation of probes

Genommic DNA probes—Simple, reproducible methods have been devised for the preparation of molecular probes. A random or specific DNA sequence (probe) can be isolated from the genomic DNA (DNA extracted from cell nuclei), using following steps : (i) extract DNA from plant or animal tissue; (ii) digest extracted DNA with a restriction enzyme (e.g. EcoRI or HindIII), which cuts DNA on specific sites or positions where a specific sequence recognized by the enzyme is found; (iii) run the digested DNA on an agarose or polyacrylamide gel for electrophoresis to separate fragments of different sizes; (iv) isolate DNA of specific fragment size from a particular band identified through Southern blots by hybridization with specific labelled mRNA or cDNA molecules, if available. If random probes are needed, no Southern blots is required; (Southern blots are DNA bands transferred from gel to nitro-cellulose membrane with the help of a buffer rising through capillary action; sec later); (v) clone this DNA in a vector (as discussed in Chapter 2,); (vi) allow chimeric vector to infect bacteria for multiplicaton, where it can make billions of copies. The steps (ii) to (vi) above are also used for preparing cDNA probes. From the transformed bacteria, the chimeric vector can be obtained and used in one of the following ways: (i) It may be used directly as probe (the presence of vector DNA will not interfere in probe assays, so that it is not really necessary to remove it). (ii) The cloned segment may be separated (or retrieved), by using the same enzyme which was used for cloning. In the latter case, cleaved chimeric vector DNA will be again electrophoresed for separating the inserted segment on the gel. This inserted segment thus retrieved can now be used as a probe. (iii) The chimeric DNA may be used for PCR, using flanking sequences as primers; the PCR product can be separated and used as a probe.

cDNA probes—A DNA sequence corresponding to a part of a specific genes) (an oligonucleotide) can be obtained by reverse transcription of mRNA. cDNA thus obtained can be cloned and used as a probe.

Synthetic oligonucleotides as probes—DNA probes with known nucleotide sequence can also be synthesized chemically using automated DNA synthesizers now available commercially. These synthetic probes will be efficient only when they are not more than 2(1 to 40 nucleotides in length).

RNA probes or riboprobes—High specific activity RNA probes or riboprobes may also be synthesized from DNA templates cloned in expression vector. This is achieved through RNA synthesized in vitro (on DNA inserted in vector) and labelled simultaneously employing specific RNA polymerise and labelled nucleotides which are supplied by commercial firms in a kit for riboprobe preparation. RNA probes offer several advantages over DNA probes, since these are single stranded, and provide improved signal or hybridization blots. However, widespread presence of ribonucleases creates some problems in their preparation and use, because these ribonucleases will cause degradation of RNA which is synthesized. Therefore, extreme care is needed in the preparation of riboprobe, keeping all glassware ribonuclease free (no glassware should be touched with bare hands, because ribonucleases may be available on our skin).

## Labelling of probes

The detection of homologous sequences after hybridization with the probe is like finding a needle in the haystack. Therefore, for the success of DNA probe assay, it is necessary to develop simple, safe and sensitive techniques for their use. As probes transmit no signal of their own, they have to be (i) either labelled with radioactive isotopes or (ii) methods are used to couple nonradioactive signal molecules to the probes without impairing the hybridization ability of these probes. These signal molecules may include fluorescent antibodies and enzymes that produce colour changes in dyes and chemiluminescent catalysts *(Fig.)*.

Radiolabelled probes—Traditionally radioactively labelled probes are used for a variety of experiments. $^{32}P$ is the most commonly used radioisotope, although other labels such as $^{3}H$, $^{35}S$ and $^{125}I$ have also been used. Conventional labelling replaces a proportion of the nucleotides in n nucleic acid sequences with $^{32}P$ derivatives (e.g. $^{32}P$dCTP) or adds $^{32}P$ to the end of the molecule. At least three methods are available for labelling of the probe. (i) nick translation involves creation of nicks in the probe DNA followed by extension of the broken ends using a labelled (BC-22)

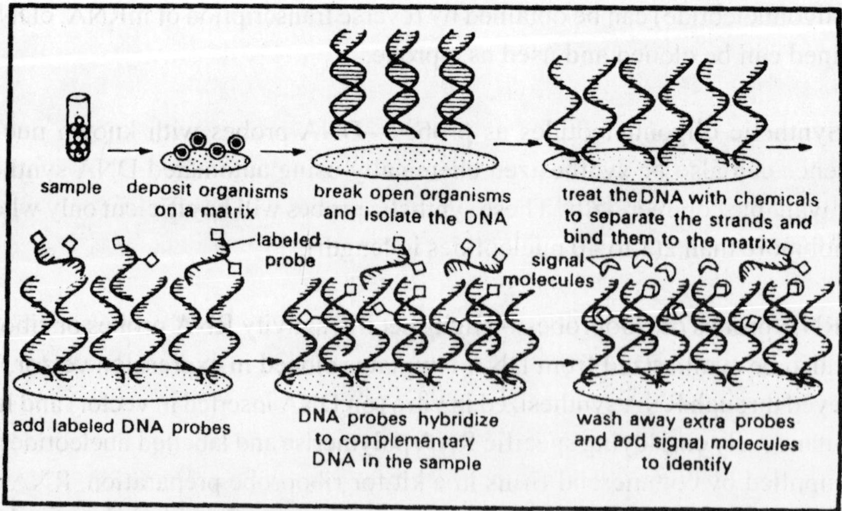


**Fig.** Use of labelled DNA probe for identification of homologous DNA sequence through DNA hybridization

deoxyribonucleotide with help of DNA polymerase I (or 'klenow fragment' of this enzyme); (ii) oligonucleotide labelling involves the use of short random oligonucleotides, which are used as primers for copying the probe DNA in presence of labelled deoxyribonucleotides; (iii) riboprobe preparation involves synthesis of labelled RNA, using DNA probe as template, in presence of a labelled ribonucleotide.

After hybridization with labelled probe, hybrids are detected by autoradiography (Fig.). $^{32}P$ has the advantage over other radioisotopes, since it has high specific activity. However, in general, radioisotopes have some disadvantages. They are difficult to handle and expensive to dispose off. Detection by autoradiography, while sensitive, may take a long time if there are few counts in the hybrid. Furthermore, radioisotopes have a short half lifes (e.g. $^{32}P$ has a half-life of 143 days) and therefore experiments should be completed preferably within one half life.

Non-radioactive probes—(a) Biotin labelled probes. Recent advances in nucleic acid technology now offer alternatives to radioactively labelled probes. One procedure that is becoming increasingly popular is biotin labelling of nucleic acids. This system exploits the affinity, which the glycoprotein °avidin' has for biotin (vitamin H). Avidin is commonly found in egg white. Biotinylated probes are prepared through a nick-translation reaction by replacing nucleotides with biotinylated derivatives. After hybridization and washing, detection of hybrids is done by a series of cytochemical reactions which finally give a blue colour whose intensity is proportional to the amount of biotin in the hybrid. There are several advantages of using biotinylated probes. For example, these assays employ non- toxic materials, whose half-life is longer. These probes can be prepared in advance in bulk and stored at -20°C for repeated uses. Detection of hybrids is much faster than by radioactive probes. A limitation of this technology is that very small probe (20 nucleotides) contains only a small number of biotinylated sites, limiting the intensity of signal obtained. It has been solved by adding long 'tails' of biotinylated nucleotides to the probes through enzymatic methods. Sometime the probe does not need to be labelled with biotin, but only coupled with a tail. Another disadvantage of biotin labelled probes is that the cytochemical visualisation reactions lead to precipitation of insoluble material which can not be removed and therefore, the filter can not be reused, while in radiolabelled probes, the filters can be used repeatedly for hybridization with a number of probes, one at a time.

(b) Digoxigenin labelled probes—Digoxigenin is another chemical derived from plants and used for non-radioactive labelling of probes. An antibody associated with an enzyme (anti-digoxigenin - alkaline phosphatase conjugate) is used for the detection of the presence of digoxigenin. The probe may be labelled with digoxigenin-11-dUTP supplied with a digoxigenin kit (these kits are available from any commercial firm, e.g. Boehringer Mannheim). The labelled and denatured probe may be used for hybridization with denatured DNA on Southern blots. After hybridization the membrane or the slide may be washed and the membrane or the slide is transferred into detection buffer containing 20 μg/ml of antidigoxigenin fluorescein and 5% (w/v) BSA (bovine serum albumin). The system is incubated for 1h at 37°C, and then the membrane/slide is washed in detection buffer three times (8 min each at 37°C). Alkaline phosphatase activity can be detected using 0 . 17 mg/ml BCIP (5-bromo-4 chloro-3 indolyl phosphate) and 033 mg/ml NBL (nitroblue tetrazolium) as dye substrate.

(c) Alternatives to biotin and digoxigenin labelling—The techniques of nonradioisotopic labellig have been further expanded and new methods have been devised for attaching other ligands (e.g. hapten determinants, 2, 4-dinitrophenol, arsenate derivatives, etc.) to nucleotides without hampering their ability to be incorporated into DNA (sec Figure for some details). These alternatives require binding of attached ligands to specific proteins that can be tagged with enzymes or fluorescent molecules. It is possible to monitor many probes simultaneously by using several different ligands since each ligand would yield a different signal. A chemiluminescent (light emitting) probe system has also been developed in which two different probes complementary to a continuous segment of DNA hybridize to adjacent segments of a gene. The first label is a chemiluminescent complex that emits light at a specific wavelength; this emission excites the label molecule on the second probe to emit light at a different wave length which can be detected using a photomultiplier device (Fig.). This process called non-radioactive energy transfer, can occur only if the two probes hybridize correctly and the two labels are close to each other. This system has great fidelity and provides basic technology for a homogeneous assay. In addition, DNA does not need to be immobilised and no washing steps are necessary which may be an additional advantage for large scale testing.

## Amplification of DNA Probe Signals

In recent years, DNA probe assays and immunoassays (using antibodies) have been competing with each other for a variety of purposes. While DNA probes have an advantage of detecting the presence of gene rather than its product, the immunoassay has the advantage of getting amplified target (the gene product) firstly due to transcription (several mRNA molecules can be synthesized on single copy of gene) and secondly due to translation (several protein molecules can be produced from same mRNA molecule). The amplified target gives an amplified signal. Therefore, if DNA probes have to compete with monoclonal antibodies for immunoassay, DNA probe signals need to be amplified. Following devices have been used for the signal amplification:

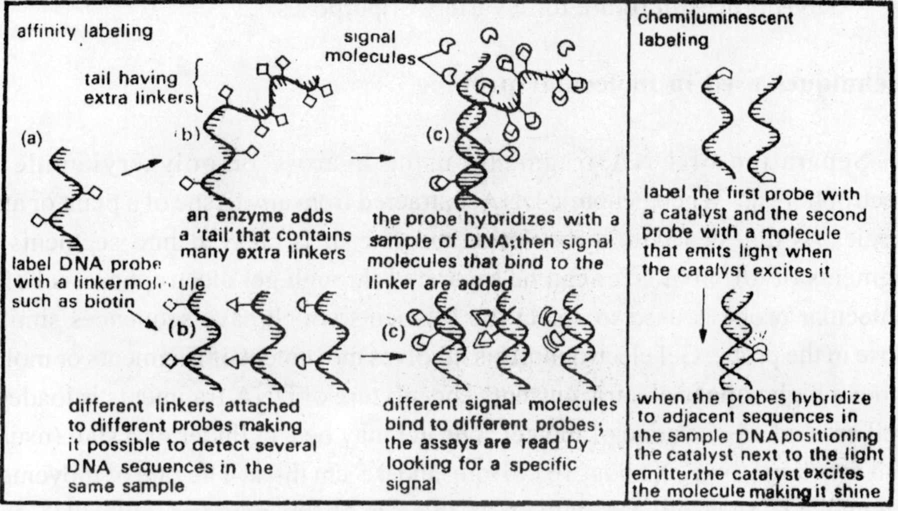


**Fig.** Non-radioactive labelling of probes, and the use of three labelled probes in DNA hybridization

(1) Probes have been developed for rRNA, since rRNA genes are present in thousands thus amplifying the target (as done by companies like Gene-Probe and Gene-Trak systems in USA).

(2) Polymerase chain reaction (PCR) technology (which utilizes a thermostable polymerase) is used to generate millions of copies of target DNA sequence, which can then be identified by standard methods (fetus Corp. in USA introduced this technique; for more details consult next Chapter).

(3) Signal generating capacity of the probe can also be increased by any one of the following methods: (i) concentrating more label at the site of target molecule by attatching multiple enzyme molecules to each of its DNA probe or (ii) by using multiple probes, or (iii) by attatching multiple enzymes to each of the multiple probes.

(4) Ampli probe technology uses multiple secondary probes (each having one or more enzyme molecules) that hybridize to multiple target specific primary

probes. The secondary probe is independent of target but helps in amplifying the signal. These approaches have been described as Christmas Tree or Christmas Forest approaches.

The above amplification systems will play a significant role in the use of DNA probes in future for a variety of purposes.

**Techniques used in molecular probing**

Separation of DNA fragments using agarose or polyacrylamide geld electrophoresis. When genomic DNA, extracted from any tissue of a plant or animal species, is digested with a restriction enzyme, it is cleaved into segments. The segments of different sizes can be separated through gel electrophoresis before a molecular probe is used to detect the segments which have sequences similar to those in the probe. Gel electrophoresis involves movement of fragments or molcules under a high voltage electric current. The mixture of DNA fragments is loaded in a well created on one edge of the gel. The gel may be a cylinder or a slab (usually a slab for cloning expt.), about 10 cm long and 0.5 cm thick. The rate of movement of fragments is inversely correlated with the size of fragments or molecules, so that heavier fragments will remain closer to the site of loading and the lighter fragments will move away (Fig.). Fragments of different sizes will appear as bands on the gel and can be examined or isolated for further study. More often agarose gels are used, but for separation of fragments differing by few base pairs, polyacrylamide gels are used. Polyacrylamide gels are more commonly used for DNA sequencing experiments.

Separation of large DNA molecules (whole chromosomes) using PFGE. The technique of gel electrophoresis is used for separation of DNA molecules of different sizes. However, DNA molecules of large size could not be handled in this technique. In recent years, a new technique called pulsed field gel electrophoresis (PFGE) has been used for separation of large sized DNA molecules, sometimes representing whole chromosomes. Using this technique, separation of DNA molecules belonging to each of the 16 individual chromosomes of yeast has become possible (Fig.). In this technique, short pulses of electricity are used in two different directions and DNA is embedded and used in the form of agarose plugs to avoid fragmentation of large DNA molecules. Using the technique of PFGE and a more refined technique CHEFE (countour clamped homogeneous electric field electrophoresis), genome of several fungi could be resolved into chromosomal bands and used for mapping of DNA sequences on specific chromosomes.

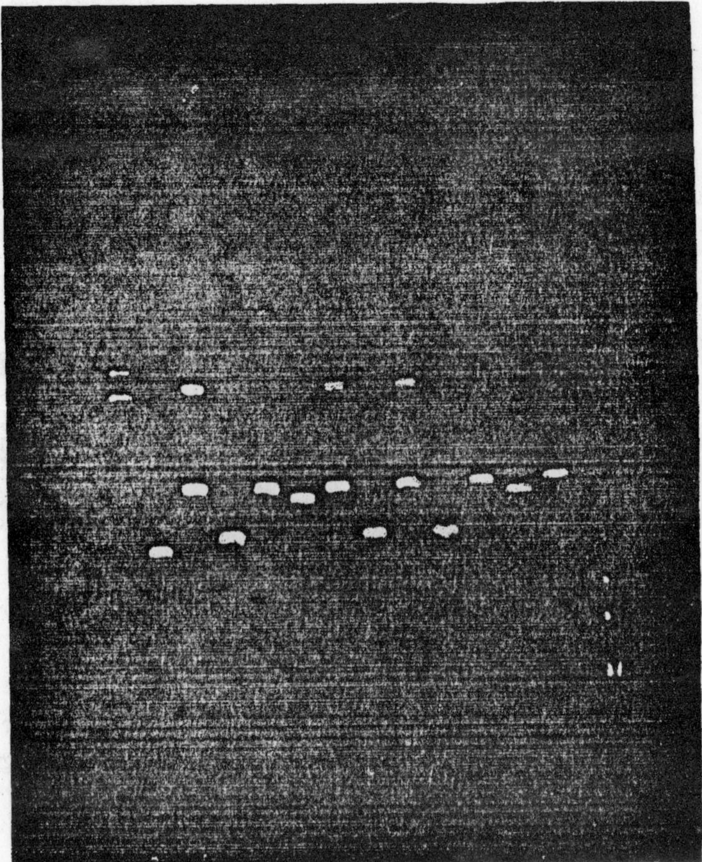

**Fig.** A gel with electrophoresed DNA in several lanes.

Southern, Northern and Western blotting—A mixture of DNA, RNA or protein fragments can be separated by gel electrophoresis and the separated bands can be stained and visualized directly in the gel. However, to confirm the identity of these bands or to find similarity of one or more of these bands with a known and available molecular probe, it is possible to hybridize these bands with a labelled probe. To facilitate this hybridization, the bands are often transferred to a nitrocellulose membbrane through a technique described as blotting. When DNA bands are thus blotted, it is called Southern blotting (after the name of E.M. Southern); when RNA bands are thus transferred it is described by the jargon term Northern blotting and similarly when protein bands are transferred, the technique is described as Western blotting.

(a) Southern blotting—For Southern blotting, DNA sample is first digested with a restriction enzyme and digested sample is gel electrophoresed. The DNA bands in

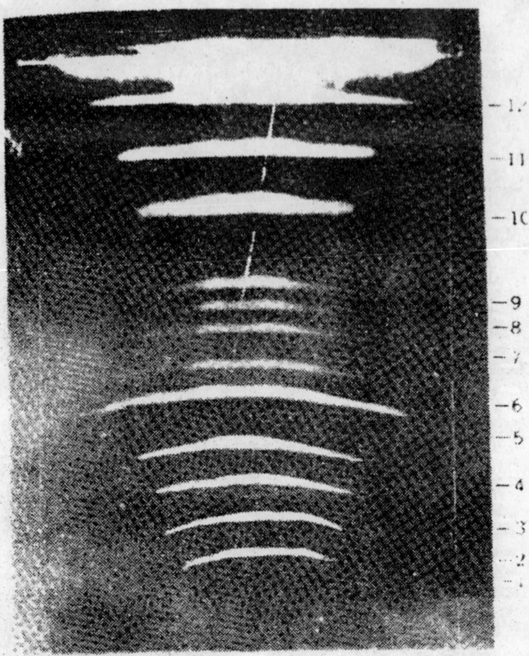


**Fig.** A gel with 12 bands representing 15 chromosomes of yeast as resolved by PFGE (the three doublet bands i.e. 5, 10 and 11 can be resolved on other gels)

the gel are denatured into single strands with the help of an alkali solution. Subsequently, the gel is large on top of a buffer saturated filter paper, placed on a solid support (e.g. glass plate), with its two edges immersed in the buffer. A sheet of nitrocellulose membrane is placed on top of the gel and a stack of many papers (paper towels) on top of this membrane. A weight of about 0.5 kg is placed on top of paper towels. The buffer solution is drawn up by filter paper wick, and passes through the gel to the nitrocellulose membrane and finally to the paper towels. While passing through the gel, the buffer carries with it single stranded DNA, which binds on to the nitrocellulose membrane, when the buffer passes through it to the paper towels. After leaving this arrangement for a few hours or overnight, paper towels are removed and discarded. The nitrocellulose membrane with single stranded DNA bands blotted on to it, is baked at 80°C for 2-3 hours to fix the DNA permanently on the membrane. This membrane now has a replica of DNA bands from agrose gel, and can be used for hybridization with radioactively labelled DNA or RNA probe. The membrane may then be washed to remove any unbound DNA and X-ray film is exposed to the hybridized membrane to get autoradiographs. The above seeps involved in Southern blotting are shown in *Figure*.

(b) Northern blotting—Initially the technique of Southern blotting used for DNA transfer from gel to the membrane, could not be used for blot-transfer of RNA. Instead, mRNA bands from the gel were blot-transferred into a chemically reactive paper, prepared by diazotization of aminobenzyloxymethyl paper. The technique being related to Southern blotting was called Northern blotting (not after the name of any scientist as in Southern blotting). Later, it was shown that mRNA bands can be blotted directly onto nitrocellulose membrane, a technique which has been widely adopted. The mRNA bands blotted onto nitrocellulose membrane can be hybridized with a labelled DNA or RNA probe . The single stranded regions of the probe are removed by nuclease (mungbean nuclease or S-1 nuclease), so that quantitative estimation of hybridized mRNA can also be made.

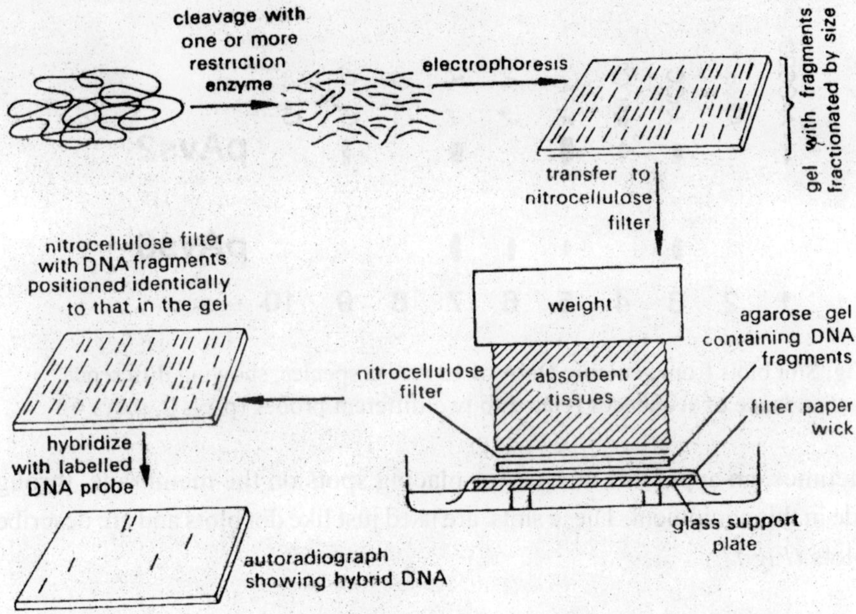


**Fig.** Southern blotting method of analyzing DNA segments that share homology with a nucleic acid probe

(c) Western blotting—This technique is used to detect proteins of particular specificity. Particularly, when a transferred gene expresses transformed cells, the translated product in the form of protein can be identified by this technique. The extracted proteins are subjected polyacrylamide gel electrophoresis (PGE) and are then transferred on nitrocellulose, to which they bind. Nitrocellulose membrane is then used for probing with a specific labelled antibody (antibody will not hybridize with protein, but bind with it). The antibody may be labelled with and the signal is defected again with autoradiography. If radioactive labelled is not used, bound antibody may be defected by a second anitbody lagged with an enzyme (as in ELISA technique).

Dot blots and slot blots—Another variety of Southern or Norther blots is dot blots or slot blots. In dot blots, cloned or pure extracted DNAs to be tested are spotted adjacent to each other on a nitrocelluloser membrane. DNA blots thus produced are immobilized and denatured as that they are bound on the membrane as single stranded DNA blots. This membrane is then hybridized with readioactively labelled probe (DNA or RNA). The dot representing sequence related to probe will light autoradiography (Fig.). The intensity of the dot will indicate the relative concentration of a sequence related to the probe in the DNA sample used for a particular dot. No steps involve in digestion with enzyme, gel electrophoresis or transfer from gel to membrane are needed in this technique. Therefore, it is considered much more convenient, when no restriction sites are needed to be studied.

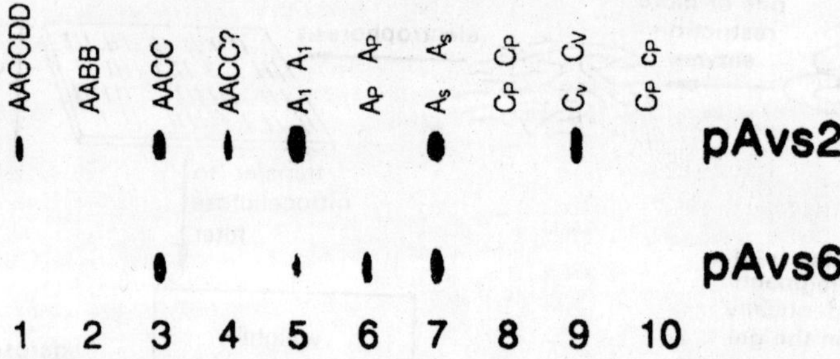


**Fig.** Slot blots from genomic DNA of 10 Avena species, showing difference in abundance of sequences related to two different probes (pAvs 2, pAvs 6).

Sometimes, an apparatus is used for placing spots on the membrane, through slots made in this equipment. These slots, are used just like dot blots and are described as slot blots *(Fig.)*.

## Applications of Molecular Probes

Molecular probes can be used for both basic and applied antibodies in the field of molecular biology and biotechnology. These uses will be discussed in detail in the different subsequent chapters of this hook at appropriate places, but will be described briefly in this section.

Use of molecular probes in restriction fragment length polymorphisms (RFLPs) and related analyses. (a) WIzal are RFLPs. When genomic DNAs from each of several individuals belonging to one or more species are digested separately with restriction enzyme, electrophoresed, blotted on a membrane, and probed with a labelled DNA clone, polymorphism in the hybridization pattern is sometimes revealed and attributed to sequence differences between the individuals *(Fig.)*. Such variation has been termed as "restriction fragment length polymorphism" (RFLP). Variation in one DNA fragment obtained with a specific enzyme is treated as one RFLP.

(b) Detection of RFLP—A RFLP can he demonstrated using the following steps: (i) Extract and purify DNA from several individuals which may differ in some respects among themselves. (ii) Digest DNA samples with a restriction endonuclease. (iii) Subject each DNA digest to gel electrophoresis separately, but on the same gel slab. (iv) The DNA fragments of different lengths resulting from digestion can be separated by gel electrophoresis, but the differences in the distribution patterns of fragments in several DNA samples due to digestion by several endonucleases can not be detected directly. This is because the number of fragments is large and the range in size is rather continuous, such that they form a continuous smear on the gel. Electrophoresis is , therefore, followed by the following steps. (v) Fragments are transferred from gel to nitrocellulose filters using the technique of Southern blotting. (vi) The filters with small fragments, fully or partially homologous to the probe, can be detected by autoradiography after hybridization (Fig.). Since DNA clones having repetitive DNA may hybridize with many fragments, they will give a smear on autoradiography. Therefore, unique DNA sequences are generally used as probes for detecting RFLPs. The DNA fragments hybridizing with the probe, can be visualized by autoradiography, and will be called RFLPs at the phenotypic level. In Figure, results of a hypothetical experiment are shown where RFLPs are detected by using four plants (which differ by point mutations as well as by insertions) and three enzymes individually and in combination.

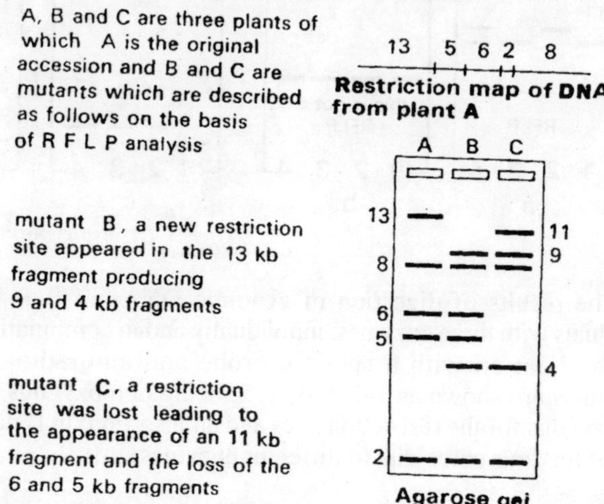


**Fig.** The technique for detection of restriction fragment length polymorphisms (RFLPs)

(c) RFLPs for evolutionary studies—The restriction fragment length polymorphisms (RFLPs) can be studied in a set of related species, using a random or a specific DNA probe. The similarities and differences can be used to infer phylogenetic relationships. This has actually been done in a number of cases, both in plants and in animals.

(d) RFLP maps and linkage of RFLPs with specific genes—RFLPs have been used to prepare chromosome maps in humans, mice, fruitfly, and in plants including maize, tomato, lettuce and rice. The use of Mendelian markers for genetic mapping is sometimes limited due to non-availability of mutants. The list of markers can be extended by using molecular markers, which are examined in the form of RFLPs.

Once a large number of RFLPs are available in a species, the parents, Fl and Ft's can be used to study their inheritance and linkage relationship and genetic linkage maps can be prepared. They have been assigned to specific chromosomes using monosomics in maize and efforts are being made to relate them with morphological and economic traits so that they can be used for practical plant breeding.

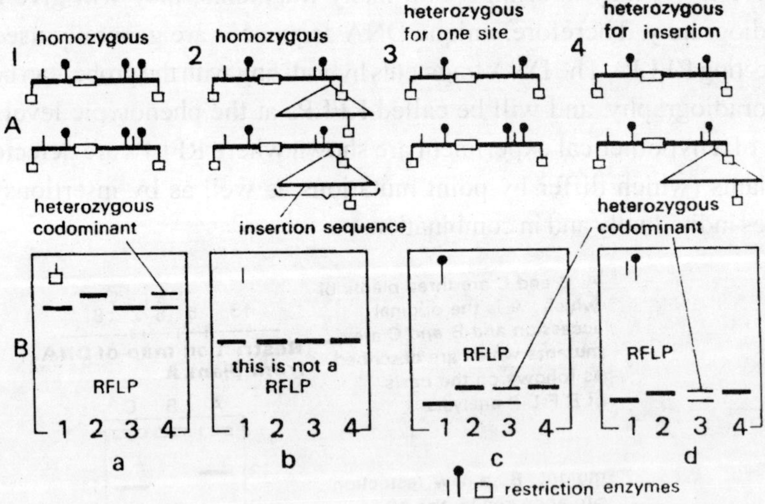


**Fig.** The results of digestion of genomic DNAs of four hypothetical individuals with three enzymes, individually and in combination, followed by hybridization with a specific probe and autoradiography (four individuals are shown as 1-4 in A, 1, 2, being homozygous and 3, being heterozygous for the restriction sites and an insertion; in B note RFLPs in three of the four cases due to different enzymes)

A difference in restriction maps between two individuals or a RFLP can be used as a genetic marker in much the same way as any other phenotypic marker.

The polymorphism at the molecular level may be due to the presence or absence of a restriction site. *Figure* shows how the presence of an additional site will produce two fragments and its absence will produce only one fragment. Although, there may be linkage between a molecular marker and a phenotypic trait, the change in molecular map (restriction map) may not affect the phenotype. In order to study linkage relation between a molecular marker and a phenotypic marker, we may examine the restriction fragment patterns and the phenotypes of two parents that differ morphologically as well is for molecular markers. By making crosses between two such parents, we can estimate the recombination frequencies between molecular markers and phenotypic markers using a test cross. In *Figure*, we used a hypothetical example showing linkage of red and white eyes with different molecular markers in Drosophilu. It can be seen that as a result of recombination, in test cross, progeny shows association of red eye with a molecular marker associated with white eye in the parent and vice versa. Therefore, it is concluded that restriction marker is 15 map units away from the marker eye colour, although it has no causal relation with eye colour. Such a linkage between molecular marker and phenotypic marker allows identification of genetic loci at the molecular level. A tight linkage of such a molecular marker with disease resistance may also be recommended as a tool for identification of disease resistant cultivars in many of our crop plants and will be increasingly utilized in future. Molecular markers are also being used for selection of plants at the seedling stage for plant breeding purposes.

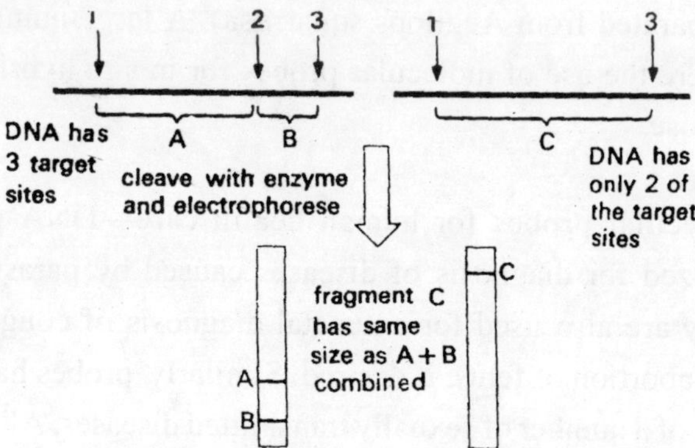


**Fig.** Detection of a change in DNA as revealed by the appearance or disappearance of a restriction site (an additional restriction site is present in the DNA shown on the left).

Use of molecular probes in molecular cytogenetics. (a) Isolation of gene or related sequence. This can be facilitated through the use of its probe carrying a known sequence. For this purpose, DNA sequences may be hybridized following any one of the following techniques: (i) Southern blots can be prepared (from genomic

DNA or cDNA) to isolate DNA corresponding to a band that hybridizes with the probe. This DNA can then be used for cloning experiment to gel a partial genomic library or cDNA library (see later in this chapter for details); (ii) The genomic or cDNA library can be screened through colony hybridization or plaque hybridization to identify and isolate the clone carrying the desired sequence; sometimes the step (i) is not used and complete genomic/eDNA libraries can be prepared and used in step (ii); (iii) The identified clone should carry the gene or sequence of interest, so that the clone can be multiplied and the sequence or gene retrieved according to standard procedure.

(b) In situ hybridization (ISH) for locution of sequences on chromosomes. A molecular probe can also be used for in situ hybridization where chromosomes in miotic or meiotic preparation are used for hybridisation with the probe (without extracting DNA). In a number of cases, rye chromosomes in wheat background have been indentified using this technique. Sat- chromosomes with NORs (nucleolar organizing regions) are recognized using probes for ribosomal DNA as done in many cases (wheat, barley, etc). Telemeteres have been identified in human and other eukaryotes using a telomeric sequence as a molecular probe, showing that telemeteres of all chromosomes carry the same sequence. In wheat and related species, chromosomes of D genome can be identified using a D ,genome specific probe (pAS1 separated from Aegilops squarrosa). A large number of such cases can be cited where the use of molecular probes for in situ hybridization served a very useful purpose.

Use of molecular probes for human health care—DNA probes are being extensively utilized for diagnosis of diseases caused by parasitic Protozoa and Helntinths. They are also used for antenatal diagnosis of congenital diseases to allow advice on abortion of fetus, if desired. Similarly, probes have been designed for the diagnosis of a number of sexually transmitted diseases. All these aspects will be discussed in some detail in Chapter.

Use of molecular probes in DNA fingerprinting—With the help of probes prepared from human minisatellite DNA dispersed throughout human genome, it has become possible to identify individual humans, since the possibility of any two humans (except identical twins) showing same pattern of fingerprinting is remote. This aspect is already playing an important role in identifying criminals like murderers and rapists.

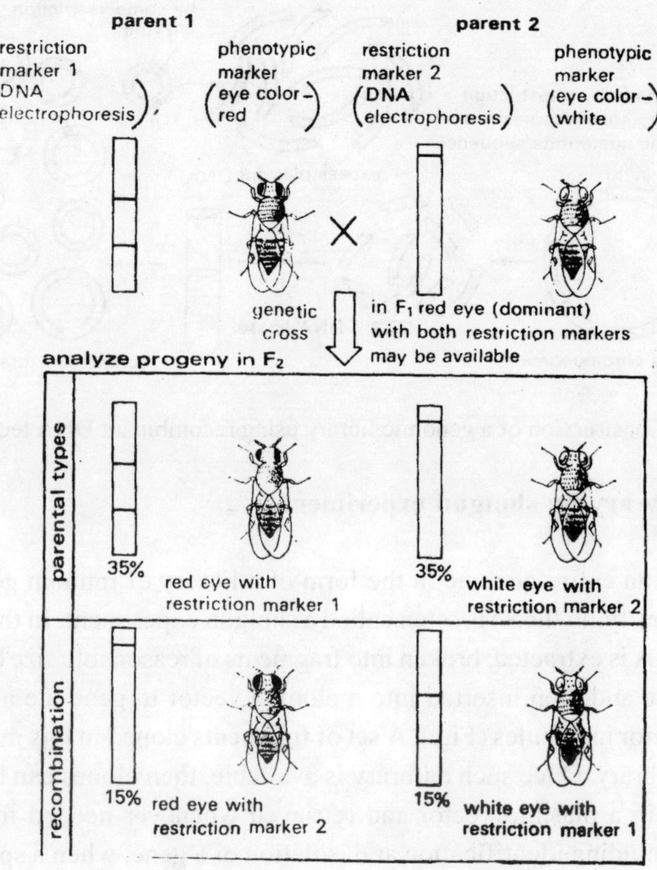


**Fig.** A diagrammatic representation showing linkage and recombination between a MIT (a molecular genetic marker in the form of restriction pattern) and a phenotypic marker (eye colour)

## IV. Genomic Libraries

In order to isolate one or more related genes from a genome, we like to prepare a mixture of clones each carrying DNA derived either from the genemic DNA or from cDNA (derived from the mRNA isolated from a specific metabolically active tissue of an organism). This mixture may contain thousands of clones, which when derived directly from the genemic DNA are collectively called a genomic library. Similarly, when these clones are derived from cDNA, they are collectively called a cDNA library. Construction and use of these libraries will be discussed in this section.

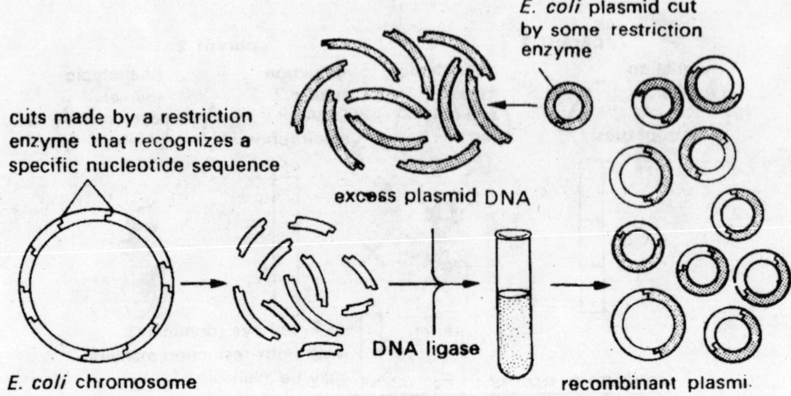


**Fig.** Construction of a genomic library using recombinant DNA technique.

## Genomic library by shotgun experiment

Cloning an entire genome in the form of a library of random genomic clones (without identifying them) is often called a shotgun experiment. In this experiment, genomic DNA is extracted, broken into fragments of reasonable size by a restriction endonuclease and then inserted into a cloning vector to generate a population of chimeric vector molecules (Fig.). A set of fragments cloned in this manner is called a genomic library. Once such a library is available, then clones can be perpetuated indefinitely in a plasmid vector and retrieved whenever needed for a variety of purposes, including identification and isolation of a gene, when a specific probe is available.

Genomic libraries can be prepared by using a number of restriction endonucleases, one at a time, so that fragments of varying sizes having cuts at different places of the genome will be available. However, this may lead to cuts at inconvenient places, including sites within a gene, so that fragments having complete genes will be difficult to obtain. In order to overcome this difficulty, we use the following strategy in the shotgun experiment: (i) We use restriction endonucleases, which have short (4 bp) recognition sequences, so that such a sequence may be frequently distributed. (ii) Conditions are used which give only partial digests, so that a particular restriction site is only occasionally cleaved, and long fragments without having any breaks on recognition sites available within a gene can be easily obtained. This technique of shotgun experiment leads to the construction of a random genomic library, in which all fragments have same fragment ends thus helping retrieval of a fragment from the vector with the help of the same enzyme.

The number of fragments representing every sequence of the genome increases with genome size. For instance, for a parobability level of 99% that all sequences are present in our library of a species, we may need 1,500 cloned fragments for E. coli, 4,600 for yeast, 48,000 for Drosophila melanogaster and 8,00,000 for a mammal like human being. Libraries reaching these desired limits have been prepared in all these cases.

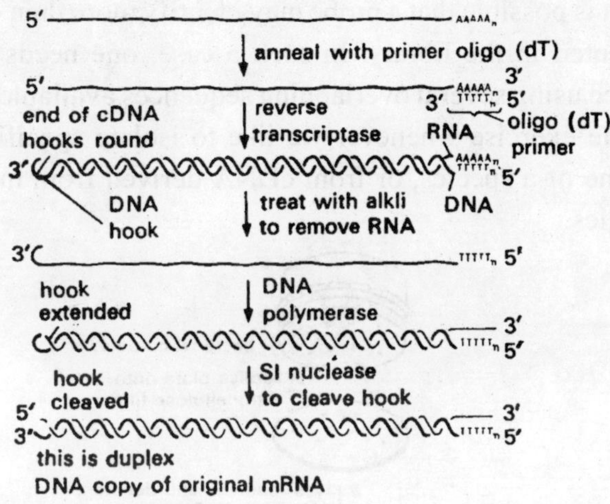


**Fig.** Synthesis of cDNA from mRNA, using reverse transcriptase enzyme

## cDNA library from mRNAs

Complementary DNA (cDNA) libraries can also be prepared by isolating mRNAs from tissues which are actively synthesizing proteins, like roots and leaves in plants, ovaries or reticulocytes in mammals, etc. The mRNA is used for copying it into cDNA through the use of reverse transcriptase. A cDNA molecule can be made double stranded, and cloned as shown in Figure and discussed later in Chapter. However cDNA clones will differ from genomic clones in lacking the introns present in split genes, but may have the advantage of being capable to be expressed in bacteria, which do not have the machinery to process the HnRNA (obtained from split genes) into mRNA.

## Colony (or plaque) hybridization for screening of libraries

Once a genomic library or cDNA library is available, we may like to use it for isolation of a gene sequence. This can be achieved by colony hybridization technique illustrated in Figure. In this technique, bacteria carrying chimeric vectors are grown into colonies, which are lysed on nitrocellulose filters. Their DNA is denatured in

.situ and fixed on the filter, which is hybridized with a radioactively labelled probe carrying a sequence related to the gene to be isolated (usually it cloned cDNA for screening of a genomic library). Colonies carrying this sequence will be identified by dark spots after aoutoradiography, in that the original chimeric vector carrying the desired gene sequence can be recovered from one or more colonies in the original master plate and used for further experiments. This technique is described as colony hybridization. It is possible that a probe may identify more than one clones or that a gene is fragmented in the library. In such a case, one needs to reconstruct the desired sequence using several overlapping sequences available in the library. This is a very routine exercise whenever we like to isolate specific DNA sequences from the genome of a species, or from cDNA derived from mRNA of a specific tissue of a species.

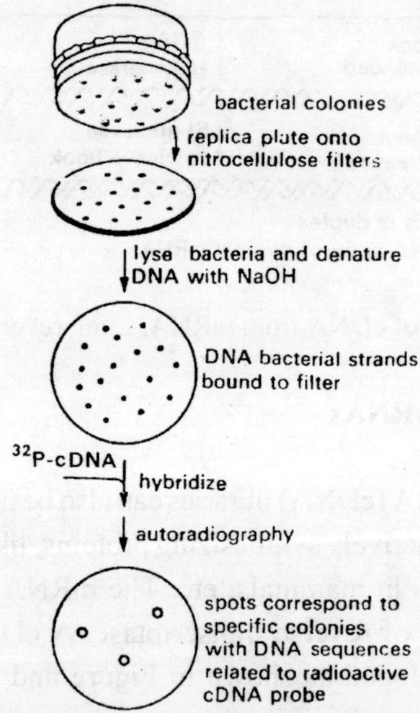


**Fig.** Colony hybridization technique for selection and isolation of a sequence complementary to a radioactively labelled probe

Sometimes, the library may be available not in the form of bacteria transformed with chimeric DNA molecules, but in the form of chimeric phage particles carrying the cloned segments. In such a situation, a bacterial lawn is infected with a mixture of chimeric phage particles (i.e. the library) and a large number of plaques develop overnight. These plaques can be treated just like the colonies in colony hybridization to identify and isolate the chimeric phage particle carrying the gene of interest. This technique is then described as plague hybridization.

**Chromosome walking and characterization of chromosome segments**

When a probe is used for the identification of a gene sequence in a genomic library, the probe may hybridize with a number of clones, each carrying a part of a large gene fragmented during preparation of genome library. If we obtain partial digests (by digesting the DNA only partially) from the genome, different genomes (from large number of cells) may give fragments which have overlapping sequences, because sites cleaved in different genomes of the same organism, will differ being random. Since none of these fragments may have entire sequence represented in the probe, overlapping sequences may be used to construct the original genomic sequence. Identification of fragments with overlapping sequences may be a key to the reconstruction or characterization of large chromosome regions. This is achieved by the technique popularly described as chromosome walking.

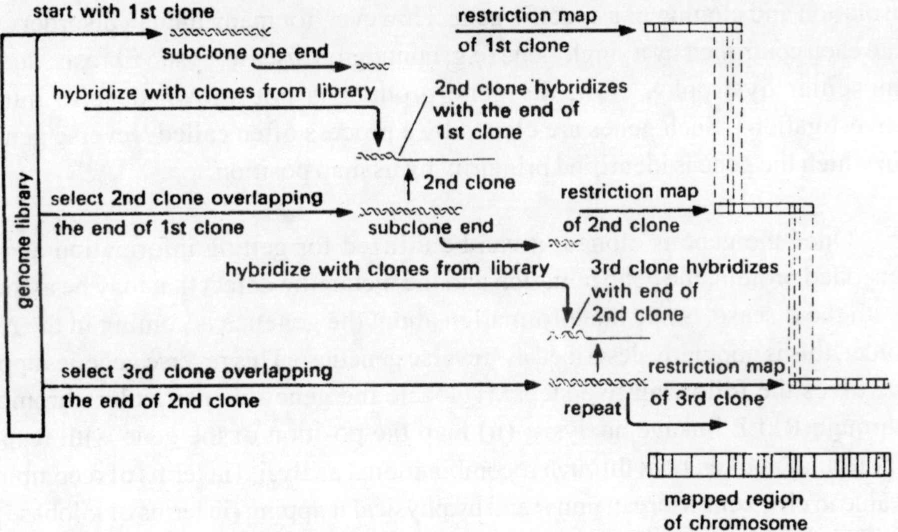


**Fig.** The technique of chromosome walking through successive hybridization between overlapping genomic clones.

The technique of chromosome walking involves the following steps: (i) from the genomic library select a clone of interest (identified by a probe) and subclone a small fragment from one end of the clone (there is a technique available to subclone a fragment form the end); (ii) the subcloned fragment of the selected clone may be hybridized with other clones in the library and a second clone hybridizing with the subclone of the first clone is identified due to presence of overlapping region; (iii) the end of the second clone is then subcloned and used for hybridisation with other clones to identify a third clone having overlapping region with the subcloned end of the second clone ; (iv) the third clone identified as above is also subcloned and hybridized with clones in the same manner and the procedure may be continued; (v) restriction map of each selected clone may be prepared and compared to know the region of overlapping as shown in Figure, so that identification of few overlapping restriction sites will amount to walking along the chromosome, or along a long chromosome segment.

Regions of chromosomes approaching 1000kb have been mapped following the above technique. Restriction maps of entire chromosomes can be prepared in this manner following the technique of chromosome walking.

**Reverse genetics and chromosome jumping (or hopping) libraries**

In most cases, we identify the genes by their protein products, which help in the isolation and cloning of a specific gene. However, for many human disorders which are each controlled by a single gene (e.g. huntington disease, cystic fibrosis, duchenne muscular dystrophy, etc.), the gene product is not known despite intensive investigations. Such genes are cloned by a process often called 'reverse genetics', in which the gene is identified primarily by its map position.

Once the gene is cloned, it can be utilized for getting information about the encoded protein, the gene mutation and the metabolic defect that may be associated with the disease. Since the information about the genetics is coming in the reverse order, this is popularly described as 'reverse genetics'. This reverse gene.ics approach involves the following two steps: (i) locate the gene to a particular chromosome through RFLP linkage analysis; (ii) map the position of the gene with respect to molecular marker, both through recombinational analysis (in terms of recombination value in cM=centiMorgan units) and by physical mapping (in terms of kilobase pairs) with pulsed - field gel electrophoresis (PFGE) and preparation of restriction map. Even if the distance is 1cM, which is the limit of resolution, this will be equivalent to 1,000 kb. Therefore, if the segment carrying the gene is to be cloned using the linked molecular marker, such a segment will be too large to be cloned. In order to bring the molecular marker close to the gene of interest, 'chromosome jumping' approach has been utilized. The technique of 'chromosome jumping' is based on the following steps shown in Figure: (i) depending upon the distance between the gene and the marker, decide about the distance of 'jumps' or 'hopsize' (e.g. 100 kb or 200 kb); (ii) genomic DNA molecules in the range of selected size (say 80 kb-130 kb in case of 100 kb 'hopsize' or 160-240 kb for 'hopsize' of 200 kb) are selected through pulsed-field gel electrophoresis; (iii) for circularization of DNA segments, ligation between two ends of each long linear DNA molecule was allowed using T4 ligase in the presence of supF ); (iv) DNA circles obtained in (iii) above are digested with EcoRI; (v) the vector Ch3A O lac, an ambermutated phage vector (supF-is also cut with EcoRI and used for cloning small DNA fragments representing the junctions of the circularized genomic. DNA molecules and carrying sup F+; (vi) the cloned DNA fragments obtained in (v) above represent the jumping library, which can be plated on a bacterial host and screened through the technique of plaque hybridization described earlier.

The above technique of chromosome jumping will help narrowing the gap between the gene and available molecular markers. After several cycles of chromosome jumping followed by cloning the regions that are close to the gene, it will be possible to approach very close to the desired gene and clone it. It is thus obvious that in the field of 'reverse genetics' using technique of chromosome jumping, it will be possible to isolate, clone and characterize genes, whose gene products are unknown. The gene for cystic fibrosis has recently been isolated using this technique. It could also be shown that 'cystic fibrosis' disease is caused by a single base substitution in this gene. This technique of 'reverse genetics' will become a very active area of research particularly in human genetics in the coming years.

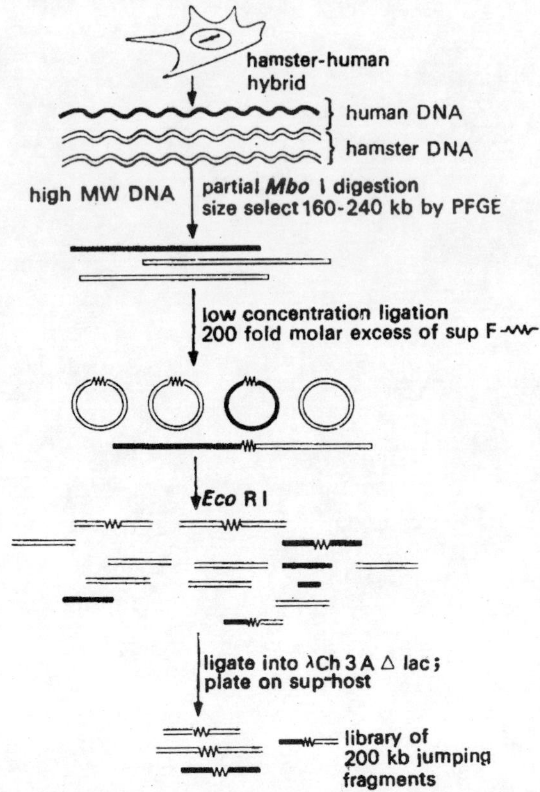


**Fig.** Steps involved in chromosome jumping

The term 'reverse genetics' has however been redefined recently. The classical genetics, sometimes described as 'forward genetics' involves a study which starts from the phenotype, and follows through the identification of the gene and finally concludes with the isolation and sequencing of the gene. In view of this, if 'reverse genetics' has to use steps in the reverse order for a genetic study, then it should start with DNA molecules or clones with unknown effect and should conclude with the determination of the phenotype -which it controls. This can be achieved by preparing random genomic DNA clones, which can be subsequently utilized for a genetic study in the reverse order. This will then be described as 'reverse genetics'. This definition of 'reverse genetics' differs from the earlier conventional meaning of this term, where 'reverse genetics' was used to describe cases of genes with unkwon products, which are isolated ( with the help of phenotype and its linkage with molecular markers) and then studied thoroughly.

# SIX

# Antibodies and Hybridomes

In post-war years, in the euphoria it seemed to many that science had effected a decisive turn in mankind's endless struggle against disease. In recent years, it is being felt that in a world with limited resources, expenditure on prevention is more productive than expenditure on cure. It is better to acquire immunity against a disease than to wait for the disease to appear, then cure it. Indeed, we can see the beginnings of preventive medicine in the Hebraic health code, the Greek dedication to personal hygiene and the pioneer work of the Romans in the field of public sanitation. Obvious examples from more recent times are statutory requirements and warnings for purity of drinking water and for the disposal of sewage, regulations concerning the preparation and adulteration of food and warnings against the hazards of tobacco, alcohol and over-eating. Consequently, immunobiology and understanding immunogenetics have opened new avenues in clinical applications and several other industrial processes such as synthesis of antibiotics and pharmaceuticals. The potential of monoclonal antibodies to revolutionize methodologies in the biosciences in general has been discussed since 1975 when Kohler and Milstein introduced the hybridoma technology. The present chapter outlines briefly the basic principles of antigen-antibody interaction, immuno-genetic engineering applications in various fields.

### Antigen Antibodies Reactions

An antigen is a proteinaceous molecule (or a polysaccharide) or a foreign substance (which is no longer true) that elicits a response in the body's immune system. In recent years, however, the definition of antigen has been broadened in that it need not be a foreign substance (a person's own tissue may itself act as antigen). Usually, the antigen has a relatively high molecular weight—10,000 daltons or higher—with a few exceptions such as insulin. A large number of substances may act as antigens including hemoglobin, bacterial toxins, bee venom, chemical components of microbial flagella, pili and capsules and even milk proteins. Certain molecules known as haptens are not directly antigenic, but these will combine with tissue proteins or polysaccharides to form antigens. The penicillin molecule is an example of a hapten that combines with tissue protein to form an antigen. Other examples of haptens include poison ivy, poison oak, certain dyes and cosmetics. Each antigen has an antigenic determinant site (analogous to coenzyme) which consists of 6-8 amino acids or polysaccharides (m.wt about 750 dalton) (Fig.). In order for a substance to be considered an antigen, it must be subject to degradation by the host's macrophage. Antigens are broadly classified on the basis of origin and action.

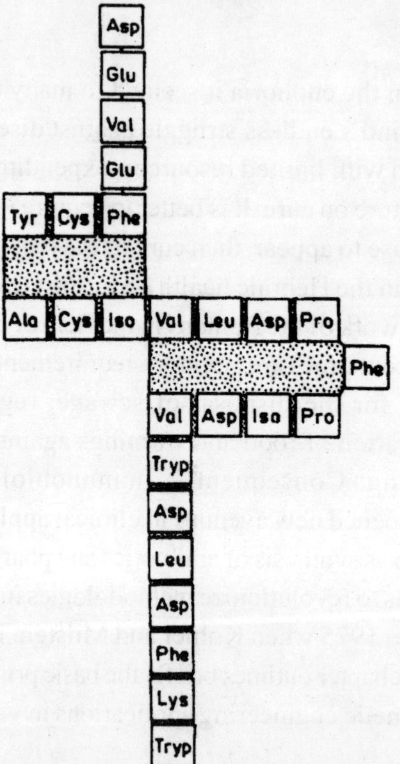


**Fig.** Molecular Configuration of an Antigen (with determinant sites dotted)

(i) Autoantigens are a person's own proteins and other biochemicals that elicit immune responses when self-tolerance breaks down.
(ii) Alloantigens-these exist in selected members (not all) of a given species.
(iii) Heterophiles-these occur in apparently unrelated specific species.

The immune system is a general term denoting a complex series of cells that respond to antigens and ultimately lend resistance to disease. The two important cells include (i) ethrythropoietic cells which are transformed to erythrocytes and (ii) lymphopoietic cells which are actually responsible for the immune system and are transformed to lymphocytes generally categorized as T-lymphocytes (Mu thymus-dependent lymphocytes) and B-lymphocytes (Bursa of Febricius-dependent lymphocytes). Immunity derives from the stimulation of B-lymphocytes by antigens contained in the macrophages B-lymphocytes that form a clone of cells called plasma cells. Their principal products are protein molecules known as antibodies. Plasma cells will continue to produce antibodies for 2-3 days or until the antigenic stimulation comes to an end. A mammalian cell can synthesize thousands of different kinds of antibodies. Each antibody is an oligomeric protein molecule which has the ability to form a complex with, and remove from the tissues or blood of the body, a particular antigen or family of antigens.

Each plasma cell is highly specialized to synthesize only one particular kind of antibody. It follows that an organism contains thousands of different plasma cell clones, each committed to the synthesis of a particular protein or groups of proteins.

Five different classes of antibodies (also known as immunoglobulins) have been recorded and designated as IgM, IgA, IgE, IgG and IgD (Table). On antigenic stimulation, IgM is the first to appear in the circulation. Every immunoglobulin, regardless of its class, contains two distinct types of polypeptide chains, one heavy and the other light, linked by disulphide bonds. The origin of numerous types of antibodies has been debated at length and the most acceptable process is the clonal selection hypothesis. Apparently there exists in the lymphoid tissue a cadre of the B-lymphocyte capable of producing each known antibody. Evidence suggests that antibodies are manufactured even without antigenic stimulation and that they may localize on the surface of B-lymphocytes. The antigenic determinant sites interact with and select those B-lymphocytes with complimentary receptor sites and stimulate them to undergo transformations leading to plasma cells. The individual clone then produces all kinds of antibodies (Fig.).

**Table**

Classes and Properties of Immunoglobulins

| *Type* | *Class* | *Molecular weight* $\times 10^{-5}$ | *Percentage of Total antibody* | *Location in body* | *Carbohydrate %* |
|---|---|---|---|---|---|
| IGG | $r_2\lambda_2$ | 1.5 | 80 | Blood lymph | 2.9 |
| Ig M | $(\lambda_2 u_2)_5$ | 9-10 | 5-10 | Blood lymph | 7.7-10.7 |
| Ig A | $(l_2\ \alpha_2)_n$ | 1.6 | 10 | Secretions<br>Body cavities | 7.5 |
| Ig E | $\lambda_2\ \varepsilon_2$ | 2 | 1-3 | Blood lymph | 10.7 |
| Ig D | $\lambda_2\ \delta_2$ | 1.7-1.8 | 0.05 | Blood<br>Lymph | 12 |

**Hybridomas and Preparation of Monoclonal antibodies**

Monoclonal antibodies are ultrapure antibodies manufactured by hybridoma cells (discussed in the following section). Hybridoma cells are fused lymphocytes and malignant myeloma cells. The normal immune system is capable of manufacturing at least a million different kinds of antibodies to combat and inactivate foreign proteins or other antigens that may invade the body. A malignant myeloma cell, however,

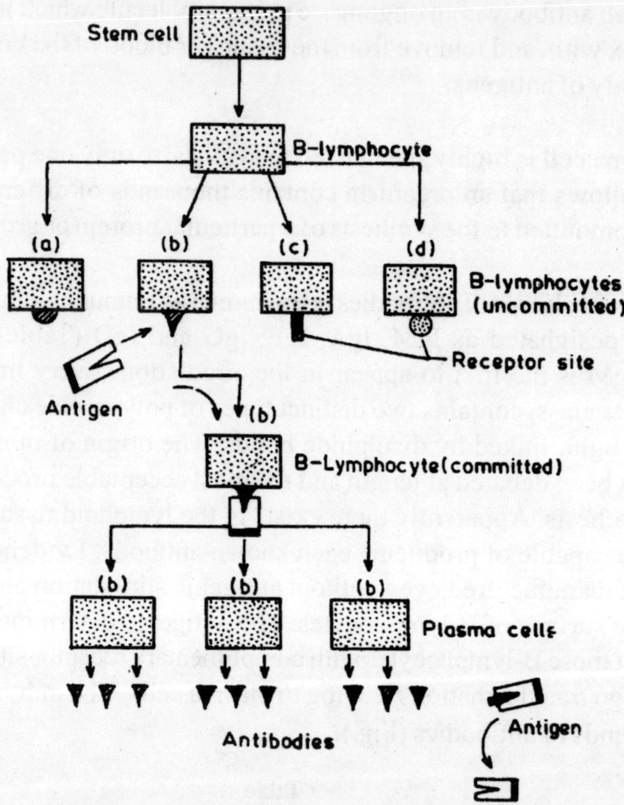


Fig. Clonal Selection Method (out lined)

synthesizes only a single type of antibody, an immunoglobulin protein that may be any one of the almost innumerable proteins possible. Myeloma cells proliferate rapidly and can be cultured indefinitely from a single cell. They cannot, however, be induced to yield antibodies to a specific antigen. That difficulty was overcome in 1975 when Cesar Milstein and his colleagues at the Medical Research Council laboratory of Molecular Biology in Cambridge, England conceived the idea of fusing mouse myeloma cells with B-lymphocytes from the spleen of a mouse immunized with a specific antigen. The resulting hybrid myeloma or hybridoma cells had the properties of both parent cells: immortality and the ability to secrete large amounts of a single specific type of antibody. This discovery was a masterpiece in the research career of Kohler who had been a Ph.D. student in Milstein's laboratory. In 1974, Kohler succeeded in producing a hybridoma by fusing P3 myeloma cells (resistant to azaguanine) and a lymphocyte from the spleen of a mouse immunized against red blood cells. Mice were immunized against these R.B.C's; then P3 myeloma cells were mixed with spleen cells from immunized mice with the help of polyethylene glycol. The hybrid cells secreted immunoglobulins and some of these cells produced antibodies against the R.B.C.'s. Clones of these cells producing a single type of antibodies (Monoclonal antibody) were isolated and maintained in culture. (Fig.).

The separated clones can be maintained indefinitely at freezing temperatures. These cells/clones could be injected into animals of same line (which produced the original cell required for fusion). These animals then develop tumours producing monoclonal antibodies, corresponding to those cells/clones, at a very high concentration (10-30 mgml1). These antibodies could be produced against small haptens, proteins, enzymes, carbohydrates and glycolipid viruses and even cell membrane components. Monoclonal antibodies, unlike the usual antiserum, can be reproduced many times. It is now possible to prepare human monoclonal antibodies and the motivation to produce these for therapeutic use is overwhelming.

The technique for production of Monoclonal antibodies could be briefly summarised as follows:

1. Production of hybridomas by hybridizing lymphocytes (antibody secreting) with myeloma cells which would secrete specific antibodies.
2. Separating different hybrid myelomas from each other by cloning.
3. Continuous propagation of hybridomas producing specific monoclonal antibody.

A suspension of immune lymphocytes are mixed with myeloma cells, the mixture is washed, concentrated and exposed under carefully controlled conditions to a fusing agent. Sendai virus was used earlier on lymphocyte hybridization. However, it is now being gradually replaced with polyethylene glycol (PEG) since this gives much better results, besides being inexpensive and easily available. The fusion is carried out by adding PEG to a concentrated suspension of lymphocyte-myeloma cells either in a tube or deposited on polycarbonate filter discs.

PEG preparations with mean molecular weights between 1000-6000 are suitable for cell fusion; outside of this range they have an unaceptably high toxicity. The optimum concentration of PEG is normally between 30-50% and the time for which cells are exposed to PEG at this concentration is usually short (I minute) to minimize the membrane damaging effects of the fusing agents. The cells are then gently resuspended, diluted and washed free of PEG. The entire operation is conducted at 37°C. It is not necessary to purify B-lymphocytes before myeloma cells selectively fuse with these cells. The frequency of formation of viable hybrid cells is low. Most cells die at an early stage and this can be indirectly damaging to the surviving cells. Hybridoma cells are sensitive to alkaline culture conditions and are, therefore, maintained in an atmosphere containing at least 5% $CO_2$. The fusion conditions are such that the vast majority of myeloma cells are not fused to spleen lymphocytes and these need to be destroyed. This is accomplished by culturing the cells in the presence of aminopterin or amethopterin drugs which inhibit folic acid metabolism and, therefore, kill the myeloma cell deficient in HGPRT (enzyme hypoxanthine guanine phosphoribosyl transferase) which are mutational variants of parent myeloma and produced for purposes of selecting the hybrid myeloma cells in the background of HGPRT minus cells. Hybrid cell lines searching specific monoclonal antibody are isolated and cloned by routine established procedure. It is technically simpler to clone cells in fluid phase media.

Production of monoclonal antibodies has been commercialized by several public and private enterprises all over the world since 1981. The price ranges between Rs. 1500-4000 per milligram. Many pharmaceutical firms have been interested in large-scale production of ultrapure monoclonal antibodies because of enormous applications of these molecules.

Milsteins work has opened a new era of experimental immunology. The problems associated with heteroantiserums (serums containing mixture of antibodies) could in principle be now circumvented. In 1980, Carlo M. Croce and his coworkers at the Wistar Institute of Anatomy and Biology in Philadelphia succeeded in generating a stable, antigen-producing intraspecific human hybridoma by fusing B-lymphocytes from a patient suffering from multiple myeloma with peripheral lymphocytes from a patient with subacute penencephalitis. The hybridoma cells of this fusion were found to secrete molecules of human immunoglobalin M specific for components of the measles virus. Although only a limited number of human cell lines lead to hybridomas that actively secrete specific antibodies, Croce's work indicates the possibility of obtaining human B-cell hybrids tnat continuously secrete human antibodies against a variety of pathogenic viruses:

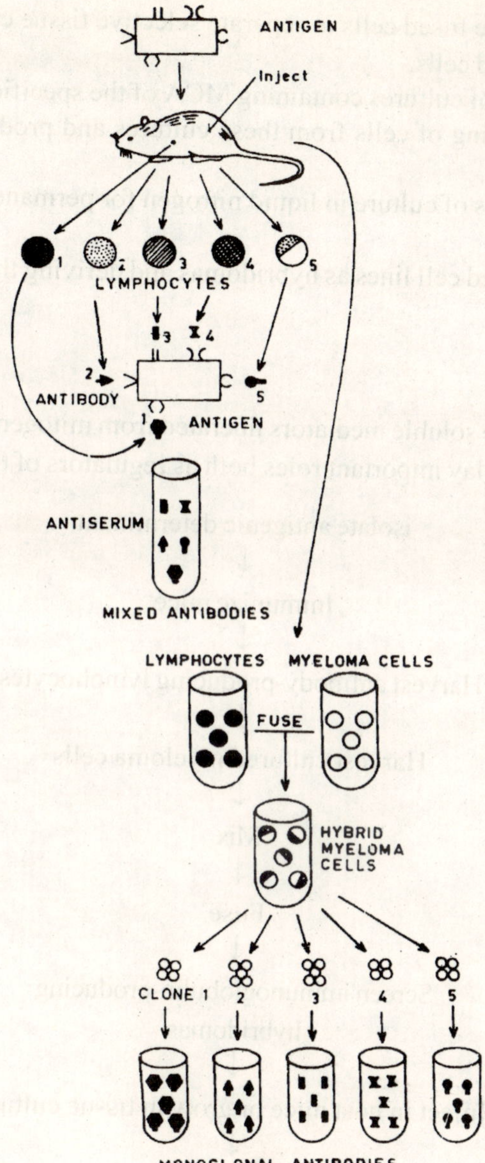


**Fig.** Various steps in production of Monoclonal Antibodies

*Various Steps for Production of MCA*

1. Immunization of mice with specific antigen required for a particular MCA.
2. Preparation of immune lymphocyte cell suspension and mixing with suitable myeloma cells.
3. Addition of a cell fusion agent (such polyethylene glycol) to the cell mixture.

4. Distributing the fused cells in separate selective tissue culture conditions to raise hybridized cells.
5. Identification of cultures containing MCA of the specific type.
6. Repeated cloning of cells from these cultures and production of particular MCA.
7. Freezing stocks of culture in liquid nitrogen for permanent preservation and utilization.
8. Growing cloned cell lines as hybridomas and deriving the MCA. (Fig.).

**Lymphokines**

Lymphokines are soluble mediators liberated from mitogen or antigen-activated lymphocytes. They play important roles both as regulators of the immune response

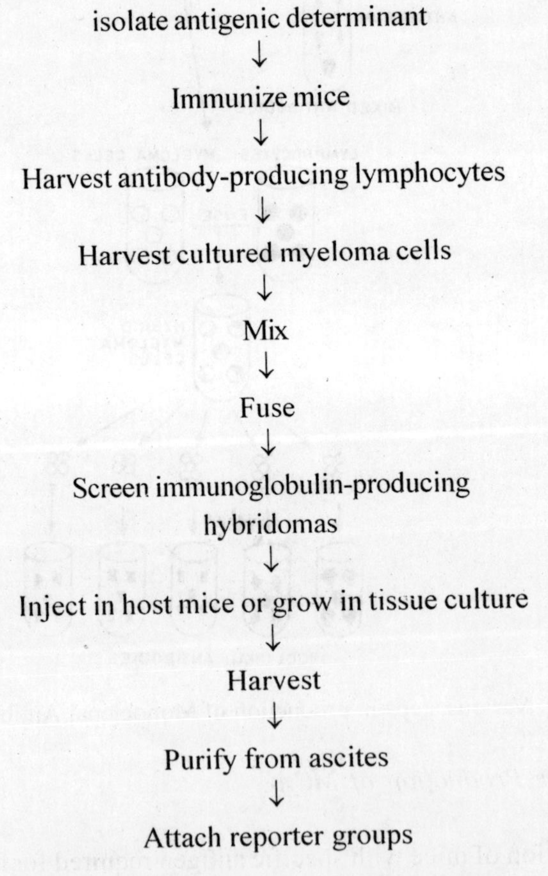


**Fig.** Production of Monoclonal Antibodies.

and as effector agents in the induction of delayed-type hypersensitivity and can roughly be divided into two groups: Cell regulatory lymphokines and inflammatory lymphokines. The lymphokines in the former group play a role in the effector mechanism of the immune response, interleukine 2 and T cell replacing factors belong to the group. The latter group includes lymphokines listed in Table and are considered to be involved in the induction of delayed type hypersensitivity.

**Table**

Some Inflammatory Lymphokines

| *Abbreviation* | *Full name* | *Major lymphokine effects* |
|---|---|---|
| MIF | Macrophage migration inhibitory factor | Inhibition of spontaneous motility of macrophages. |
| MAF | Macrophage activating factor | Increases of glucose oxidation, $O_2^-$ production, microbicidal capacity and tumor cell cytotoxicity of macrophages. |
| MCF | Macrophage chemotactic factor | Accumulation of macrophages at the sites of delayed-type skin lesions |
| VPF | Vascular permeability factor | Vascular change at the sites of the immune reaction |
| SRF | Skin reactive factor | Local inflammation at the sites of delayed-type skin lesions |
| MF | Mitogenic factor | Activation and proliferation of lymphocytes |
| LT | Lymphotoxin | Cell lysis |
| CSF | Colony stimulating factor | Growth of monocyte and granulocyte colonies |

Inflammatory lymphokines which react with macrophages and other inflammatory cells, or with the vascular endothelium are considered to play important roles as effector agents in the development of delayed hypersensitivity. Although each inflammatory lymphokine exhibits very interesting biological activity in vitro, as shown in Table, it is very difficult to prove that these molecules actually play an important role in vivo, because activated lymphocytes produce a great mixture of inflammatory lymphokines in very tiny amounts (approx 1-10 ng from $10^6$ activated lymphocytes). Also, their limited availability has hampered the chemical characterization of inflammatory lymphokines and presented several controversial problems concerning the molecular differences between them (e.g. migration inhibitory factor and macrophage activating factor; macrophage chemotactic factor and skin reactive factor). To solve these problems and facilitate medical applications of these physiologically interesting substances, effective methods for their large-scale production and purification need to be devised.

Recently, Toshiaki Osawa from the University of Tokyo has developed an effective method for their production. T cell hybridomas are expected to be excellent sources for the production of large amounts of inflammatory lymphokines. Human acute lymphatic leukemia cells, treated with an irreversible protein synthesis inhibitor (emetine) and an irreversible RNA synthesis inhibitor (actinomycin D), are fused with mitogen activated human peripheral blood lymphocytes. The monoclonal human T cell hybridomes thus established have been used for identifying and differenciating between several inflammatory lymphokines.

Fusion of human peripheral blood lymphocytes (activated by a T cell mitogen PHAP or Con A) under the optimum conditions for the production of inflammatory lymphokines with the above emetine and actinomycin D-treated leukemia cells, results in the formation of human T cell hybridomas secreting inflammatory lymphokines (Fig.). The optimum concentrations of emetine and actinomycin D for the treatment of leukemia cells vary depending on the nature of the leukemia cell lines and are determined for each individual cell line. However, the great advantage of Osawa's method is a the higher growth rate of T cell hybridomas compared with previous methods.

Apparently, preservation of functional stable human T cell hybridomas and their culture in a serum-free medium is vitally importance for the largescale preparation of human lymphokines and their possible medical application.

*Applications and utility of Monoclonal antibodies*

— The utility of monoclonal antibodies has been envisaged for the labelling and precise identification of specialized cells such as neurons.

— Monoclonal antibodies have been useful in determining the structure of cell membranes. These studies have been made possible since 1977 when Milstein and coworkers isolated several antibodies specific for mouse cell membrane surface antigens. These antibodies could be used as standard reagents to identify molecules present in cell membranes and to distinguish between cell populations.

— Currently monoclonal antibodies are being used directly in serotherapy as well as in the preparation of very specific vaccines active against viral strains.

— Monoclonal antibodies are now being seriously considered to neutralize the action of lymphocytes responsible for the rejection of grafts and destroy the autoantibodies produced in auto-immune diseases.

— Recently monoclonal antibodies are now finding application in tumour and cancer therapy. The medical literature is full of several recent trials and results in this connection and the detailed description is beyond the scope of the text.

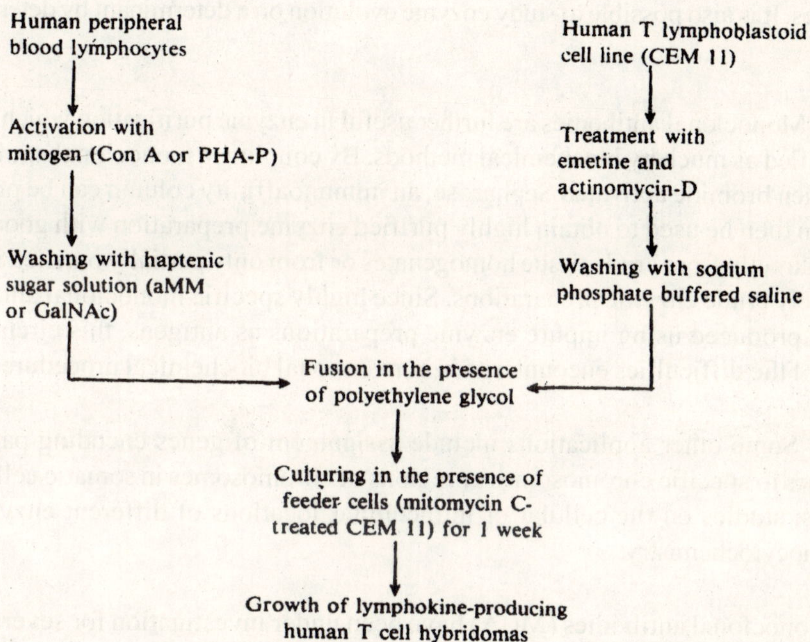


**Fig.** Production of Lymphokine-Producing Human T Cell Hybridomas by the Emetine-Actinomycin D Selection Method

— Monoclonal antibodies are useful in identification and classification of major histocompatibility complex products for tissue typing in transplantation and population genetic studies.

— For serogenetic classification of infectious microorganisms and protozoans and metazoan parasites.

— It is envisaged that application of a monoclonal antibody based assay will replace conventional tissue section immuno florescence assays (as employed in clinical laboratory test). Besides, it is likely that such assay will be very useful in forensic sciences.

—The potential of monoclonal antibodies as tools in enzyme genetics is developing fast as a discipline of immunogenetics. Different monoclonal antibodies can be raised against a given enzyme protein, each recognizing a small specific region on the molecular surface (antigenic determinant or opitome as described earlier). It is thought that the specificity of such a determinant is usually determined by only a few amino acids that are not necessarily sequential on a polypeptide chain. Thus, by using a panel of monoclonal antibodies it becomes possible to distinguish, on a determinant by determinant basis, between structurally distinct though genetically closely related enzymes. It is also possible to study enzyme evolution on a determinant by determinant basis.

—Monoclonal antibodies are further useful in enzyme purification which cannot be purified as much by biochemical methods. By coupling a monoclonal antibody to cyanogen bromide activated Sepharose, an immunoaffinity column can be prepared that can then be used to obtain highly purified enzyme preparation with good yield, either directly from crude tissue homogenates or from only partially purified but still relatively crude enzyme preparations. Since highly specific monoclonal antibodies can be produced using impure enzyme preparations as antigens, this circumvents many of the difficulties encountered in conventional biochemical procedures.

— Some other applications include assignment of genes encoding particular enzymes to specific chromosomes or regions of chromosomes in somatic cell hybrid studies; studies on the cellular or intracellular locations of different enzymes by immunocytochemistry.

Monoclonal antibodies (MCA) have been under investigation for several years in both the public sector and industry with a view to developing useful reagents for either monitoring clinical outcomes or elucidating clinical conditions. At times, it has appeared that exploitation of these new technologies has proceeded for reasons other than the merits of the project itself. At other times, suitable broad applications have been tantalizingly close. Today, it is possible to discern areas in which diagnosis using these approaches may be possible; they provide clinicians with information which genuinely benefits and has an impact on patient management.

There are questions which can be decided, if not answered with precision; if one asks whether products derived from biotechnology will contribute to health care and company profits, the answer is undoubtedly yes-even if the precise products cannot be named. It is also possible to identify laboratory uses of such products. The above figure describes in general terms some of the areas where the first in-vitro tools of the new technologies are likely to be used and also indicates which areas seem best suited to each tool; it also shows that there are areas where monoclonal antibodies have procedural or technical advantages, and an area of overlap between them.

# SECTION - 3

# SIGNIFICANCE FOR ANIMALS

# SEVEN

# Cell and Tissue Culture

## History and Scope

Animal cell and tissue culture came into light in the beginning of the present century, when cells could be shown to survive and divide in vitro (Jolly, 1903). The true beginning of animal cell and tissue culture was, however, made a little later, when using frog (a cold blooded animal = poikilotherm) as a source of tissue, Ross Harrison (1907) developed a reproducible technique for tissue culture and later Alexis Carrel (1912) used tissue and embryo extracts as culture media. They successfully demonstrated that animal cells can be grown indefinitely in vitro, just like protozoa and microorganisms. Although initially cold blooded animals like frog were used, so that no incubation was required, but later warm blooded animals like chick and rodents were used and eventually mammals became a favourite material. For about 50 years, mainly tissue explants rather than cells were used for culture techniques, although later after 1950's, mainly dispersed cells in cultures were utilized. As in plants, the terms 'cell culture', 'tissue culture' and 'organ culture' are commonly used for animal material also. However, for animals if the explant maintains its structure and function in culture, it is described as an 'organotypic culture' and if cells in culture reassociate to create a three dimensional structure, irrespective of the tissue from which it was derived, it is described as a 'histotypic culture'.

For many years, the lower vertebrates and the invertebrates were not used for tissue culture studies, even though some aspects of their development, like tissue regeneration in amphibia and metamorphosis in insects, made them attractive systems for tissue culture studies. More recently, a resurgence of interest in tissue culture in these groups has also been witnessed even though it still remains a very specialized area. Consequently, the bulk of interest in animal cell and tissue culture utilises avian (birds) and mammalian tissues. The techniques, therefore, are largely devleoped for the culture of cells and tissues from mammals. The technical advances in tissue culture became possible mainly due to the following reasons: (a) commercial supply of reliable media and sera (singular 'serum'), (b) greater control of contamination with the use of antibiotics and that of clean air equipment (laminer air flow chamber).

A vareity of areas of research in biotechnology depend largely on tissue culture and, therefore, these areas have given the desired impetus to the development of tissue culture methods in animals. These areas include the following: (i) production of antiviral vaccines, which required the standardization of cell lines for the multiplication and assay of viruses; (ii) cancer research, which required the study of uncontrolled cell division in cultures; (iii) cell fusion techniques, (iv) genetic manipulation, which is possible in cells or organs in culture; (v) monoclonal antibodies whose production requires cell lines in culture; (vi) production of pharmaceutical drugs using cell lines; (vii) chromosome analysis of cells derived from womb; (viii) study of the effects of toxins and pollutants using cell lines; (ix) use of artificial skin; (x) study of the function of nerve cells (even though neurons can not be propagated in vitro, without resorting to the use of transformed cells).

## Advantages and Disadvantages of Tissue Culture

Tissue culture offers a number of advantages for a variety of studies and applications, but there are also disadvantages, so that tissue culture should not be used indiscriminately. These advantages and disadvantages are listed in Table.

**Table**

Advantages and disadvantages of the use of animal tissue culture

| *Advantages* | *Disadvantages* |
|---|---|
| 1. Controlled physiochemical environment (pH, temperature, osmotic pressure, $O_2$, $CO_2$, etc. | 1. Expertise is needed, so that behaviour of cells in culture can be interpreted and regulated. |
| 2. Controlled and defined physiological conditions (constitution of medium, etc.) | 2. Ten times more expensive for same quantity of animal tissue; therefore reasons for its use should be compelling. |
| 3. Homogeneity of cell types (achieved through serial passages) | 3. Unstable aneuploid chromosome constitution |
| 4. Economical, since smaller quantities of reagents are needed than in vivo. | |
| 5. Legal, moral and ethical questions of animal experimentation are avoided. | |

## Laboratory Facilities

The laboratory facilities for animal culture (as also for plant tissue culture) should allow for the following: (i) sterile handling; (ii) incubation (iii) preparation of glassware, media and tissue; (iv) wash-up; (v) sterilization and (iv) storage. The clean area for handling should be available at one end of the room and the wash up and sterilization at the other end, with preparation, incubation and storage in between, Tissue culture facilities required in any laboratory are listed in Table.

**Table**

Tissue culture facilities (modified from Freshney, 1987)

| *Minimum requirements (essential)* | *Desirable features (beneficial)* | *Useful additions* |
|---|---|---|
| 1. **Sterile area:** clean and quiet, no thorough traffic, separate from animal house and microbiological labs | 1. Filtered air (air conditioning) | 1. Piped COQ and compressed air |
| 2. **Preparation area** | 2. Hot room with temperature recorder | 2. Store room for bulk plastic |
| 3. **Storage areas:** | 3. Microscope room | 3. Containment room for biohazard work. |
| (a) *liquids*-ambient, 4°—30° | 4. Dark-room | 4. Liquid $N_2$ storage tank (—500 1) |
| (b) *glassware* (shelving) | 5. Service bench adjacent to culture area | |
| (c) *plastics* (shelving) | 6. Separate preparation room. | |
| (d) *small items* (drawers) | 7. Separate sterilizing room | |
| (e) *specialized equipment* (slow turnover), cupborard (s) | 8. Cylinder store | |
| (f) chemicals-ambient, 4°-20°C (share with liquids but keep chemicals in sealed container over desiccant) | | |

The introduction of laminar flow cabinets or hoods to provide a clean bench has greatly facilitated the maintenance of aseptic and sterile conditions, so that no separate room for sterile handling is needed and a laminar flow cabinet can be kept in an unspecialized laboratory space. Sometimes, however, aseptic rooms are disigned, where Filtered air is supplied from the ceiling, so that the whole room is regarded as a sterile working area. Incubation is generally done in incubators or thermostatically controlled hot rooms. In hot rooms, racks are designed for the placement of cultures (tubes, Petri plates, flasks, etc.).

The media may be purchased ready made or may be prepared in the laboratory, the latter being cheaper, if needed in large quantities. The washing and sterilization facility (autoclave, etc.) are often kept outside the tissue culture laboratory, since they would generate moisture (increased humidity) and heat. Facilities should also be available for the storage of the following: (i) sterile liquids (media, serum, etc.) at 4°C, -20°C or -70°C; (ii) sterile glassware and plasticware (bottles, flasks, pipettes, Petri dishes, syringes); (iii) screw caps, filter tubes, stoppers, etc.; (iv) gloves, disposable bags, etc.; (v) other equipment (filters, large receiver flasks). For a detailed account of the design and layout of tissue culture facilities, the readers are advised to consult the book 'Culture of Animal Cells: A Manual of Basic Techniques' written by R. Ian Freshney and published by Alan R. Liss Inc., New York.

**Table**

Tissue culture equipment (modified from Freshney, 1987)

| ***Minimum requirements (essential)*** | ***Desirable features (beneficial)*** | ***Useful additions*** |
|---|---|---|
| 1. Incubator | 1. Laminar flow hood(s). | 1. -70°C freezer |
| 2. Sterilizer (autoclave, pressure cooker, oven) | 2. Cell counter | 2. Glassware washing machine |
| 3. Refrigerator | 3. Vacuum pump | 3. Closed-circuit TV for inverted microscope(s) |
| 4. Freezer (for -20°C storage) | 4. $CO_2$ incubator | 4. Colony counter |
| 5. Inverted microscope | 5. Coarse and fine balance | 5. High-capacity centrifuge (6 × 1 liter) |
| 6. Soaking bath or sink | 4. | 6. Cell seer (e.g., Coulter ZB series) |
| 7. Deep washing sink | 6. pH meter | 7. Time-lapse cinemicrographic equipment |
| 8. Pipette cylinder (s) | 7. Osmometer | 8. Interference-contrast microscope |
| 9. Pipette washer | 8. Phase-contrast and fluorescence microscope(s) | 9. Polythene bag scaler (for packaging sterile items of long-term storage) |
| 10. Water purifier | 9. portable temperature recorder | 10. Controlled-rate cooler (for cell freezing) |
| 11. Bench centrifuge | 10. Permanent temperature recorders on sterilizing oven and autoclave | 11. Filing for freezer records and catalogues |
| 12. Liquid $N_2$ freezer (351, 1,500-3,000 ampules) | 11. Roller racks for roller bottle culture | 12. Centrifugal elutriator centrifuge and rotor |
| 13. Liquid $N_2$ storage flask (251) | 12. Magnetic stirrer racks for suspension cultures | 13. fluorescence-activated cell sorter |
| | 13. Pipette drier | 14. Densitometer |
| | 14. Pipette plugger | 15. Density meter (for density gradient cell separation) |
| | 15. Trolleys for collecting soiled glassware and redistributing fresh supplies | |
| | 16. Pipette aid | |
| | 17. Autopipette or other form of automatic dispensor, dilator | |
| | 18. separate sterilizing oven and drying oven | |

## Substrate on which Cells Grow

The majority of vertebrate cells cultured in vitro have been grown as monolayers on an artificial substrate. Spontaneous growth in suspension is restricted to hemopoietic cell lines, rodent ascites tumour cell lines and a few other selected cell lines. With these exceptions, most cells need to spread out on a substrate inorder to proliferate. Inadequate spreading will inhibit cell proliferation. Cells shown to require attachment for growth are said to be anchorage dependent.

However, cells which have undergone transformation , frequently become anchorage independent and can grow in suspension. (Note: In microbiology and plant cell culture, transformation means uptake of new genetic material but in animal cell culture, particularly mammalian cell culture, transformation means phenotypic modification of cells in culture, without involving the uptake of new genetic material, which is described as transfection to distinguish it from transformation). Whenever, a substrate is needed, it may be adhesive (glass, plastic, palladium, etc. or a metallic surface; these surfaces may sometimes be variously modified) or nonadhesive (agar, agarose, methocel) and will be briefly discussed in this section.

## Glass as a Substrate

Glass is the most commonly used substrate, in the form of slides, coverslips, flasks and test tubes, which may or may not be variously modified. It is cheap, easily washed without losing its growth supporting properties, can be readily sterilized by dry heat (oven) or moist heat (autoclave) and is optically clear for microscopic examination of cultures. Treatment with strong alkali renders glass unsatisfactory for culture unless it is neutralized by acid wash.

## Disposable Plastic as a Substrate

Polystyrene is often used as the most common and cheapest substrate to be discarded after single use. This plasticware is supplied sterile and is not autoclavable. However, it is hydrophobic and needs to be treated by gamma irradiation or some chemical or with electric arc to produce a charged surface, which thus becomes wettable. One may, however, have to determine the plating efficiency and growth rate of cells, since great variation occurs in the plasticware manufactured by different commercial companies.

Other plastics used for tissue culture include the following: polyvinylchloride (PVC), polycarbonate, polytetrafluoroethylene or teflon (PTFE), thermanox (TPX), etc. In several of these other plastics, one should try to grow a regular monolayer and then attempt to culture or clone cells on it with and without pretreatment. PTFE films are available as displosble Petri dishes or as membranes to be incorporated in an autoclavable reusable culture vessel. These dishes have two other advantages: (i) substrate is permeable to O, and CO., and (ii) plastic is thin, and is suited to sectioning for microscopy.

Substrates involving polystyrene, sephadex and polyacrylamide are also available in the form of beads for propagation of anchorage dependent cells in suspension.

## Palladium and Metallic Surface as Substrates

For 90% of animal cell and tissue cultures, glass or plastic is employed as a substrate. However, other substrates have been used for specialized applications. For instance, palladium deposited on agarose (using electron microscopy shadowing equipment) was used as a substrate for growth of fibroblasts and glia. Stainless steel discs and other metallic surfaces were also used.

## Treatment of Substrate Surfaces

The substrate surfaces described above can be variously treated to improve cell attatchment. The treatments with culture medium meant for another culture, or with purified fibronectin (lng/ml) added to the medium or with collagen are commonly used. Coating with collagen can be achieved, using rat tail collagen or other commercially available forms, and pouring the collagen solution over the surface of the dish, draining the excess, and permitting the residue to dry. After coating with collagen, the substrate may be sterilized under UV and used either immediately or may also be stored. Sometimes on the inert coating of collagen (or gelatin or poly - D-lysine), a monolayer of an appropriate cell type may be used to provide a matrix for the maintenance of some specialized cells.

## Feeder Layers on Substrate

Sometimes, on the substrate surface a monolayer of special kind of cells is used, partly to supplement the culture medium for growth and partly for conditioning of the substrate by cell products. These monolayers are called feeder layers, since they also feed the growing cultures. The feeder layer may consist of mouse embryo fibroblasts, normal fetal intestine, glial cells, etc. and has been effectively used for selective growth of breast and coelomic epithelium, central and peripheral neurons, and also for growing cells used for the production of transgenic animals through transfection.

### Gas Phase for Tissue Culture

The animal cell and tissue cultures also require for their growth a gas phase consisting of $O_2$ and $CO_2$ even though cultures (cell culture *vs* organ culture) may vary in their oxygen requirement. While cell cultures, may grow on an atmospheric oxygen concentration or even lower $O_2$ concentration, some organ cultures (e.g. late embryo, newborn or adult) may require upto 95% $O_2$. While $O_2$ is directly available to cells in cell cultures, in organ cultures there may be a problem of diffusion. Since there is a trend towards the use of serum free media, $O_2$ concentration may become an important factor in tissue cultures in future. $CO_2$ concentration , which is a complex factor, also needs to be manipulated, since it also influences the pH and the $HCO_3^-$ ion concentration of the culture. At very low concentration of cells, it is necessary to add $CO_2$ in gas phase of sealed flasks of most cultures. However, in high concentration of cells, $CO_2$ is liberated from the cultured tissue or cells. Therefore, it is not necessary to add $CO_2$ in gas phase of such sealed flasks with high concentration of cells; but instead sometimes we may need to slacken the cap of a culture flask and allow excess $CO_2$ to escape.

## Culture Media

Culture medium is one single most important factor for culturing cells and tissues. It provides (i) the optimum conditions of factors like, pH, osmotic pressure, etc. and (ii) the chemical constituents, which the cells or tissues are incapable of synthesizing (unlike microorganisms, which can synthesize them from simple inorganic substances). Animal cells need either a completely natural medium or an artificial medium supplemented with some natural products. Sometimes the natural medium is also preferred, since it is the cheapest and most convenient.

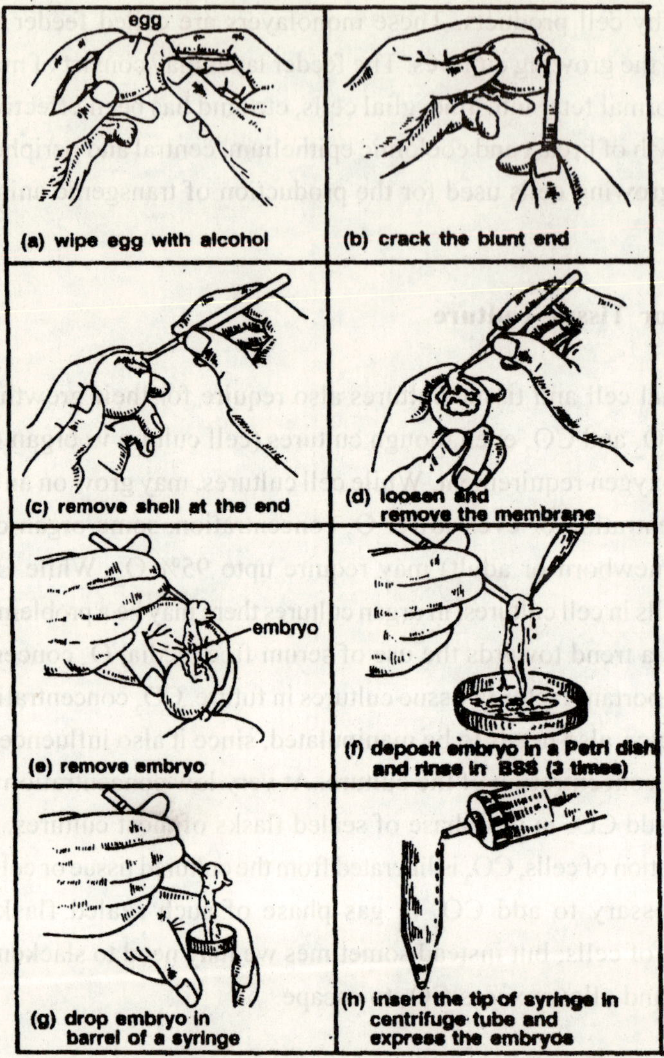


Fig. Different steps involved in the preparation of embryo extract

## Natural Media

The natural media used to promote cell growth fall in following three categories: (i) coagula, such as plasma clots; (ii) biological fluids, such as serum and (iii) tissue extracts of which embryo extract is the most common.

Plasma clots—Plasma clots have been in use for long time and are now available commercially either (i) as liquid plasma in siliconed ampules or (b) as lyophilized plasma to be reconstituted by the addition of distilled water saturated with $CO_2$.

However, one may also prepare plasma in the laboratory, usually from a male fowl, where efforts are made to prevent coagulation of blood, often by adding heparin as an anticoagulant.

Biological fluids (e.g. serum)—The most commonly used biological fluid is serum, which is commonly obtained from human adult blood, placental cord blood, horse blood or calf blood. Of these, human placental-cord serum and fetal calf serum seem to be particularly satisfactory. The serum is obtained as exuded liquid from blood undergoing coagulation, and is filtered using millipore filters. The serum is tested for sterility and toxicity before using, and is stored at low temperature. Testing may be done by incorporating the serum in a growth medium and checking (i) the presence of contaminants and (ii) the plating efficiency of a standard strain (e.g. fiborblast strains). According to some workers, toxicity of sera can be reduced by heat inactivation.

Besides serum, other biological fluids used as natural media include the following: (i) amniotic fluid, (ii) ascitic and pleural fluid, (iii) aqueous humour (from eye), (iv) serum ultrafiltrate, (v) insect haemolymph, (vi) coconut milk (more often used for plant tissue culture).

Tissue extracts—Embryo extract is the most commonly used tissue extract, although effective substitutes have now been developed to replace the embryo extract as a natural medium. The most important component of the substitutes is a mixture of amino acids. Embryo extract is often prepared from chick or bovine embryos of different ages (upto 10 days). The detailed procedure for the preparation of embryo extract is shown in Figure, and can be found in any advanced text book on animal tissue culture (e.g. Paul, 1975).

Other tissue extracts have also been employed, e.g. spleen, liver, bone marrow, leucocytes and tumours. However, all tissue extracts deteriorate rapidly at room temperature, and within a week even in refrigerator. At 20°C, these can be stored satisfactorily.

**Defined Media (for warm blooded animals)**

Media with serum—Although natural media are very useful and convenient for a wide range of uses, they suffer from disadvantage of poor reproducibility due to lack of knowledge of the exact constitution. For this reason, synthetic media have been designed for a variety of tissues and purposes, like the following: (i) media for immediate survival, (ii) media for prolonged survival, (iii) media for indefinite growth and (iv) media for specialized functions. For immediate survival, a 'balanced salt solution' with defined osmotic pressure and pH is needed (Table). This requirement is met by combination of certain inorganic ions, salts and glucose. For longer survival, serum may be used or the balanced salt solution may be supplemented

**Table**

Constitutions of different balanced salt solutions

| *Component* | *Earle's balanced salt solution; g/1* | *Dulbecco's phosphate buffered saline (solution A) (PBSA): g/1* | *Hanks' balanced salt solution; g/1* | *Spinner salt solution (Eagle); g/1* |
|---|---|---|---|---|
| **1. Inorganic sells** | | | | |
| $CaCl_2$ (anhyd.) | 0.02 | — | 0.14 | — |
| KCl | 0.04 | 0.20 | 0.40 | 0.40 |
| $KH_2PO_4$ | — | 0.20 | 0.06 | — |
| $MgCl_2 . 6H_2O$ | — | — | 0.10* | — |
| $MgSO_4 . 7H_2O$ | 0.20 | — | 0.10* | 0.20 |
| NaCl | 6.68 | 8.00 | 8.00 | 6.80 |
| $NaHCO_3$ | 2.20 | — | 0.35 | 2.20 |
| $Na_2HPO_4.7H_2O$ | — | 2.16 | 0.09** | — |
| $NaH_2PO_4.H_2O$ | 0.14++ | — | — | 1.40 |
| **2. Other components** | | | | |
| D-glucose | 1.00 | — | 1.00 | 1.00 |
| Phenol red | 0.01 | — | 0.01^^ | 0.01 |

* MgCl2.6H2O added to original formula.

++ Original formulation calls for 0.15 mg/1 NaH2PO42H2O, In vitro, 9. #6 (1974)

** Original formulation calls for 0.06 g/1 Na2HPO4.H2O. Proc. Fap. Biol. and Med.. 71[1949]

^^ Original formulation calls for 0.02 g/1. Proc. Soc. Exp. Not. and Med., 71[1939.

with amino acids, oxygen, vitamins, and serum protein. One such medium was developed and modified by Eagle (1955), which is described as minimum essential medium (MEM). This contains essential amino acids, vitamins and salts. Other more complex synthetic media include the following: (i) 199 (Morgan et al., 1950), (ii) CMRL 1066 (Parker et al., 1957), (iii) RPMI 1640 (Moore et al., 1967), and (iv) F12 (Ham, 1965). Common constituents of some of these media for animal tissue culture are listed in Table (for more details, consult Freshney's book published in 1987). All these defined media are usually supplemented with 5-20% serum.

**Table**

Different commonly used media (with serum), for animal cell and tissue culture, and their constituents (all quantities in mg/1)

| Component | *Eagle's MEM* | *Dulbecco's modiftcation* | *Ham's F12* | *CMRL 1066* | *RPMI 1640* | *199* | *L15* | *Fischer's* | *Waymouth MB 752* |
|---|---|---|---|---|---|---|---|---|---|
| **1. Amino acids** | | | | | | | | | |
| L-alanine | — | — | 8.90 | 25.0 | — | 25.0 | 225 | — | — |
| L-arginine (free base) | — | — | — | — | 200 | — | 500 | — | — |
| L-arginine-HCI | 126 | 84.0 | 211 | 70.0 | — | 70.0 | — | 15.0 | 75.0 |
| L-asparagine | — | — | — | 50.0 | — | — | — | — | — |
| L-asparagine-$H_2O$ | — | — | 15.0 | — | — | 260 | 11.4 | — | — |
| L-aspartic acid | — | — | 13.3 | 30.0 | 30.0 | 10.0 | — | — | 60.0 |
| L-cysteine (free base) | — | — | — | — | — | — | 120 | — | 61.0 |
| L-cystine | 24 | 48.0 | — | 20.0 | 50.0 | — | — | — | 15.0 |
| L-cystine.2Na | — | — | — | — | 23.7 | — | 23.7 | — | — |
| L-cysteine. HCI.120 | — | — | 35.1 | 260 | — | 0.0987 | — | — | — |
| L-glutamic acid | — | — | 14.7 | 75.0 | 20.0 | 66.8 | — | — | 150 |
| L-glutamine | 292 | 584 | 146 | 100 | 300 | 100 | 300 | 200 | 350 |
| Glycine | — | 30.0 | 7.5O | 50.0 | 10.0 | 50.0 | 200 | — | 50.0 |
| L-histidine (free base) | — | — | — | — | 15.0 | — | 250 | — | 128 |
| L-histidine HCl.$H_2O$ | 42.0 | 42.0 | 21.0 | 20.0 | — | 21.9 | — | 81.1 | — |
| L-hydroxy-proline | — | — | — | 10.0 | 20.0 | 10.0 | — | — | — |
| L-isoleucine | 52.0 | 105 | 3.94 | 20.0 | 50.0 | 20.0 | 125 | 75.0 | 25.0 |
| L-leucine | 52.0 | 105 | 13.1 | 60.0 | 50.0 | 60.0 | 125 | 30.0 | 50.0 |
| L-lysine. HCI | 73.1 | 146 | 36.5 | 70.0 | 40.0 | 70.0 | 93 | 50.0 | 240 |
| L-methionine | 15.0 | 30.0 | 4.48 | 15.0 | 15.0 | 15.0 | 75.0 | 100 | 50.0 |
| L-phenylalanine | 33.0 | 66.0 | 4.96 | 25.0 | 15.0 | 25.0 | 125 | 67.0 | 50.0 |
| L-proline | — | — | 34.5 | 40.0 | 20.0 | 40.0 | — | — | 50.0 |
| L-serine | — | 42.0 | 10.5 | 25.0 | 30.0 | 25.0 | 200 | 15.0 | — |
| L-threonine | 48.0 | 95.0 | 11.9 | 30.0 | 20.0 | 30.0 | 300 | 40.0 | 75.0 |
| L-tryptophan | 10.0 | 16.0 | 2.04 | 10.0 | 5.0 | 10.0 | 20.0 | 10.0 | 40.0 |
| L-tyrosine | 36.0 | 72.0 | 5.40 | 40.0 | 20.0 | — | — | — | 40.0 |
| L-tyrosine. 2Na | — | — | — | — | — | 49.7 | 373 | 74.6 | — |
| L-valine | 47.0 | 94.0 | 11.7 | 25.0 | 20.0 | 25.0 | 100 | 70.0 | 65.0 |
| **2. Vitamins** | | | | | | | | | |
| L-ascorbic acid | — | — | — | 50.0 | — | 0.050 | — | — | 17.5 |
| Biotin | — | — | 0.0073 | 0.010 | 0.200 | 0.010 | — | 0.010 | 0.02 |
| D-Ca-pantothenate | 1.00 | 4.00 | 0.480 | 0.010 | 0.250 | 0.010 | 1.00 | 0.500 | 1.00 |
| Calciferol | — | — | — | — | — | 0.100 | — | — | — |
| Choline chloride | 1.00 | 4.00 | 14.0 | 0.500 | 3.00 | 0.500 | 1.00 | 1.50 | 250 |
| Folic acid | 1.00 | 4.00 | 1.30 | 0.010 | 1.00 | 0.010 | 1.00 | 10.0 | 0.40 |
| i-inositol | 2.00 | 7.20 | 18.0 | 0.050 | 35.0 | 0.050 | 2.00 | 1.50 | 1.00 |
| Nicotinamide | 1.00 | 4.00 | 0.04 | 0.025 | 1.00 | 0.025 | 1.00 | 0.50 | 1.00 |
| Pyridoxal. HCI | 1.00 | 4.00 | 0.062 | 0.025 | — | 0.025 | — | 0.50 | — |
| Riboflavin | 0.10 | 0.40 | 0.038 | 0.010 | 0.20 | 0.010 | — | 0.50 | 1.00 |
| Thiamin. HCI | 1.00 | 4.00 | 0.34 | 0.010 | 1.00 | 0.010 | — | 1.00 | 10.0 |
| Vitamin Bu | — | — | 1.36 | — | 0.005 | — | — | — | 0.20 |
| Pyridoxine HCI | — | — | 0.062 | 0. 025 | 1.00 | 0.025 | — | — | 1.00 |

*(Table Contd.)*

| Component | Eagle's MEM | Dulbecco's modiftcation | Ham's F12 | CMRL 1066 | RPMI 1640 | 199 | L15 | Fischer's | Waymouth MB 752 |
|---|---|---|---|---|---|---|---|---|---|
| Cholesterol | — | — | — | 0.200 | — | — | — | — | — |
| Para-amino-benzoic acid | — | — | — | 0.050 | 1.00 | 0.050 | — | — | — |
| Nicotinic acid | — | — | — | — | — | 0.025 | — | — | — |
| Menaphthone sodium bisulphate $3H_2O$ | — | — | — | — | — | 0.019 | — | — | — |
| DI-α tocopherol $PO_4.2Na$ | — | — | — | — | — | 0.01 | — | — | — |
| Vitamin A acetate | — | — | — | — | — | 0.115 | — | — | — |
| Riboflavin $PO_4.2Na$ | — | — | — | — | | — | 0.10 | — | — |
| Thiamin mono $PO_4$, $_2H_2O$ | — | — | — | — | — | — | 1.00 | — | — |
| **3. Inorganic salts** | | | | | | | | | |
| $CaCl_2$ (anhyd.) | 200 | 200 | 200 | — | — | — | — | — | — |
| $CaCl_2.2H_2O$ | — | — | 44.0 | — | — | 186 | 186 | 91.0 | 120 |
| $Fe(NO_3)_3.9H_2O$ | — | 0.10 | — | — | — | 0.10 | — | — | — |
| KCl | 400 | 400 | 224 | 100 | 400 | 400 | 400 | 400 | 150 |
| $KH_2PO_4$ | — | — | — | — | — | 60.0 | 60.0 | — | 80.0 |
| $MgCl_2.6H_2O$ | — | — | 122 | — | — | — | — | — | 240 |
| $MgSO_4.7H_2O$ | 200 | 200 | — | 200 | 100 | 200 | 400 | 121 | 200 |
| NaCl | 6.800 | 6.400 | 7.599 | 6.799 | 6.000 | 8.000 | 8.000 | 8.000 | 6.000 |
| $NaHCO_3$ | 2.200 | 3.700 | 1.176 | 2.200 | 2.200 | 350 | — | 1.125 | 2.240 |
| $Na_2H_2PO_4.H_2O$ | 140 | 125 | — | 140 | — | — | — | 78.0 | — |
| $Na_2HPO_4$ (anhyd) | — | — | — | — | — | 47.5 | 190 | 60.0 | — |
| $Na_2HPO_4.7H_2O$ | — | — | 268 | — | 1.512 | — | — | — | 506 |
| $CuSO_4.5H_2O$ | — | — | 0.00249 | — | — | — | — | — | — |
| $FeSO_4.7H_2O$ | — | — | 0.834 | — | — | — | — | — | — |
| $ZnSO_4.7H_2O$ | — | — | 0.863 | — | — | — | — | — | — |
| $CaNO_3.4H_2O$ | — | — | — | — | 100 | — | — | — | — |
| **4. Other components** | | | | | | | | | |
| D-glucose | 1.000 | 4.500 | 1.802 | 1.000 | 2.000 | 1.000 | — | 1.000 | 5.000 |
| D-galactose | — | — | — | — | — | — | 9.000 | — | — |
| Lipoic acid | — | — | 0.21 | — | — | — | — | — | — |
| Phenol red | 10.0 | 15.0 | 12.0 | 20.0 | 5.00 | 17.0 | 10.0 | 5.00 | 10.0 |
| Sodium pyruvate | — | 110 | 110 | — | — | — | 550 | — | — |
| Hypoxanthine | — | — | 4.10 | — | — | 0.30 | — | — | — |
| Linoleic acid | — | 0.084 | — | — | — | — | — | — | 25.0 |
| Putrescine 2HCl | — | — | 0.161 | — | — | — | — | — | — |
| Thymidine | — | — | 0.73 | 10.0 | — | — | — | — | — |
| Cocarboxylase | — | — | — | 1.00 | — | — | — | — | — |
| Coenzyme A | — | — | — | 2.50 | — | — | — | — | — |
| Deoxyadenosine | — | — | — | 10.0 | — | — | — | — | — |
| Deoxycytidine HCl | — | — | — | 10.0 | — | — | — | — | — |
| Deoxyguanosine | — | — | — | 10.0 | — | — | — | — | — |
| Diphosphopyridine nucleotide.$4H_2O$ | — | — | — | 7.00 | — | — | — | — | — |
| Ethanol for solubilizing lipid components | — | — | — | 16.0 | — | — | — | — | — |

*(Table Contd.)*

| *Component* | *Eagle's MEM* | *Dulbecco's modification* | *Ham's F12* | *CMRL 1066* | *RPMI 1640* | *199* | *L15* | *Fischer's* | *Waymouth MB 752* |
|---|---|---|---|---|---|---|---|---|---|
| Flavine adenine dinucleotide | — | — | — | 1.00 | — | — | — | — | — |
| Glutathione (reduced) | — | — | — | 10.0 | 1.00 | 0.05 | — | — | 15.0 |
| S-methyl-deoxy-cytidine | — | — | — | 0.10 | — | — | — | — | — |
| Sodium acetate.3$H_2O$ | — | — | — | 83.0 | — | — | — | — | — |
| Sodium glucuronate.$H_2O$ | — | — | — | 4.20 | — | — | — | — | — |
| Triphosphopyridine nucleotide | — | — | — | 1.00 | — | — | — | — | — |
| Tween 80 | — | — | — | 5.00 | — | 5.00 | — | — | — |
| Uridine triphosphate. 4$H_2O$ | — | — | — | 1.00 | — | — | — | — | — |
| Adenine SO4 | — | — | — | — | — | 10.0 | — | — | — |
| 5'AMP | — | — | — | — | — | 0.20 | — | — | — |
| ATP-2Na | — | — | — | — | — | 10.0 | — | — | — |
| Cholesterol | — | — | — | — | — | 0.20 | — | — | — |
| 2-Deoayribose | — | — | — | — | — | 0.50 | — | — | — |
| Guanine HCl | — | — | — | — | — | 0.30 | — | — | — |
| D-ribose | — | — | — | — | — | 0.50 | — | — | — |
| Na acetate | — | — | — | — | — | 36.7 | — | — | — |
| Thymine | — | — | — | — | — | 0.30 | — | — | — |
| Uracil | — | — | — | — | — | 0.30 | — | — | — |
| Xanthine | — | — | — | — | — | 0.30 | — | — | — |
| 5. $CO_2$ (gas phase) | 5% | 10% | 5% (pH 7.0) 2% (pH 7.4) | 5% | 5% | Ambient | Ambient | 5% (pH 7.0) 2% (pH 7.4) | 5% |

* Note : For use in a gas phase of air in sealed containers, flanks' salts may be substituted in Eagle's MEM

Serum free media—The use of serum in culture media has several disadvantages, which led to the development of many serum free media. Some of the disadvantages include the following: (i) serum varies from batch to batch and deteriorates within one year; (ii) changing serum batch requires fresh testing, so that the replacement is as close to the previous batch as possible; (iii) if more than one cell types are used, each may require different serum batch, so that a number of batches may need to be maintained and co-culturing may be difficult; (iv) the demand of serum usually exceeds the supply for a variety of reasons; (v) when cell culture is used for downstream processing to recover cell products, the presence of serum is an obstacle to purification; (vi) serum increases the cost of the medium, since its cost is ten times the cost of its constituents if used as a substitute; (vii) serum may stimulate undesirable growth and may even inhibit growth in some cases.

The major advantage of serum free media is the ability to make a medium selective for a particular cell type, since each cell type appears to require a different recipe. However, the removal of scrum also requires (i) high degree of purity of reagents and water and (ii) high degree of cleanliness of all apparatus. This is necessary, since removal of serum removes the protective and detoxifying action, which some of the serum proteins have.

The serum free media may be developed using any of the following two approaches: (i) A known recipe for a related cell type may be used, and individual constituents may be altered, until the medium is optimized, as has been done by Ham (1984). This is time consuming, and sometimes takes as many as three years for developing a new medium. (ii) In the second, approach, an existing medium like RPMI 1640 or Ham's F12 or DMEM (Dulbecco's modified minimal essential medium) is taken and a shorter list of constituents like selenium, transferrin, albumin, insulin, androgen, hydrocortisone, estrogen, etc. is used for manipulation. Using these approaches, many recipes for serum free media have been prepared and are available for commercial use or can be made up in the laboratory (Table). However, despite several advantages associated with the use of serum free media, the serum free media will not be put to very common use, till better, convenient, cheap and less

**Table**

Examples of serum-free media

(basal formulae*); (modified from Freshney,1987)

| *Component* | *MCDB110* | *MCDB202* | *MCDB402* | *MCDBl53* | *Iscove's* | *LHC* |
|---|---|---|---|---|---|---|
| 1. Amino acids | | | | | | |
| (All as L-enantiomers) | | | | | | |
| Alanine | 1.0E-4 | 1.0E-4 | — | 1.0E-4 | 2.8E-4 | 1.0E-4 |
| Arglnine HCl | 1.0E-3 | 3.0E-4 | 3.0E-4 | 1.0E-3 | 4.0E-4 | 2.0E-3 |
| Asparagine | 1.0E-4 | 1.0E-3 | 1.0E-4 | 1.0E-4 | 1.0E-4 | 1.0E-4 |
| Aspartic acid | 1.0E-4 | 1.0E-4 | 1.0E-5 | 3.0E-5 | 2.3E-4 | 3.0E-5 |
| Cysteine HCl | 5.0E-5 | 2.0E-4 | — | 2.4E-4 | — | 2.4E-4 |
| Cystine | — | 2.0E-4 | 4.0E-4 | — | 2.9E-4 | — |
| Glutamic acid | 1.0E-4 | 1.0E-4 | 1.0E-5 | 1.0E-4 | 5.1E-4 | 1.0E-4 |
| Glutamine | 2.5E-3 | 1.0E-3 | 5.0E-3 | 6.0E-3 | 4.0E-3 | 6.0E-3 |
| Glycine | 3.0E-4 | 1.0E-4 | 1.0E-4 | 1.0E-4 | 1.0E-4 | 1.0E-4 |
| Histidine . HCl | 1.0E-4 | 1.0E-4 | 2.0E-3 | 8.0E-5 | 2.0E-4 | 1.6E-4 |
| Isoleucine | 3.0E-5 | 1.0E-4 | 1.0E-3 | 1.5E-5 | 8.0E-4 | 3.0E-5 |
| Leucine | 1.0E-4 | 3.0E-4 | 2.0E-3 | 5.0E-4 | 8.0E-4 | 7.0E-3 |
| Lysine . HCl | 2.0E-4 | 2.0E-4 | 8.0E-4 | 1.0E-4 | 8.0E-4 | 2.0E-4 |
| Methionine | 3.0E-5 | 3.0E-5 | 2.0E-4 | 3.0E-5 | 2.0E-4 | 6.0E-5 |
| Phyenylalanine | 3.0E-5 | 3.0E-5 | 3.0E-4 | 3.0E-5 | 4.0E-4 | 6.0E-5 |
| Proline | 3.0E-4 | 5.0E-5 | — | 3.0E-4 | 3.5E-4 | 3.0E-4 |
| Serine | 1.0E-4 | 3.0E-4 | 1.0E-4 | 6.0E-4 | 4.0E-4 | 1.2E-3 |

*(Table Contd.)*

| *Component* | *MCDB110* | *MCDB202* | *MCDB402* | *MCDBl53* | *Iscove's* | *LHC* |
|---|---|---|---|---|---|---|
| Threonine | 1.0E-4 | 3.0E-4 | 5.0E-4 | 1.0E-4 | 8.0E-4 | 2.0E-4 |
| Tryptophan | 1.0E-5 | 3.0E-5 | 1.0E-5 | 1.5E-5 | 7.8E-5 | 3.0E-5 |
| Tyrosine | 3.0E-5 | 5.0E-5 | 2.0E-4 | 1.5E-5 | 4.6E-4 | 3.0E-5 |
| Valine | 1.0E-4 | 3.0E-4 | 2.0E-3 | 3.0E-4 | 8.0E-4 | 6.0E-4 |
| **2. Vitamins** | | | | | | |
| d-Biotin | 3.0E-8 | 3.0E-8 | 3.0E-8 | 6.0E-8 | 5.3E-8 | 6.0E-8 |
| Folic acid | — | — | — | 1.8E-6 | 9.1E-6 | 1.8E-6 |
| Folinic acid | 1.0E-9 | 1.0E-8 | 1.0E-6 | — | — | — |
| DL-a-lipoic acid | 1.0E-8 | 1.0E-8 | 1.0E-8 | 1.0E-6 | — | 1.0E-6 |
| Niacinamide | 5.0E-5 | 5.0E-5 | 5.0E-5 | 3.0E-7 | 3.3E-5 | 3.0E-7 |
| D-pantothenate l/2 Ca | 1.0E-6 | 1.0E-6 | 5.0E-5 | 1.0E-6 | 1.7E-5 | 1.0E-6 |
| Pyridoxal | — | — | — | — | 2.0E-5 | — |
| Pyridoxine -HCl | 3.0E-7 | 3.0E-7 | 1.0E-4 | 3.0E-7 | — | 3.0E-7 |
| Riboflavin | 3.0E-7 | 3.0E-7 | 1.0E-6 | 1.0E-7 | 1.1E-6 | 1.0E-7 |
| Thiamin-HCl | 1.0E-6 | 1.0E-6 | 1.0E-4 | 1.0E-6 | 1.2E-5 | 1.0E-6 |
| Vitamine B12 | 1.0E-7 | 1.0E-7 | 1.0E-8 | 3.0E-7 | 9.6E-9 | 3.0E-7 |
| **3. Other organic constituents** | | | | | | |
| Acetate | — | — | — | 3.7E-.3 | — | 3.7E-3 |
| Adenine | 1.0E-5 | 1.0E-6 | 1.0E-6 | 1.8E-4 | — | 1.8E-4 |
| Choline chloride | 1.0E-4 | 1.0E-4 | 1.0E-4 | 1.0E-4 | 2.9E-5 | 2.0E-4 |
| D-glucose | 4.0E-3 | 8.0E-3 | 5.5E-3 | 6.0E-3 | 2.5E-2 | 6.0E-3 |
| i-Inositol | 1.0E-4 | 1.0E-4 | 4.0E-5 | 1.0E-4 | 4.0E-5 | 1.0E-4 |
| Linoleic acid | — | 2.0E-7 | 3.0E-7 | — | — | — |
| Putrescine-2HCl | 1.0E-9 | 1.0E-9 | 1.0E-9 | 1.0E-6 | — | 1.0E-6 |
| Na Pyruvate | 1.0E-3 | 5.0E-4 | 1.0E-3 | 5.0E-4 | 1.0E-3 | 5.0E-4 |
| Thymidine | 3.0E-7 | 3.0E-7 | 1.0E-6 | 3.0E-6 | — | 3.0E-6 |
| **4. Major inorganic salts** | | | | | | |
| $CaCl_2$ | 1.0E-3 | 2.0E-3 | 1.6E-3 | 3.0E-5 | 1.5E-3 | 1.1E-1 |
| KCl | 5.0E-3 | 3.0E-3 | 4-0E-3 | 1.5E-3 | 4.4E-3 | 1.5E-3 |
| $KNO_3$ | — | — | — | — | 7.5E-3 | — |
| $MgCl_2$ | — | — | — | 6 0E-4 | — | 2.2E-2 |
| $MgSO_4$ | 1.0E-3 | 1.5E-3 | 8.0E-4 | — | 8.1E-4 | — |
| NaCl | 1.1E-1 | 1.2E-1 | 1.2E-1 | 1.2E-1 | 7.7E-2 | 1.0E-1 |
| $Na_2HPO_4$ | 3.0E-3 | 5.0E-4 | 5.0E-4 | 2.0E-3 | 1.0E-3 | 2.0E-3 |
| **5. Trace elements** | | | | | | |
| $CuSO_4$ | 1.0E-9 | 1.0E-9 | 5.0E-9 | 1.1E-8 | — | 1.0E-8 |
| $FeSO_4$ | 5.0E-6 | 5.0E-6 | 1.0E-6 | 5.0E-6 | — | 5.4E-4 |
| $H_2SO_4$ | 3.0E-8 | 3.0E-8 | 1.0E-8 | 3.0E-8 | 1.0E-7 | 3.0E-8 |
| $MnSO_4$ | 1.0E-9 | 5.0E-10 | 1.0E-9 | 1.0E -9 | — | 1.0E-7 |
| $Na_2SiO_3$ | 5.0E-7 | 5.0E-7 | 1.0E-5 | 5.0E-7 | — | 5.0E-7 |
| $(NH_4)_6Mo_7O_{24}$ | 1.0E-9 | 1.0E-9 | 3.0E-9 | 1.0E-9 | — | 1.0E-9 |
| $NH_4VO_3$ | 5.0E-9 | 5.0E-9 | 5.0E-9 | 5.0E-9 | — | 5.0E -9 |
| $NiCl_2$ | 5.0E-10 | 5.0E-12 | 3.0E-10 | 5.0E-10 | — | 5.0E-10 |
| $SnCl_2$ | 5.0E-10 | 5.0E-12 | — | 5.0E-10 | — | 5.0E-10 |
| $ZnSO_4$ | 5.0E-7 | 1.0E-7 | 1.0E-6 | 5.0E-7 | — | 4.8E-7 |
| **6. Buffers and indicators** | | | | | | |
| Hepes | 3.0E-2 | 3.0E-2 | — | 3.8E-2 | 2.5E-2 | 2.3E-2 |
| $NaHCO_3$ | — | — | 1.4E-2 | 1.4E-2 | 3.6E-2 | 1.2E-2 |
| Phenol red | 3.3E-6 | 3.3E-6 | 3.3E-5 | 3.3E-6 | 4.0E-5 | 3.3E-6 |
| $CO_2$ | 2% | 2% | 5% | 5% | 10% | 5% |

* Computer style notation used for concentration, e.g., 3.0 E- 2 = 30 mM. Sufficient NaOH is added to give pH 7.3-7.4 at 37°C in correct gas phase

temperamental serum free media are developed. Following factors discourage the use of these serum free media: (i) simplicity of media with serum; (ii) lack of reliable source of most serum free media and (iii) the considerable investment in time, effort and resources needed for developing serum-free media. In view of this, one would have to make a judicious choice of a medium with or without serum depending upon the type of cells or a cell line. Some of the cell lines with the corresponding suitable selective media (including media with and without serum) are listed in Tables.

**Table**

Examples of selective media (modified from Freshney, 1987)

| | *Cells of cell line* | *Medium* | *Reference* |
|---|---|---|---|
| 1. | Fibroblasts | MCDB 202 | McKeehan and Ham. 1977 |
| 2. | Fibroblasts | MCDB 110 | Bettger, 1981 |
| 3. | Keratinocytes | MCDB 153 | Tsao et al., 1982 |
| 4. | Bronchial epithelium | LHC | Lechner and Laveck, 1985 |
| 5. | Mammary epithelium | MCDB 170 | McKeehan et al., 1984 |
| 6. | Prostate epithelium (rat) | WAJC 404 | McKeehan et al., 1982 |
| 7. | Prostate epithelium (human) | WAX 404 | McKeehan et al., 1984 |
| 8. | Small cell lung cancer | HITES | Carney et al., 1981 |
| 9. | Endothelium | MCDB 130 | Knedler and Ham. 1983 |

**Table**

**Selecting a suitable medium * (modified from Freshney, 1987)**

| | | *Medium with serum* | | *Serum-free medium* |
|---|---|---|---|---|
| | *Cells or cell line* | *Medium* | *Serum* | |
| 1. | Chick embryo fibroblasts | Eagle's MEM | CS | MCDB 202 |
| 2. | Chinese hamster ovary (CHO) | Eagle's MEM | CS | MCDB 301 |
| 3. | Chondrocytes | F12 | FB | Adolphe et. at., 1983 |
| 4. | Continuous cell lines | Eagle's MEM | CS | 199, Waymouth. 1984 MB752/1. MD7505/1 CMRL1066 |
| 5. | Endothelium | DME, 199, MEM | FB | MCDB 130 |
| 6. | Fibroblasts | Eagle's MEM | CS | MCDB 110, 202, 402 |
| 7. | Glial cells | MEM. SF12 | FB | 113 Michler et al., 1984 |
| 8. | Glioma | MEM. SF12 | FB | |
| 9. | HeLa cells | Eagle's MEM | CS | Blaker et al., 1971 |
| 10. | Hemopoietic cells | RPMI 1640, Fischer's | FB | Iscove. 1984 |
| 11. | Human diploid fibroblasts | Eagle's MEM | CS | MCDB 110, 202 |
| 12. | Human leukemia | RPMI 1640 | FB | Breitman et al., 1984: Iscove. 1981 |
| 13. | Human tumours | SF12. L15. RPMI 1640. DME | FB | |

*(Table Contd.)*

| | Cells or cell line | Medium with serum: Medium | Medium with serum: Serum | Serum-free medium |
|---|---|---|---|---|
| 14. | L cells (L929, LS) | Eagle's MEM | CS | Birch and Pirt, 1970, 1971; Higuchi, 1971 |
| 15. | Lymphoblastoid cell lines (human) | RPMI 1640 | FB | Iscove, 1984 |
| 16. | MDCK dog kidney epithelium | DME, F10/DME | FB | Taub, 1984 |
| 17. | Melanoma | MEM. SF12 | FB | Barnes and Sato. 1980 |
| 18. | Mouse embryo fibroblasts | Eagle's MEM | CS | MCDB 402 |
| 19. | Mouse leukemia | Fischer's RPMI 1640 | FB HoS | Murakami, 1984 |
| 20. | Mouse erythroleukemia | SF12. RPMI 1640 | FB HoS | Iscove, 1984 |
| 21. | Mouse myeloma | DME. RPMI 1640 | FB | Murakami. 1984 HB101 |
| 22. | Mouse neuroblastoma | DME, F12/DME | FB | MCDB411 |
| 23. | Skeletal muscle | DME. F12 | FB HoS | F12 |
| 24. | Syrian hamster fibroblasts BHK21 | MEM, GMEM. DME | CS | Pardee et. al., 1984 |
| 25. | 3T3 cells | MEM.DME | CS | MCDB402 |

* Abbreviations as in Table; SF12 is Ham's F12 plus Eagle's essential amino acids and non-essential amino acids as in DME (available as 100 times stock from Flow Labs., Gibco. etc.). Further recommedations on the choice of media can be found in McKeehan. 1977, Barnes et al., 1984, Maurer, 1986.

**Table**

Sterilization of equipment and apparatus (modified from Freshney, 1987)

| Item | Sterilization |
|---|---|
| Apparatus containing glasss S: silicone tubing | Autoclave* |
| Disposable tips for micropipettes | Autoclave |
| Dispenser tubing for compupet | Autoclave |
| Filter-Millipore, Sartorius | Autoclave-do not prevac. or postvac. |
| Glassware | Dry heat** |
| Glass bottles with screw caps | Autoclave |
| Glass coverslips | Dry heat |
| Glass slides | Dry heat |
| Instruments | Dry heat |
| Magnetic stirrer bars | Autoclave |
| Pasteur pipettes-glass | Dry heat |
| Pipettes-glass | Dry heat |
| Screw caps | Autoclave |
| Silicone tubing | Autoclave |
| Stoppers-rubber, silicone | Autoclave |
| Test tubes | Dry heat |

* Autoclave = 100 kPa (15 lb/in$^2$); 121° C for 20 min.

** Dry heat = 160°C/1 hr.

**Table**

Sterlization of liquids (modified from Freshney, 1987)

| *Solution* | *Sterilization* | *Storage* |
|---|---|---|
| Agar | Autoclave* | Room temperture |
| Amino acids | Filter+ | 4°C |
| Antibodies | Filter | –20°C |
| Bacto-peptone | Autoclave | Room temperature |
| Bovine serum albumin | Filter-use stacked filters (see text) | 4°C |
| Carboxylmethyl cellulose | Steam-30 min** | 4°C |
| Collagenase | Filter | –20°C |
| DMSO | Self-sterilizing, aliquot into sterile tubes | Room temperature-keep dark; avoid contact with rubber or plastic. |
| EDTA | Autoclave | Room temperature |
| Glucose-20% | Autoclave | Room temperature |
| Glucose-l-2% | Filter | Room temperature |
| Glutamine | Filter | –20°C |
| Glycerol | Autoclave | Room temperature |
| HEPES | Autoclave | Room temperature |
| HCl 1N | Filter | Room temperature |
| Lactalbumin hydrolysate | Autoclave | Room temperature |
| Methocel | Autoclave | 4°C |
| $NaHCO_3$ | Filter | Room temperature |
| NaOH 1N | Filter | Room temperature |
| Phenol red | Autoclave | Room temperature |
| Salt solutions (without glucose) | Autoclave | Room temperature |
| Serum | Filter-use stacked filters | –20°C |
| Sodium pyruvate 100 mm | Filter | –20°C |
| Transforrin | Filter | –20°C |
| Tryptose | Autoclave | Room temperature |
| Trypsin | Filter | –20°C |
| Vitamins | Filter | –20°C |
| Water | Autoclave | Room temperature |

* Autoclave=100 kPa (15 Win) 121° C for 20 min.

\+ Filter =0.2 Et m pore size.

** Steam =100°C for 30 min.

## Culture Procedures

Culture procedures involve preparation and sterilizaiton of glassware, equipment, reagents and media. The apparatus and liquids meant for tissue culture should be reserved for that purpose alone and washing of glassware should also be separate, since the rquirement of washing for tissue culture is higher than that for general glassware.

**Preparation and sterilization of apparatus**

For an effective washing of glassware, following procedure and precautions should be followed: (i) do not allow soiled glassware to dry, and always collect soiled glassware in water with sodium hypochlorite, so that microbial contamination may be prevented; (ii) a detergent effective for available tap water of your area should be used and it should be nontoxic; (iii) after washing, glassware should be thoroughly rinsed in tap water and then in deionized or distilled water ; (iv) it should be dry inverted and then sterilized by dry heat (in oven at 160°C for 1 hr). Automatic washing machines are also used for the glassware in tissue culture laboratories. Sterilization procedure for equipment and apparatus is given in Table and that for liquids in Table.

**Preparation and sterilization of reagents and media**

Reagents and media should be produced in pure form. Most reagents and media can be sterilized either by moist heat i.e. by autoclaving (if heat stable) or by membrane filtration (if heat labile). During autoclaving (mosit heating by steam), the container may be kept sealed (if borosilicate or polycarbonate) or the caps may be left slack (if soda glass) to avoid breakage. The liquid layer in container may go down due to evaporation and should be restored with sterile distilled water after autoclaving. Culture media also may be available in autoclavable form. For instance, autoclavable versions of Eagle's MEM and other media are available and the autoclavine is much less labour intensive, less expensive and has much lower failure rate than those sterilized by filtration.

**Preparation of Animal Material**

Contamination should also be avoided during experimental manipulations. This can be achieved by collecting tissue aseptically or sterilizing it after removal. While some tissues are free from contamination (e.g. embryo or other contents of uterus), in other adult tissues bacterial contamiantion may be common. In the former case (embryo), the tissue needs to be removed carefully under aseptic conditions eliminating the possibility of contamination, but in the latter case (adult tissue), the tissue may have to be washed in a solution having high concentration of antibiotic. For this purpose, a useful antibiotic mixture is a 'balanced salt solution' containing 1000 units of penicillin and 0.5 mg of streptomycin or neomycin per ml. Further handling of tissue should be carefully done under fully aseptic conditions, which can be created either through preparing an aseptic room or through the use of 'laminar flow cabinet' or 'hood'. The use of 'hoods' has greatly facilitated the handling of material under aseptic conditions.

To develop primary cultures and cell lines, a variety of tissues, and disaggregation methods are used to give good yield of separate. It is, therefore, obvious that for any exercise involving animal cell tissue culture, tissue needs to be obtained under aseptic conditions, although it may or may not be cultured, depending upon whether or not it is immediately subjected to disaggregation to give yield of cells. However, even for obtaining cell cultures, often the tissue is cultured using 'primary explantation technique' developed by Hart (1907), Carrel (1912) and others. This technique is used for cultivation of pieces of fresh tissue derived from the organism, and this was the exclusive technique used for animal tissue culture till about 1945. Different forms of primary explantation techniques are still widely and will continue to be used for a very long time. These lecher differ only in the type of vessels (flasks, test tubes. etc.) used for growing the tissue, but are uniform in principle. The primary explanation tech is also used for embryo and organ culture, but are variously modified specialized techniques. The different explanation technique classified into the following: (i) slide cultures, (ii) carrel flask cud and (iii) roller test tube cultures. These three techniques of tissue culture and the techniques of organ culture (including tissue engineering) are discussed in this chapter.

## Tissue Culture

### Slide or Coverslip Cultures

In this technique, slides or cover slips are prepared by placing a fragment of tissue (explantation) onto a coverslip, which is subsequently inverted over the cavity of a depression slide. This is the oldest method of tissue culture and is still quite widely used. This has a number of advantages and disadvantages listed in Table. From this Table, it is obvious that the application of slide culture is limited. It may be very useful for morphological studies through the use of lime-lapse cinemicrographic investigations. There are several general methods for preparation of slide culture; three of these methods will be briefly described in this section.

**Table**

Advantages and disadvantages of slide coverslip cultures of tissue

| *Advantages* | *Disadvantages* |
|---|---|
| 1. It is simple and relatively inexpensive. | 1. Supply of oxygen and nutrients is rapidly exhausted, so that the medium quickly becomes acidic and requires transfer of rapidly growing tissue. |
| 2. Cells in living state are spread out in a manner suitable for microscopy and photography. | |
| 3. Cells grow directly on covet-slip and can be fixed and stained to make permanent slides. | 2. Sterility can not be maintained for a long period. |
| | 3. Only very small amounts of tissue can be cultured. |

Single coverslip with plasma clot—This technique developed by Harrison (1907) has been most commonly used during the last more .than fifty years and includes the following steps (Fig.): (i) prepare medium in two parts, one containing 50% plasma in BSS (balanced salt solution) and the other containing 50% embryo extract

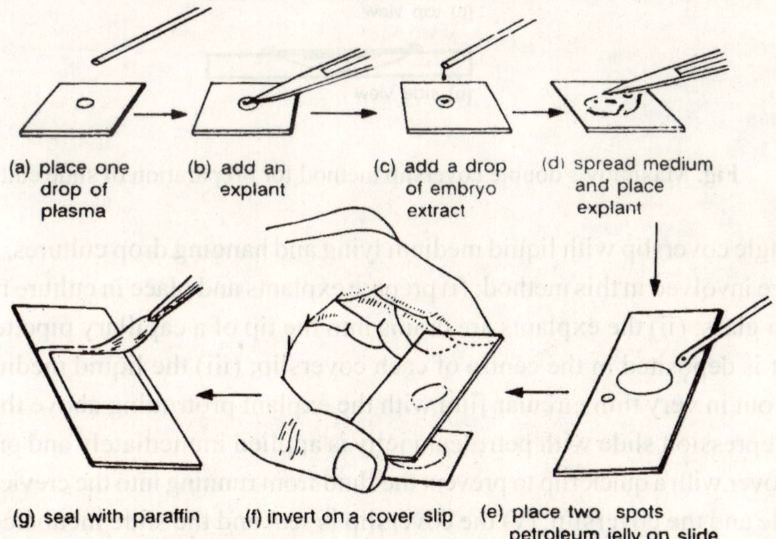


**Fig.** Preparation of single coverslip cultures

in serum; (ii) under sterile ax aseptic conditions, using a capillary pipette, place one drop of plasma containing solution in the centre of each of one, or more coverslips (22mm) (iii) transfer a fragment (one or two pieces) of tissue (called explant) to this drop without crushing the tissue; (iv) add the embryo extract containing, solution and mix thoroughly before clotting starts and then locate the explant; (v) place two small spots of petroleum jelly (using a glass rod) see the concavity of a depression slide and invert this slide over the coverslip apply gentle pressure, so that jelly sticks to coverslip; (vi) allow culture medium to clot; (vii) turn over the slide and seal the margins of coverslip with paraffin; (viii) label and incubate at 37°C.

Double coverslip with plasma clot—This technique was developed by Maximov and resembles the single coverslip method described above. It includes the following steps: (i) a small drop of BSS is placed on large coverslip (40mm); (ii) a square or round coverslip (22mm) is placed over BSS in the centre of large coverslip. These two sleps are then followed by the steps listed above for single coverslip method. A large depression silde is used and the entire preparation is attached to it by petroleum jelly and wax in such a way that the small coverslip is not in contact with the slide at any point (Fig).

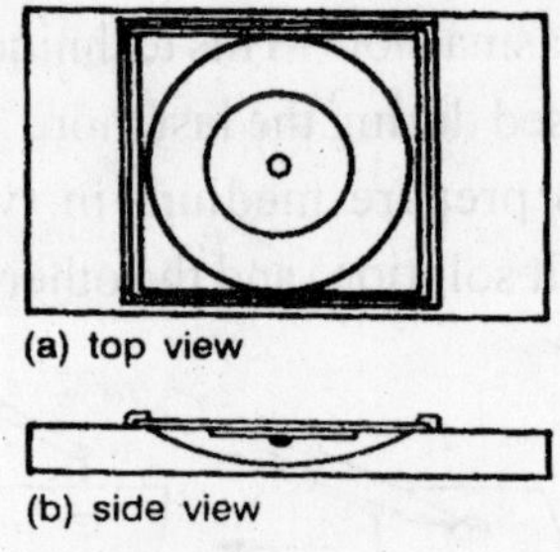


**Fig.** Maximov's double coverslip method for preparation of slide cultures

Single coverslip with liquid medium lying and hanging drop cultures, following steps are involved in this method: (i) prepare explants and place in culture medium in a watch glass; (ii) the explants are drams into the tip of a capillary pipette, and one explant is deposited in the centre of each coverslip; (iii) the liquid medium can be spread out in very thin circular film with the explant protruding above the surface; (iv) a depression slide with petroleum jelly is applied immediately and preparation turned over with a quick flip to prevent the fluid from running into the crevice between the slide and the coverslip; (v) the coverslip is seal and the slide incubated at 37°C, upright or inverted; the tissue grows on the coverglass.

After-care of slide cultures—Single coverslip cultures are very useful for short term studies but they are difficult to handle subsequently except by a process of transfer. Therefore, double coverslip method is recommended, whenever it is desirable to leave the explant in its original location to obtain long term tissue cultures. This would require washing, feeding, patching and transfer of cultures.

(a) Washing and feeding double coverslip cultures—Washing and feeding involves the following steps: (i) remove seal using razor blade and remove large coverslip with small one still attatched and flip it over, so that culture is uppermost in orientation; (ii) detatch small coverslip from the large coverslip using needle 'and forceps and transfer it to a Columbia staining dish (watch glass or Petri dish may also be used) containing balanced salt solution (BSS); (iii) remove small cover slip treated as above (one at a time) and place it on a large coverslip with culture up; while removing small coverslip from dish with BSS, the amount of BSS brought with it may be controlled by the rate of removal (too much liquid will interfere in feeding operation and too little will allow air bubbles); (iv) feed the culture by adding a drop of feeding solution (e.g. BSS : serum : 50% embryo extract = 1:1:1) to the small coverslip; (v) attatch a clean depression slide, using petroleum jelly, as done earlier.

(b) Patching the plasma clot in slide cultures—If there is evidence of liquifaction, plasma clot should be patched as follows: (i) wash small coverslip with culture as above; (ii) in a watch glass take two drops of a mixture of plasma and BSS and to this add two drops of a mixture of serum and embryo extract; mix this quickly and place a drop on each washed coverslip having a culture; (iii) a clean depression slide with petroleum jelly is then attatched, as usual.

(c) Transfer of slide culture—The coverslip cultures may need to be transferred using the following steps: (i) remove and wash coverslip with culture, and using a Bard-Parker knife, cut through the outgrowth; (ii) the square tissue may be cut into two or four pieces, each transferred to new coverslip and treated as a new explant.

## Flask Cultures

The main use of flask cultures is in the establishment of a strain from fresh explants of tissue. A good Carrel flask has excellent optical properties for microscopic examination, even though polystyrene culture flasks can also be used, provided they have a wide neck for handling the explants. The flask technique has the following advantages: (i) tissue can be maintained in the same flask for months or even years; (ii) large number of cultures can be easily prepared and large amount of tissue can be grown with large amount of medium.

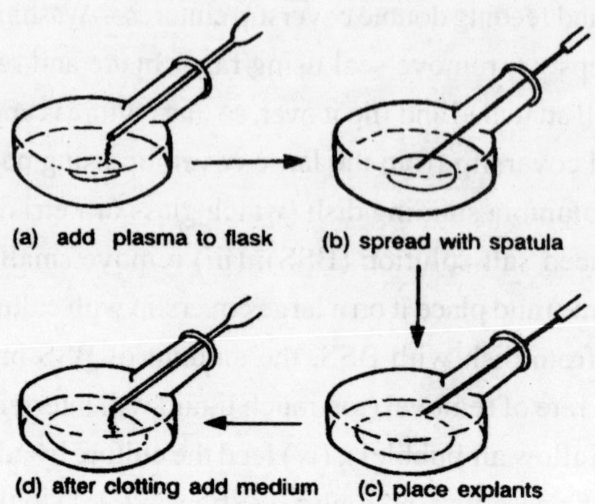


**Fig.** Preparation of flask cultures

There are two types of flask techniques: (i) thick clot culture which allow rapid growth suitable for short term cultures and (ii) 'the clot cultures', which can be maintained for a considerable period of time.

Preparation of flask cultures—Following steps are involved in the preparation of flask cultures (Fig.): (i) place upto six D3.5 Carrel Flasks in a rack with their necks flamed and pointing to the right; place a drop of plasma on the floor of flask and spread this plasma out in a circle; (iii) with the help of spatula, transfer the desired number of explants to the plasma and allow clotting to occur; (iv) after the plasma clots and explants fused in position, add extra medium; for thick old 1.2ml of dilute plasma and for thin clots 1.2ml of dilute serum is added instead of plasma; the whole thing is left for clotting; (v) the flasks are gassed with gas phase (5% $CO_2$, in air).

Renewal of medium—The medium may be periodically replaced (task cultures as follows: (i) old fluid is drawn off by means of a piper (ii) 1.2ml of fluid medium is added as replacement, (iii) the flask is gassy as above.

Transfer of culture. The culture grown in a flask culture need to be removed and cut into pieces, when there is a need for transfer. The pieces are then used for replantation as usual.

## Test tube cultures

Test tubes are cheap and convenient vessels for tissue culture and can be used for preparing a large number of cultures, which can be placed on stationary racks or roller drums. However, this technique has disadvantages like poor optical property for microscopy, difficulty in quantitation due to the curvature and high risk of contamination. Cultures on plasma clots test tube are prepared just like those in flasks, but tissues may be grown technique on the wall of test tube without a plasma clot. Test tubes may also be used for developing suspension cultures. The feeding, patching and transfer of culture is done as in other primary explantation techniques.

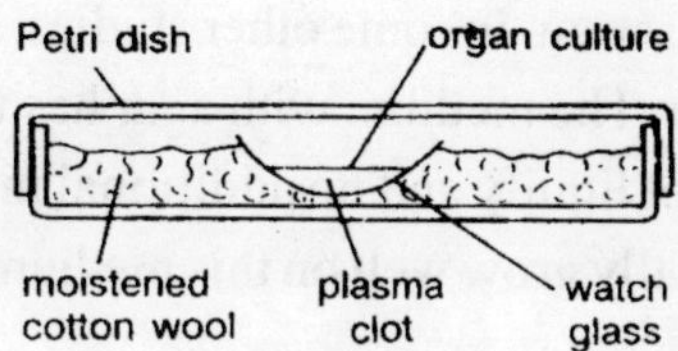


**Fig.** The classical organ culture technique

## Organ Culture

Organ culture usually implies culturing pieces of an organ (not necessarily whole organ) in vitro, and its objective is to maintain the architecture of the tissue and direct it towards normal development such as occurs in vivo. In order to achieve this aim, it is essential that the tissue should never be disrupted or damaged and this requires careful handling. In view of this, organ culture techniques generally demand more careful manipulation than tissue culture techniques discussed earlier in this chapter. Media used for growing organ cultures are generally the same as those used for tissue culture. The techniques of organ culture can be divided into (i) those employing a solid medium and (ii) those employing liquid medium.

### Culture of embryonic organs

It is easier to culture embryonic organs than to culture organs from adult animals and the techniques used differ in two cases. Embryo organs can be cultured, using any of the following three techniques.

Organ cultures on plasma clots—The organ cultures can be prepared on plasma clots using the following steps (Fig.): (i) prepare a plasma clot by mixing 15 drops of plasma with five drops of embryo extract in a watch glass; (ii) place the watch glass on a pad of cotton wool in a Petri dish; cotton wool is kept moist to prevent excessive evaporation from the dish; (iii) place a small, carefully dissected piece of tissue on top of the plasma clot in the watch glass. The technique has now been modified, and a raft of lens paper or rayon net is used on which the tissue is placed. Transfer of the tissue can then he achieved by moving the raft easily. This also facilitates feeding of the culture, so that the net may be removed. Excess fluid is removed and the net with the tissue placed again on fresh pool of medium.

Organ cultures on agar—Media solidified with agar are also used for organ culture and these media consist of 7 parts 1% agar in BSS, 3 parts chick embryo extract and 3 parts horse scrum. In some other studies, defined media with or without serum are used with agar. The medium with agar has the advantage that the support for organ culture does not liquify and no additional mechanical support is required. Embryonic organs generally grow well on this medium, but tumours from adults will not survive.

Organ cultures on fluid media—Organ cultures also make use of liquid media. In this method, often a raft made of lens paper, or cellulose acetate net or perforated metal gauge is used.

**Culture of Adult Organs**

The culture of organs or parts from adult animal is more difficult due to their greater requirement of Off. A variety of adult organs (e.g. liver) have been cultured using special media (e.g. T8; see Paul, 1975) with a special apparatus (Towell's Type II culture chamber). Since serum was found to be tonic, serum free media were used, and the special apparatus permitted the use of 95% $O_2$.

**Whole Embryo Culture**

The effect of metabolic inhibitors on embryonic development was studied by Spratt (1956, 1957), using embryo cultures. In this technique, 40-hours old embryos were used and the embryo development could be followed for another 24-48 hours in vitro before the embryo dies. This involves the following steps: (i) prepare a suitable defined medium or a synthetic medium (consult a manual), and 1ml aliquots of the medium are added to sterile watch glasses, placed on moist absorbent cotton wool pads in Petri dishes (as for organ culture); (ii) incubate hen's eggs at 38°C. for 40-42 hours to provide about a dozen embryos; (iii) the shell is wiped with alcohol and broken into a sterile evaporating dish containing 50m1 chick saline or BSS; (iv) a circular cut is made (using scissors) into the vitelline membrane around the blastoderm and the latter is transferred to a Petri dish containing BSS; (v) the adherent vitelline membrane is removed with the aid of forceps and the embryo is examined under the microscope to determine the stage of development; (vi) the blastoderm is transferred to the top of the medium in the watch glass prepared in (i) above; (vii) the blastoderm is spread on agar gel (ventral side down) and the excess BSS is removed; (viii) culture is incubated at 37.5°C.

**Culture of mammalian embryos or ova**

The mammalian embryos (rabbit or mouse), as young as 2-8 celled fertilized ova, have been cultured. It was shown in these cases that considerable development can really take place in vitro, and mouse embryos were shown to develop upto blastocysts. Such cultured embryos or ova can be reimplanted and give rise to healthy animals. The media used for this purpose varied from 100% serum through the common types of plasma-embryo extract media to simple Krebs-Ringer solutions supplemented with 7% of thin egg white or bovine albumin. The embryo can be cultured either in the pre-implantation stage or at the post-implantation stage. In the latter case, special care is needed to remove the embryo in an undamaged form.

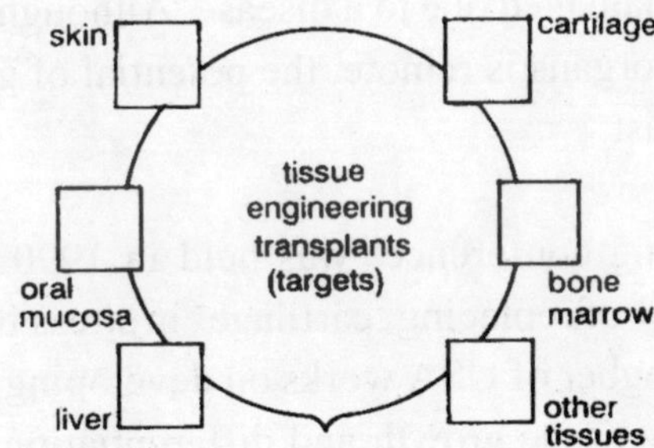


**Fig.** Different tissues, which are the targets for tissue engineering, starting from stromal cells of human tissue.

## Tissue Engineering

As disscussed in the last chapter and earlier in this chapter, a variety of animal tissues can now be cultured. The number of types of these human tissues is more than 200. During the last decade, the tissue culture work in animals demonstrated that virtually any human tissue or organ can be grown in culture (Fig.). This has become possible when it was found that the ability of cultured cells to undergo differentiation can be restored. These developments led to the emergence of a new science called 'tissue engineering'. Some of the goals of tissue engineering are listed in Table. 'Skin' was the first organ, which could be cultured in artificial media and could be used successfully for transplantation, after serious skin burns. Eugene Bell, who is the pioneer of artificial skin development feels that eventually a tissue engineer will be able to reconstitute body parts for (i) use in grafting, and to be used as (ii) better models for drug delivery and action.

**Table**

Some of the technological goals of tissue engineering

1. Growth of cells in three-dimensional systems
2. Delivery systems for protein therapeutics
3. Cell cultivation methods for culturing ' recalcitrant cells'
4. Transgenic protein expression in transplantable cells
5. Vehicles for delivering transplantable cells
6. Avoiding immunogenicity in transplantation systems
7. Development of markers for tracking transplanted cells
8. Developing in vivo and ex vivo biosensors for monitoring cell behaviour during tissue production

It is believed by tissue engineers like Robert (Bob) Nerem of USA, that after the success of artificial skin, the next step is the growth of functional artificial organs like pancreas and liver damaged due to a disease. Although the clinical application of these other artificial organs is remote, the potential of growing and using these artificial organs does exist.

A 'tissue engineering conference' was held in 1990 at Keystone, Colorado (USA), where possibility of replacing 'cartilage' in stress bearing joints was found to be a reality. Donald Ingber of USA works on developing 'extra cellular matrices (ECMs)' that will allow tissue growth and differentiation in vitro, as a means to organ replacement. However, tissue engineering would need a knowledge of physics and chemistry along with the knowledge about cell and tissue culture. For example, one will like to know, how the cellular environment affects three dimensional structure of tissue grown in culture. One will also have to know the fine structure of carbohydrate aggregates surrounding the cells. Detailed knowledge on these aspects will allow the induction of stem cells to differentiate in required manner. It is also believed that stem cells exist for all tissue types, and if these can be recovered, any human tissue can he induced to reform.

**Artificial skin**

There are three biotech companies in USA, which are developing artificial skin to the stage of clinical trials (i) ATS (Advanced Tissue Science), located at La Jolla,

California which is now merged with Neomorphics (Lexington, MA), (ii) Biosurface Technology (BTI, Cambridge, MA) and (iii) Organogenesis. These three companies use different technological approaches, which are really the basis of all future tissue engineering,

The real breakthrough in culturing artificial skin was made in 1970s at Harvard. It was demonstrated that when an irradiated 3T3 fibroblast cell line is grown with skin cells (keratinocytes, which make 90% of the skin epidermis), 3T3 cells stimulated growth of skin cells and allowed their differentiation into epidermis. 3T3 cells are believed to secrete factors which help in growth and differentiation. This technique was used by BTI, where postage stamp sized skin biopsies are taken and expanded 10,000 times within three to four weeks. This gives only the cultured epidermis, which when applied to burnt area, helps regeneration of dermis underneath. Since in severely burned patients, skin biopsies may not be available from the patients, following two approaches were used to overcome the problem: (i) Skin from cadavers (dead bodies) are taken to cover and protect against infection and desiccation, until their own skin is grown. (ii) Neonatal foreskin (loose skin from the lip of penis of newly born child ) is cultured, which creates artificial dermis layer. This leads to allogeneic grafts stimulating rapid healing. New born skin cells are used, since they grow relatively more vigorously. ATS company uses a synthetic polymer, PGA, for growing artificial skin from neonatal foreskin. The use of synthetic polymers allows cells to grow into skin without scars, so that it is believed that this technology will be the basis for future organ development.

A further advancement was made by the company 'Organogenesis' which claims to have created a 'living skin equivalent (LSE) ' comprising of both a lower dermis layer and an upper epithelial layer. Collagen matrix is used as a support for the growth of LSE. This LSE when used for grafting may be replaced by host cells eventually, hut is not rejected. More details about the skin culture and transplantation are discussed in Chapter.

## Artificial cartilage

Artificial cartilage is another candidate for tissue replacement, since it is not vascularized and therefore is not rejected due to immunogenic response. Immediate application of artificial skin involves apart related injuries and trauma, in much the same way as the patients predisposed to arthritis. For this purpose, the precursors of cartilage i.e. the chondrocyte cells are combined with materials (e.g. collagen or PGA) that have mechanical properties, and are grown in culture.

Experiments have been conducted with rabbis using artificial cartilage, and encouraging results have already been obtained, even though normal condition is not restored. Like artificial skin, artificial cartilage can also be used for providing a microenvironment that promotes healthy growth. For instance, insertion of a hyaline cartilage in a defect prevents the growth of scar-like fibrous cartilage. Future tissue engineering research will revolutionize.the culturing of skin, cartilage, hone, liver, and other organs and tissues.

## Disaggregation

We may obtain a primary cell culture either (i) by allowing cells to migrate out from the tissue, which is adhering to a substrate or (ii) by disaggregating the tissue mechanically or enzymatically to produce a suspension of cells. Most normal untransformed cells survive and proliferate to produce 'primary culture' when attached to a substrate, but these cells need to be obtained by disaggregation. The disaggregated cells can also grow in suspension to produce the primary culture. Disaggregation is achieved by any one of the following three methods: (i) physical disruption, (ii) enzymatic digestion and (iii) treatment with chelating agents. Physical disruption, which may involve cutting the tissue into pieces, is often combined with other methods, but chick embryo suspension may also be obtained by expressing (squeezing) the embryo through the nozzle of a syringe.

Of the many enzymes that are used for disaggregation, trypsin and pronase are the most commonly used. Other enzymes used successfully include collagenase, elastase, mucase, pancreatin pappain, etc. Similarly, tissues like epithelium ( which needs Ca + + and Mg + +ions for its integrity) can be treated with chelating agents, such as citrate and ethylene- diaminetetra-acetic acid (EDTA). Chelating agents are mainly used for production of cell suspensions from establised cultures of epithelial type.

### Isolation of Tissue

Before working with animal or human tissue, it is essential to be sure that your work does not violate medical-ethical rules or the current legislation. You may sterilize the site with 70% alcohol and remove the tissue aseptically and transfer it to balanced salt solution (BSS) or to a suitable culture medium. The tissue may also be stored in a refrigerator before transferring it to BSS or to a culture medium, because viable cells may be recovered from chilled tissue several days after explantation.

Different protocols are available for isolation of tissues like mouse embryo, hen's egg, human biopsy material, etc. Readers are advised to consult a book on animal tissue culture to obtain protocols (e.g. Freshney, 1987).

## Enzymatic disaggregation

Although dissociated cells can also be obtained by the primary explant technique to be discussed in the next chapter, mechanical or enzymatic disaggregation gives a much higher yield of representative cells in a short duration. Further, the embryonic tissue disperses more readily and gives higher yield than the newly born or adult, since in the latter case there are fewer undifferentiated cells, more fibrous connective tissue and extracellular matrix. In several tomours, the tissue (being fragile) is mostly destroyed during enzyme treatment and viable cells are obtained with difficulty.

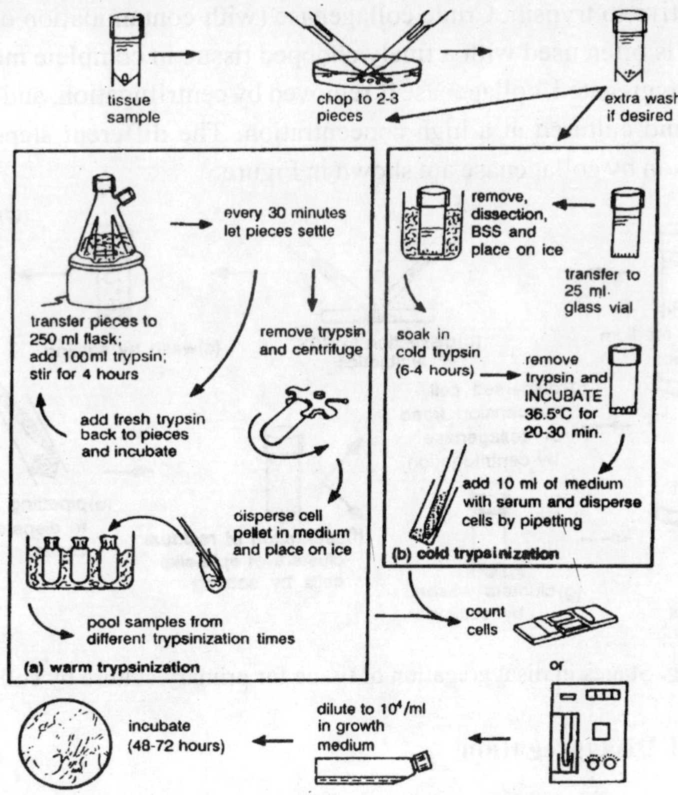


**Fig.** Preparation of primary culture by disaggregation in trypsin method

Disaggregation by trypsin (for embryonic tissue)—Crude trypsin is the most common enzyme used for disaggregation for the following reasons: (i) it is tolerated by a variety of cells, (ii) it is effective for many tissues and (iii) its residual activity is neutralized by serum of the medium or by a trypsin inhibitor (e.g. soyabean trypsin inhibitor) in case of a serum free medium. To facilitate exposure of cells to the warm enzyme (36-5°C) for a minimum period, dissociated cells are collected every half an hour and the trypsin is removed by centrifugation after 3-4 hours required for complete disaggregation. Alternatively, cold trypsinization may be used which involves soaking of tissue in trypsin at 40°C ( to allow penetration of enzyme), followed by incubation at 365°C for a shorter period. Different steps of warm trypsinization and cold trypsinization are shown in Figure.

Disaggregation by collagenase (for embryonic, normal and malignant tissues). Disaggregation by trypsin can be damaging (e.g. epithelial cells) or ineffective (in Fibrous tissues). Since intracellular matrix contains collagen, collagenase has proved effective for disaggregation of several normal and malignant tissues, which may be rather sensitive to trypsin. Crude collagenase (with contamination of nonspecific protceases) is often used with a finely chopped tissue in complete medium. When tissue is disaggregated, collagenase is removed by centrifugation, and the cells may be seeded and cultured at a high concentration. The different steps involved in disaggregation by collagenase are shown in Figure.

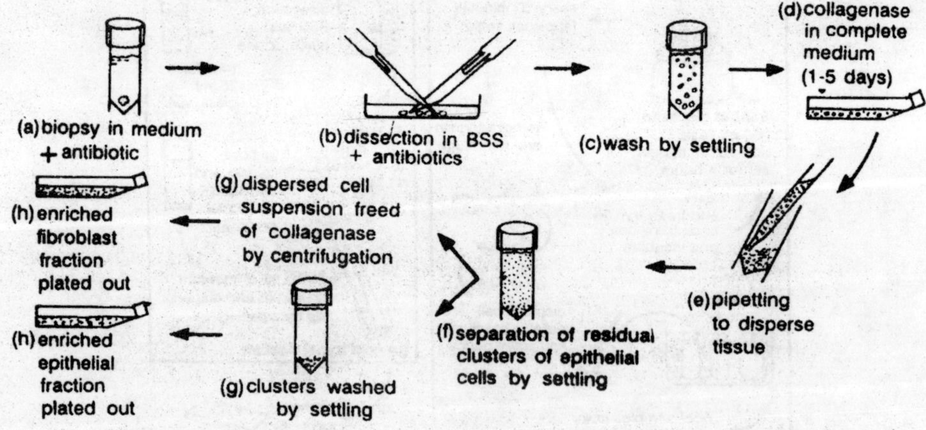


**Fig.** Stages in disaggregation of tissue for primary culture by collagenase

## Mechanical Disaggregation

Enzymatic disaggregation is labour intensive and involves damage of cells. Therefore, mechanical disaggreation of cells is sometimes preferred. In this method, tissue is carefully sliced and the cells which spill out are collected. Alternatively, the cells are either (i) pressed through the sieves of gradually reduced mesh, or (ii) forced through a syringe and needle, or even (iii) repeatedly pipetted. Although the method may cause mechanical damage, the cell suspension is more quickly obtained than in the enzymatic disaggregation. Therefore, when the availability of tissue is no limitation and the efficiency of yield unimportant, mechanical disaggregation may be used to obtain good yield of cells in a shorter time, but at the expense of very much more tissue. Steps involved in one of the methods of mechanical disaggregation are shown in Figure.

## Separation of viable and nonviable cells

The dissociated cells obtained as above, usually described as primary cells, grow well when seeded on culture plates at high density. These are adherent primary cultures, but primary cultures can also be maintained in suspension. In the first case

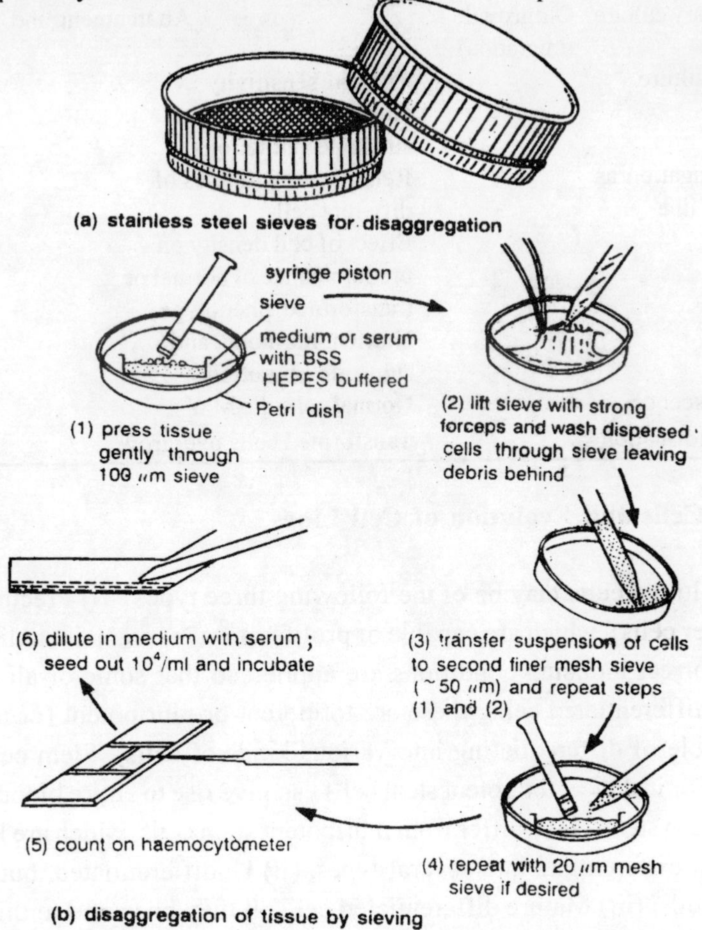


**Fig.** (a) Stainless steel sieves suitable for disaggregation of tissue; (b) different steps involved in disaggregation of tissue by sieving

(adherent culture), nonviable cells will be removed at the first change of medium. In suspension, on the other hand, non-viable cells are gradually diluted out, when cell proliferation starts. However, non-viable cells can also be removed from primary disaggregate by centrifuging the cells on a mixture of Ficoll and sodium metrizoate, when viable cells are collected from the interlace after centrifugation.

**Table**

Elements of selection in the evolution of cell line

| | | *Factors influencing selection* | | |
|---|---|---|---|---|
| | *Stage* | *Primary explant* | | *Enzymatic disaggregation* |
| 1. | Isolation | Mechanical damage | | Enzymatic damage |
| 2. | Primary culture | Outgrowth (migration) | | Attatchment and spreading |
| 3. | Subculture | | Trypsin sensitivity<br>Nutrient, hormone and substrate limitations. | |
| 4. | Propagation as a cell line | | Relative growth rates of different cells;<br>Effect of cell density on predominance of normal or transformed phenotype;<br>Nutrient hormone, and substrate limitation | |
| 5. | Senescence, transformation | | Normal cells die MW transformed cells overgrow | |

## Cultured Cells and Evolution of Cell Lines

The cultured cells may be of the following three types : (i) Precursor or stem cells (master cells), which are capable of proliferation, but remain undifferentiated until the correct inducing conditions are applied so that some or all of the cells mature to differentiated cells; these are totipotent or pluripotent (or multipotent) cells (capable of differentiating into various kinds of cells). Stem cells differ in sternness. For instance, totipotent stem cells can give rise to entire blood producing, and immune systems; they differ from pluripotent stem cells, which are less general, but can still differentiate into several types. (ii) Undifferentiated, but committed precursor cells. (iii) Mature differentiated cells. It may be useful to think of a cell culture as being in equilibrium between multipotent stem cells, undifferentiated but committed precursor cells and mature differentiated cells. This equilibrium may shift according to the environmental condition. Following are some of the examples:

(i) High serum, growth factors, low cell density (i.e. routine serial passage) will promote cell proliferation. (ii) Low serum, appropriate hormones and high cell density will promote differentiation (Fig.).

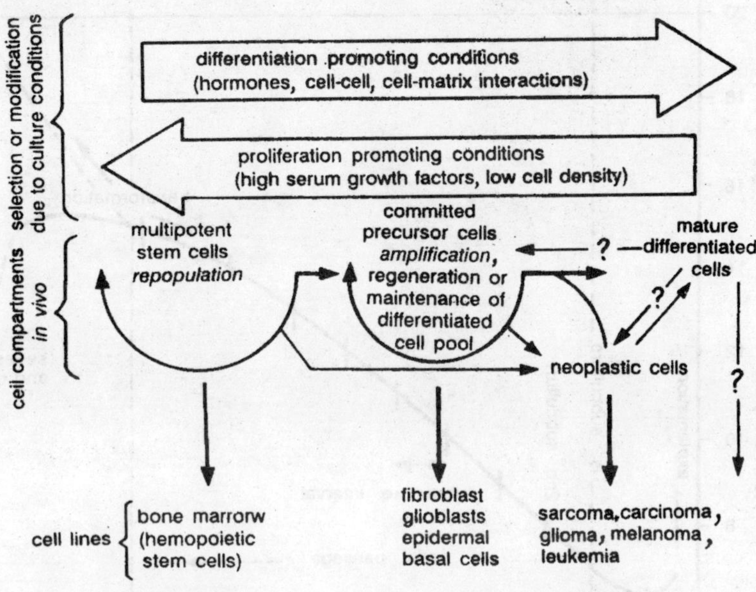


**Fig.** Effect of culture conditions on cell type and origin of cell lines.

The source of the culture will also determine which of the above three types of cells will be present in the culture. Following are some examples: (i) Cell lines derived from the embryo may contain more stem cells (or precursor cells), which will be capable of greater self renewal than the cultures from adults. (ii) The cultures from tissues, which are undergoing continuous renewal in vivo (e.g. epidermis, intestinal epithelium, hemopoietic cells) will still contain stem cells, even though they are derived from adults. These may survive indefinitely. (iii) Cultures from tissues, which renew only under stress (fibroblasts, muscle, glia), may only contain committed precursor cells with a limited culture life span. From any of the three kinds of cells derived from a primary explant, cell lines may be developed (Table). After the first subculture, the primary culture becomes a cell line and may be propagated and subcultured several times. With each successive subculture, the component of the population with the ability to proliferate most rapidly will gradually predominate and nonproliferating or slowly proliferating cells will be diluted out. Although some selection and phenotypic drift will continue, by the third passage the culture becomes more stable and is characterized by rapidly proliferating cells (Fig.)

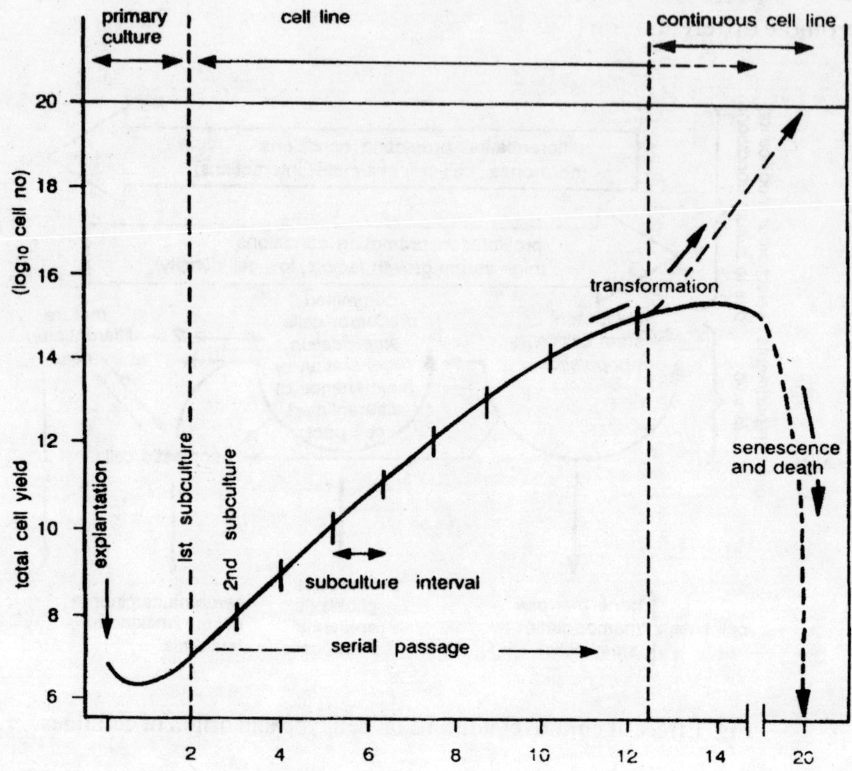


**Fig.** Evolution of a cell line

The cell lines, as above, may be propagated in an unaltered form for a limited number of cell generations, beyond which they may either die out or give rise to continuous cell lines. The continuous cell lines are often aneuploid and have larger variation in chromosome number, while the finite cell lines are often cuploid, with little variation, if any in the chromosome numbers (Fig.). The alteration in a culture giving rise to a continuous cell line is commonly called in vitro transformation, which may be spontaneous or chemically or virally (Epstein Barr Virus) induced.

The term 'transformation' has been applied to the process of formation of a continuous cell line, not only because the culture undergoes morphological and kinetic alterations, but also because the formation of a continuous cell line is often accompanied by an increase in tumorigenicity exhibiting the following properties of malignant transformations: (i) reduce serum requirement, (ii) reduced density, (iii) limitation of growth, (iv) growth in semisolid media and (v) aneuploidy.

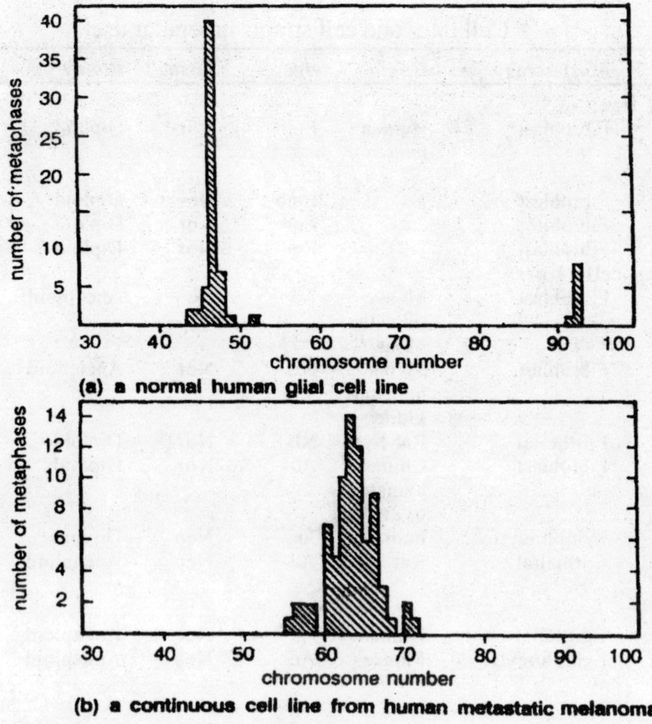


**Fig.** Chromosome numbers in a finite cell line and a continuous cell line

## Maintenance of Cultures-Cell Lines

When a primary culture produced on a substrate or in suspension discussed above has increased to the extent that all the available substrate is occupied or the medium largely consumed, there arises the need to subculture it. From a very heterogeneous primary culture containing many types of cells derived from the original tissue, during subculturing (passages or transfer) a more homogeneous 'cell line' emerges (as discussed above). The culture now called a cell line, can be propagated, characterized and stored. The potential increase in cell number and uniformity of cells, open up a much wider range of possibilities. The term cell line implies the presence of several cell lineages either similar or distinct. Among the cell lineages, if a particular cell lineage has specific properties, which are identified in bulk of the cells of that culture, it is described then as a 'cell strain'. A "cell line" or 'cell strain' may be finite or continuous, depending upon whether it has limited culture life-spans or it is immortal in culture. Finite cell lines grow upto 20-80 population doublings before extinction. Some commonly used cell lines and cell strains are listed in Table and a comparison of the properties of finite and continuous cell lines is presented in Table.

**Table**

Cell lines and cell strains in regular use

| Name | Morphology | Origin | Age+ | Tissue++ | Ploidy | Characteristic |
|---|---|---|---|---|---|---|
| **1. Finite cell lines** | | | | | | |
| MRCS | Fibroblast* | Human lung | Emb | Nor | Diploid | Susceptible to human viral infection |
| MRC9 | Fibroblast | " | Emb | Nor | Diploid | " |
| W138 | Fibroblast | " | Emb | Nor | Diploid | " |
| IMR90 | Fibroblast | " | Emb | Nor | Diploid | " |
| **2. Continuous cells lines** | | | | | | |
| A9 | Fibroblast | Mouse subcutan Mous | Ad | Neo | Aneuploid | HGPRT-ve: deriv L929 |
| BHK21 C13 | Fibroblast | Syrian hamster kidney | NB | Nor | Aneuploid | Transformable by polyoma virus |
| BR13A | Epithelial | Rat liver | NB | Nor | Diploid | Produce MSA |
| CHOK1 | Fibroblast | Chinese hamster ovary | Ad | Nor | Diploid | Simple karyotypic |
| EB 3 | Lymphocytic | human | Ju | Neo | Diploid | EB virus + ve |
| GH1, GH3 | Epithelial | Rat | Ad | Neo | Aneuploid | Produce growth hormone |
| HeLa | Epithelial | Human | Ad | Neo | Aneuploid | G6PD Type A |
| 11210 | Lymphocytic | Mouse | Ad | Neo | Anucploid | Rapidly growing Suspension |
| L5178Y | Lymphocytic | Mouse | Ad | Neo | Anueploid | |
| L929 | Fibroblast | Mouse | Ad | Neo | Aneuploid | Clone of L-cell |
| L5 | Fibroblast | Mouse | Ad | Neo | Ancuploid | Grow in suspension deriv. L929 |
| MC17 | Epithelial | Human breast pleural effusion | Ad | Neo | Aneuploid | Estrogen recep. +ve |
| P388D | Lymphocytic | Mouse | Ad | Neo | Aneuploid | Grow in suspension |
| 5180 | Fibroblast | Mouse | Ad | Neo | Aneuploid | Cancer chemotherapy. screening |
| 3T31.1 | Fibroblast | Mouse swiss | Emb | Nor | Aneuploid | Adipose diff. [Greenet. at., 978] |
| 3T3A31 | Fibroblast | Mouse BALB/c | Emb | Nor | Aneuploid | Contact inhibited: readily transformed |
| NRR49F | Fibroblast | Rat kidney | Ad | Nor | Aneuploid | Induction of suspension growth by transforming growth factors |
| Vero | Fibroblast | Monkey kidney | Ad | Nor | Aneuploid | Viral substrate and essay |

* Fibroblast epithelial lymphocytic refer to appearance and not lineage.
\+ Emb = embryonic; Ad = adult; NB = newborn; Ju = juvenile
++ Nor = from normal tissue, neo = from neoplastic tissue.

**Table**

Advantages and disadvantages of finite and continuous cell lines

| *Features* | *Finite* | *Continuous* |
|---|---|---|
| 1. Ploidy | Diploid<br>Euploid | Heteroploid<br>Aneuploid |
| 2. Transformation | Normal | Transformed |
| 3. Tomourigenicity | Non-tumourigenic | No Tunuourigenic |
| 4. Anchorage dependence | Yes | No |
| 5. Contact inhibition | Yes | No |
| 6. Density limitation of growth | Yes | No (or less so) |
| 7. Mode of growth | Monolayer | Monctlaycr of suspension |
| 8. Maintenance | Cyclic | Steady state possible |
| 9. Serum requirement (in simple media) | High | Low |
| 10. Cloning efficiency | Low | High |
| 11. Markers | May be tissue specific | Chromosomal, enzymic |
| 12. Special functions (e.g., virus susceptibility, differentiation) | May be retained | Often lost |
| 15. Growth rate | Slow (24-96 hr. doubling time) | Rapid (13-24 hr doubling time) |
| 16. Yield | Low ( $< 10^6$ cells/ml,$< 10^5$ cells/$cm^2$) | High ( $> 10^6$ cells/ml. $> 10^5$ cells/$cm^2$) |
| 17. Control features | Generation number; in vivo markers | Strain characteristics |

It will be noticed that every cell line is given (i) it code (e.g. NHB = normal human brain), (ii) a cell line number, when several cell lines are derived from the same source (e.g. NHBI, NHB2), (iii) number of population doublings, the cell line has already undergone (e.g. NHB2/2 means two doublings). For publication, each cell line should also be prefixed with a code designating the laboratory, where it was derived, e.g. WI = Wistar Institute (Federoff, 1975).

## Maintenance of adherent cell lines

Most primary cultures or continuous cell lines grow as monolayers. These cultures will need a periodic change of medium (removing old medium and adding fresh medium), whether or not the cells are proliferating. In cultures, where cells are proliferating, the usual practice in subsculturing an adherent cell line involves the following steps: (i) romoval of the medium and (ii) dissociation of the cells in the monolayer with trypsin or other enzymes. Intervals between change of medium and subculturing vary from one cell line to another depending upon the rate of growth and metabolism. Rapidly growing cell lines like HeLa are subcultured once per week and the medium changed four days later. More slowly growing cell lines are required to be subcultured every 2, 3 or 4 weeks and the medium changed weekly between subcultures. Different steps involved in the maintenance of coil lines are given in Figure. There are several factors which may be used to assess the need for replacement of medium. These include (i) drop in pH, (ii) cell concentration,

(iii) cell type and (iv) cell morphology. A more detailed discussion on these factors is available in the book by Freshney (1987).

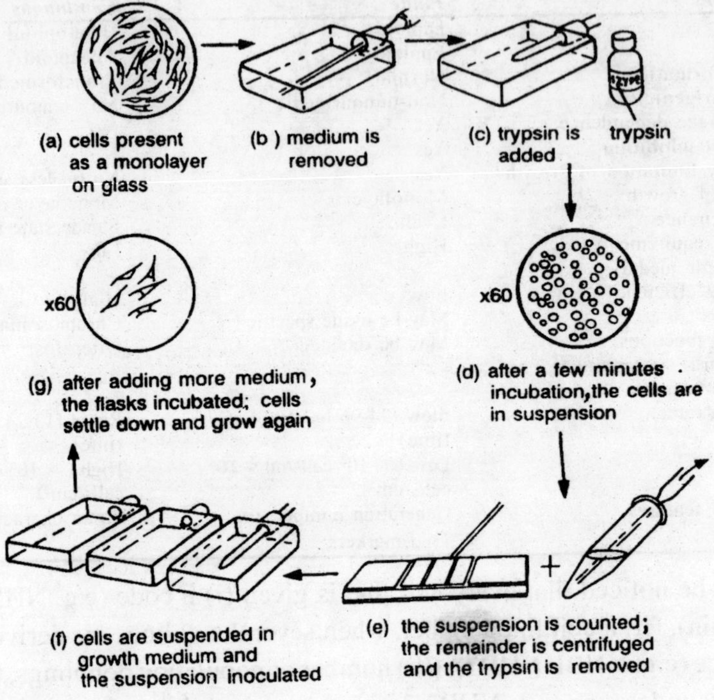


**Fig.** Maintenance of a cell line

### Maintenance of cell lines in suspension

Cells which grow continuously in suspension, either because they are nonadhesive (e.g. many leukaemias and murine ascite tumours) or because they have been kept in suspension mechanically may be subcultured like microorganisms. No enzyme treatment is required in this case and the subculturing is done by dilution of cell density, rather than by replacement of the medium, since the latter will require centrifugation of the suspension culture to obtain a pellet of cells to be added to a fresh quantity of the medium.

## Cloning of Cell Lines

From heterogeneous cultures, pure cell lines can be obtained by cloning, but since many cultures have each a finite life span, the cloning will be successful only in isolating variants from continuous cell lines. These variants may include biochemical mutants and cell strains with marker chromosomes.

**Dilution cloning**

The method of dilution cloning was proposed first by Sanford and then by Puck and Marcus (1955). Trypsinized individual cells are used for preparing suspensions and front the suspension cells are seeded at low density, in multi-well dishes, Petri dishes or plastic bottles. These seeded cells are incubated, until colonies are; formed, which are isolated and then propagated as usual. The different steps involved in this method are shown in Figure.

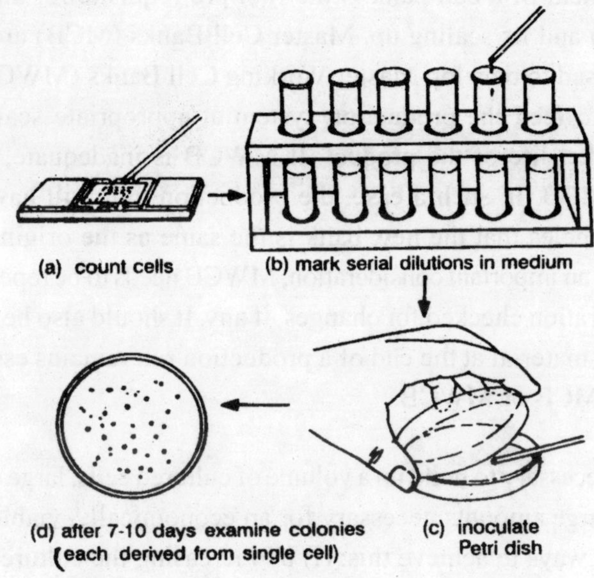


**Fig.** Dilution technique for production of colonies from single cells

## Large Scale Cell Cultures

Large scale animal cell cultures have been extensively utilized in recent years and they form the basis of a multimillion dollar industry in Europe, North and South Americas, Africa, Japan and in India. This industry, in future, will grow and will be a major area in biotechnology. Availability of genetically modified cells for the production of recombinant compounds, has increased the utility of large scale cell cultures in industry. Specific promoters like human metallothionein IIA promoter and baculovirus IEI promoter have been utilized to provide high expression of foreign genes, so that the large scale cell cultures may become commercially profitable.

Recombinant baculoviruses with foreign useful genes have also been designed, so that the cells infected by these viruses may be used in culture for the commercial production of certain chemicals. For this purpose, various 'perfusion systems' have been developed, where cells are retained in a bioreactor, while conditioned medium is constantly replaced with fresh medium, so that cell density and reactor productivity may be increased.

**Cell banking and scaling up of cell culture**

Establishment of a cell bank is the first pre-requisite for any large scale cell culture system and its scaling up. Master Cell Banks (MCB) are first established and they are used to develop Master Working Cell Banks (MWCB), which should be sufficient to feed the production system at appropriate scale for at least the predicted market life of the product. If MWCB is inadequate, one may have to make a new MCB. In such a case, the production unit will have to prove to the regulatory agencies that the new bank is the same as the original one. Since cell stability is also an important consideration, MWCB needs to be repeatedly subcultured and each generation checked for changes, if any. It should also be checked, whether or not cellular material at the end of a production run remains essentially the same as that in the MCB or MWCB.

It is also necessary to build up a volume of cultured cells, large enough to produce a product in large amounts necessary for an economically, viable production unit. There are two ways to achieve this: (i) by increasing the culture volume or (ii) by increasing the concentration of cells in a reactor by continuous perfusion of fresh medium, so that the cells keep on increasing in number without the dilution of the medium. Perfusion has a dual effect of supplying nutrients and removing potentially toxic metabolites.

**Equipment used for large scale cell culture**

The different systems available for large scale animal cell culture are listed in Table. For any production planning, several different types should be tested and compared in terms of case of operation, ease of scale-up, susceptibility to contamination, suitability for suspension or anchorage-dependent cells and the need for special culture media. Basec on all these factors, a list of pros and cons of commercially available bioreactors has been compiled, which is given in Table.

**Table**

Systems available for large scale animal cell cultures

| *System* | *Adherent (A) ! suspensions (S)* |
|---|---|
| 1. Fermenter (homogeneous) | |
| (a) Stirred tank (i) Suspension | S |
| (ii) Microcarrier | A |
| (b) Airlift | A/S |
| 2. Membrane (non-homogeneous) | |
| (a) Hollow fibres | A/S |
| (b) Flat bed | A/S |
| 3. Fused bed (non-homogeneous) | |
| (a) Ceramic matrix reactor | A/S |
| (b) Plate heat exchanger | A |
| (c) Glass bead reactor | A |
| 4. Fluidized bed (homogeneous) | |
| Porous matrix system | A/S |

**Table**

Camparison of bioreactor systems Stem

| *Bioreactor* | *Operation* | *Scale-up* | *Productivity* |
|---|---|---|---|
| 1. Stirred-tank batch reactor | Simple | Yes | Low |
| 2. Airlift batch reactor | Simple | Yes | Low |
| 3. Hollow fibre reactor | Complex | Not known | High |
| 4. Ceramic matrix | Complex | Not known | Intermediate |
| 5. Glass bead reactor | Complex | Not known | Intermediate |
| 6. Fluidized bed reactor | Complex | Not known | High |
| 7. Membroferm | Complex | Not known | High |
| 8. Stirred tank/airlift reactor | Complex | Yes | High |

## Downstream processing

In a large scale production unit, after completion of reaction, cell biomass must be separated from supernatant, which usually contains the desired product. Cell separation is usually achieved by centrifugation or by crossflow filtration. The product may then be purified by using suitable techniques.

## Large scale cell cultures as sources of valuable products

The large scale cell cultures discussed above have already been put to commercial use. Animal cell products are mainly high molecular weight proteins. They include several enzymes, hormones, animal vaccines, monoclonal antibodies, interferons, etc. Some important enzymes obtained from animal cells are asparaginase, collagenase, hyaluronidase, pepsin, rennin, typsin, tyrosine hydroxylase and urokinase. Similarly, four important hormones obtained from cultured cells include luteinizing hormone, follicle-stimulating hormone, chrionic hormone and erythropoietin. Some important products obtained from animal cell lines are shown in Table and are discussed under the following four subheads : (i) vaccines, (ii) interferons, (iii) monoclonal antibodies. (iv) other products like t-PA.

Vaccines—Production of FMD vaccines is the most important example of the use of large scale cell cultures (Table). There are several other vaccines including polio vaccine, bovine leukaemia virus (BLV) vaccines, rabies vaccines, etc., which are produced at commercial scale using cell cultures.

During vigorous purification procedures, the integrity and immunogenicity of viruses produced in cultured cells are largely destroyed. Consequently, most whole virus vaccines are relatively impure in comparison with other products, similarly derived from cultured cells. Despite this, whole virus vaccines are derived from large scale animal cell cultures. However, there are examples, where it is not possible either to culture the whole viruses or to make a safe product from them. Therefore and otherwise, using recombinant DNA technology it became possible to

**Table**

Some animal cell lines and the products obtained from them

| *Cell line* | *Product* |
|---|---|
| 1. Human tumour | Angiogenic factor |
| 2. Human leucocytes | Interferon |
| 3. Mouse fibroblasts | Interferon |
| 4. Human kidney | Urokinase |
| 5. Transformed human kidney cell line, TCL-598 | Single chain urokinase-type plasminoge activator (scu-PA) |
| 6. Human kidney cell (293) | Human protein (HPC) |
| 7. Dog kidney | Canine distemper vaccine |
| 8. Cow kidney | tool and Mouth Disease (FMD) vaccine M |
| 9. Chick embryo fluid | Vaccines fen influenza, measles and mumpe |
| 10. Duck embryo fluid | Vaccines for rabies and rubella |
| 11. Cell lines of mouse, rat or human origin | Monoclonal antibodies |
| 12. Chinese hamster ovary (CHO) cells. | (i) Tissue-type plasminogen activator (t-Pa)<br>(ii) β & γ interferons<br>(iii) factor VIII |

obtain viral subunits from animal cell lines. Vaccines derived from these proteins are purer, safer and more efficacious and are already being used for clinical trials. Similarly, Vaccine virus (orginally used for small PO vaccines) may be used as a vector for foreign viral genes, which may express on infection of host cultured cells by Vaccinia. This system has been tried and needs to be tried in future for live vaccine production. Using this system, recombinant 'Vaccinia' virus vaccines have actually been developed . One such rabies virus vaccine has been field tested in Europe and U.S.A. (using cattle).

Interferons—Due to heavy demands for a veriety of interferons for cancer therapy, recently interferon α and interferon β have been made in quantities in Europe and Japan. For production of interferon ∝, 8000 litre cell culture vessels have been used by Wellcome Foundation. Similarly, interferon-β was produced by cells grown on microcarriers scales of greater than 1000 litres.

Genetic engineering has also been used for interferon production. For instance, interferon genes were inserted in cultured mouse fibroblasts. In one case, these genes inserted in fibroblasts expressed constitutively for the production of glycosylated human interferon b. Special bioreactor have been used for the culture of these fibroblasts on suitable microcarriers and yields as high as 5000 units of interferon b/ ml/day can be obtained.

Monoclonal antibodies—Production of monoclonal antibodies with the help of cultured cells has also achieved—significance in industry, due to the increasing utility of these antibodies in (i) in vitro diagnostics, (ii) in vivo imaging, (iii) therapy in humans and animals and (iv) industrial applications like immunopurification. Ten to twenty years ago, antibodies were available in milligram amounts, today due to increasing demand, antibodies are needed in multikilograms per year.

Two approaches have been followed for the production of monoclonal antibodies (also see Chapter): (i) in vivo production using ascites tumours in mice or rats, (ii) in vitro production through cell culture technique. The second method is economically viable in view of the increasing demand. It has following additional advantages: (i) it can be scaled up ; for instance, one kilogram of antibody (which can be produced through cultured cells in a single fermenter), would require 20,000 mice, if made by ascites tumours; (ii) the risk of contamination is greatly reduced.

Hybridoma cells used for the production of antibodies can be grown either in airlift Cermenter or in stirred fermenter. These cells may also be immobilized and may then involve perfusion type systems. To increase reliability and decrease labour costs, high level of automation is also used in the process, and computer devices are used to monitor factors like pH, temperature, dissolved $O_2$, etc. After fermentation, cells and debris are removed by centrifugation and supernatant is used for isolation of antibodies.

In recent years, the need for hybridoma cells has been obviated and recombinant DNA techniques are being utilized for the production of either the whole antibodies or only a fragment thereof, which is equally useful. The genes or parts of genes are cloned, and inserted in cells which are cultured for the production of monoclonal antibodies.

Tissue-type plasminogen activator (t-PA) and other molecules—A number of other proteins have also been produced using cultured cells. The most important of these proteins is the tissue-type plasminogen activator (t-PA). Clinical trials have shown that recombinant t-PA is safe and effective for dissolving blood clots in patients with heart diseases and thrombotic disorders. This protein is now being synthesized using cell cultures. Several other recombinant molecules that can be produced in large quantities using animal cell culture, include the following: factor VIII, factor IX, protein C and immunoglobulins. Similarly, several hormones and related molecules such as erythropoietin, relaxin, kallikerin, and human growth hormome, can also be synthesized using cell cultures. A number of antigens that are used in viral vaccines are also being synthesized using cell cultures.

## Somatic Cell Fusion

Fusion of different types of cells has been successfully achieved in animals to produce hybrid cells. These hybrid cells can be used for a variety of purposes including the following: (i) study of the control of gene expression and differentiation, (ii) study of the problem of 'malignancy', (iii) viral replication, (iv) gene mapping, (iv) production of hybridomas for antibody production (see chapter for details), etc. In view of this, cell fusion in animals has been studied in some detail. Although there are examples of cell fussion within the body (e.g. myoblasts fusing to form multinucleate muscle fibres), they can also be fused in vitro to produce binucleate heterokaryons (cells fuse, but two types of nuclei do not fuse) or uninucleate hybrid cells.

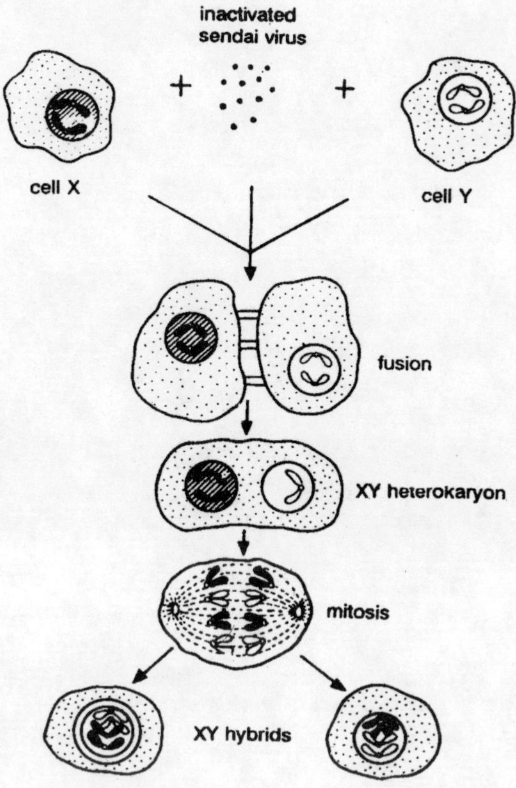


**Fig.** Different steps for production of somatic cell hybrids, using Sendai Virus

In 1960s, in France, hybrid cells were obtained from mixed cultures of two different cell lines derived from mouse. Later, many successful attempts of cell fusion using widely different taxa were made. Sendai virus is often used to induce such a cell fusion (Fig.). Polyethylene glycol (PEG) and certain other chemicals can also induce fusion of cells. Removal of surface carbohydrates may sometimes be necessary for successful cell fusion.

# EIGHT

# Mapping of Genomes

With the advent of recombinant DNA and PCR technologies recently, successful efforts have been made to prepare genetic, cytogenetic and physical maps of animal genomes. These maps utilize molecular markers (described in Chapters) in the form of RFLPs (restriction fragment length polymorphisims), RAPDs or 'rapids' (random amplified polymorphic DNAs), minisatellites or VNTRs (variable number of tandem repeats) and microsatellites or SSRs (simple sequence repeats). Preparation of physical maps is also facilitated by the preparation of contiguous sets of overlapping clones (called 'contigs') in cosmid and YAC (yeast artificial chromosome) vectors. Chromosome walking, chromosome jumping, in-situ hybridization (IS H) and several other techniques have also been used for preparation of physical maps. Various aspects of these molecular maps of animal genomes are briefly described in this chapter.

## Molecular Markers

### Restriction fragment length polymorphisms (RFLPs)

RFLPs as molecular markers, and the procedure utilized for their detection, have been described in Chapter, while discussing the uses of molecular probes. In eukaryotes, the repeated sequences being common, often only unique DNA sequences are used as RFLP probes. These DNA sequences may be obtained either as cDNA (complementary DNA) derived from mRNA (extracted from any metabolically active tissue) or as genomic DNA. For the purpose of preparing RFLP maps, generally probes are prepared from genomic DNA using the following steps : (i) DNA is extracted from cultured cells, or tissues; (ii) DNA is digested with a restriction enzyme like Pst I, which is methyl sensitive, so that repetitive sequences are not cleaved (repetitive sequences are heavily methylated); (iii) digested DNA is fractionated using a sucrose gradient (10%-40%) and DNA fragments in the size range of 500-4000bp are separated; (iv) the size fractionated fragments are cloned in a plasmid (pUC18 or pBluescript), and used for transforming efficient E. coli cells; (v) transformed bacteria are plated and those with inserts (DNA cloned in plasmid) are selected through white/blue colour of colonies (white colonies indicate presence of insert due to disruption of lac Z gene; blue colonies have no inserts); (vi) selected bacterial colonies are rechecked for the presence of the inserts by cleaving the plasmids and subjecting them to electrophoresis; (vii) the selected clones may be checked, if they all represent unique sequences; this can be achieved by hybridizing these clones with labelled genomic DNA using colony hybridization or dot blots, so that only those clones will represent unique sequences , which show weak signal or no signal; (viii) the selected clones may be radioactively labelled and used for hybridization of Southern blots derived from genomic DNA digested with a restriction enzyme like BamHI; only those clones are now selected, which give only one or two bands on autoradiograms, and those giving many bands are rejected; (ix) from these selected clones, one can select those clones, which give polymorphism in the parents used for RFLP mapping.

## Random amplified polymorphic DNAs (RAPDs)

In recent years, polymerase chain reaction (PCR) had an enormous impact on molecular biology research, which is unparalleled. PCR provides selective amplification of a specific DNA segment from an organism, if some information about target DNA is available, so that oligonucleotides may be synthesized to be used as primers at the two ends of the DNA segment to be amplified (consult Chapter for PCR). If primers with arbitrary sequences (random primers) are used for amplification, DNA segments to be amplified will be selected at random. This will provide a truly random sample of DNA markers, which are described as random amplified polymorphic D1VAs (RAPDs = pronounced as 'rapids'). RAPDs have several advantages over RFLPs and can be used as their substitute in several studies. The main advantages include the following : (i) Since some primers with arbitrary sequences can be used for different species, no species specific probes are needed for different species as required in RFLPs. For instance, primers that provide polymorphic markers in maize, also provided polymorphic markers in soybean, Ambidopsis, humans, Neurospora, etc. (ii) Collection of data using RAPDs proceeds much more quickly than those using RFLPs, since there are fewer steps involved (Table). It has been shown that generation of RAPD data is five times quicker than RFLPs. Important properties of RFLPs and RAPDs are presented in Table for comparison.

**Table**

Different steps involved in RFLP and RAPD analysis.

| *RFLP* | *RAPD* |
|---|---|
| 1. Extraction of DNA | 1. Extraction of DNA |
| 2. DNA digestion with restriction enzyme(s) | 2. Amplification of DNA by PCR using random primers |
| 3. Separation of DNA fragments on agarose gel | 3. Separation of amplified DNA on agarose gel |
| 4. Transfer of DNA fragments-Southern transfers | 4. Visualization of markers on the gel and photography of the gel |
| 5. Hybridization of Southern transfers with probe (s) | |
| 6. Detection of probe using autoradiography or using a non-radioactive system | |

RAPD techniques may also be combined with RFLPs to increase the efficiency of molecular markers for genetic analysis. A RAPD marker, known to be polymorphic, may be labeled and used as a probe for RFLP analysis, thus eliminating the need for recombinant DNA cloning of probes in bacteria.

**Table**

Properties of restriction fragment length polymorphisms (RFLPs) and randomly amplified polymorphic DNAs (RAPDs)

| *Properties* | *RFLPs* | *RAPDs* |
|---|---|---|
| 1. Distribution | Ubiquitous | Ubiquitous |
| 2. Inheritance | Simple mendelian, stable, co-dominant | Simple mendelian. stable, dominant/null |
| 3. Level of polymorphism | High | High |
| 4. Detect allelic variant | Yes | No |
| 5. No. of loci detected | 1—3 | 1—10 |
| 6. Part of genome surveyed | Generally low-copy. coding regions | Whole genome |
| 7. Detectable in all tissues | Yes | Yes |
| 8. Technical difficulty | Intermediate | Low |
| 9. Reliability | High | Intermediate/high (occasional spurious products) |
| 10. Quality of DNA required | Relatively pure | Crude |
| 11. Quantity of DNA required | 2—10 μg | 10–50 ng |
| 12. Use radioisotopes | Yes | No |
| 13. Type of probes required | Species-specific low copy gDNA or cDNA \ clones | Kandon 9 – or 10 mer oligonucleotides |
| 14. Recurring cost | High | High |

## Minisatellites or VNTRs (variable number of tandem repeats) and microsatellites or SSRs (simple sequence repeats)

It has also been recently emphasized that polyallelic markers will be very useful for mapping both the simple Mendelian traits as well as polygenic traits in segregating populations. The most widespread of these polyallelic markers are the minisatellites or 'variable number of tandem repeat' (VNTR) loci, which are uncovered by locus specific probes and exhibit highly polyallelic fragment length variation. An interesting and common example of VNTRs involves the number of tandem repeat loci associated with rRNA genes in rDNA concentrated at nucleolar organizing regions (NORs) of specific chromosomes of an organism. Like rRNA genes, most VNTR loci are concentrated in proterminal regions of human chromosomes (but not mouse chromosomes) and thus may not be able to provide the desired density of markers. This problem is largely overcome in microsatellite or SSR loci described in the following text.

In 1989, several laboratories also used PCR to demonstrate a high level of polymorphism or allelic variation in the repeat number (10-50) of SSRs (simple sequence repeats, usually I-6bp long), known as 'microsatellites'. Microsatellites occur frequently and randomly everywhere in all eukaryotic DNAs examined except yeast, and are so named by analogy with the larger minisatellite arrays or VNTRs as above. The VNTR arrays, though highly polymorphic, are less common than microsatellites and have larger sequence motifs (upto 1kbp), making them less amenable to PCR analysis. Microsatellites, on the other hand, are not only amenable to PCR analysis, but are also easy to clone and characterize. They also display considerable stable polymorphism due to variation in the number of repeat units. In view of these attributes, microsatellites are ideal markers for constructing high resolution genetic molecular maps (consult Chapter; Fig.). In 1990, the DNA committee of HGM 10.5 (Human Gene Mapping 10.5) listed 368 microsatellites from human genomc (each having motifs of $\geq$ 4bp and total length $\geq$ 20 bp) available from several databases. They estimated the occurrence of one microsatellite every 6kbp among 745 kbp of human genomic DNA sequences.

Majority of published microsatellite sequences are actually developed from randomly cloned fragments of total genomic DNA and characterized. Genomic DNA is digested with a frequent cutting enzyme (Sau3AI, or a combination of AluI, RsaI and HaeIII) and fragments are cloned in sequencing vectors. Genomic libraries generated thus are screened for the presence of microsatellites by hybridization with simple repetitive oligonucleotide probes, e.g. (AC)n. (AAG)n, etc. Positive clines are then sequenced to find out the flanking sequences to be used as primers for PCR analysis. For libraries of larger insert size, a sequencing primer may be annealed directly to the repeat to determine one of the flanking sequences.

To be useful for construction of linkage maps, microsatellites should generally have 4-5 alleles and should also have PIC (polymorphic information content) > 0 . 7. PIC is calculated as follows:

$$PIC = 1 - \left( \sum_{i=1}^{n} p_i^2 \right) - \sum_{i=1}^{n} \sum_{j=i+1}^{n} 2 p_i^2 p_j^2$$

(where pi and pj are the population frequencies of the ith and jth alleles).

When PIC is >0.7, parents are generally heterozygous at the microsatellite locus and segregation of its alleles (also linkage of this locus with other loci) can be studied unambiguously in the progeny. In a recently published comprehensive genetic linkage map of the human genome (Science, October 2, 1992), a total of 339 microsatellite loci were included among 1416 loci in the map.

## Genetic Maps

### Approaches used for mapping of molecular markers

Molecular markers available in different forms (including those representing transcribed genes) have been mapped in a variety of animals, the most extensively studied among these being humans, mouse and Drosophila. Several approaches have been used, which also vary depending upon the nature of the organism. These approaches include the following : (i) Segregation and linkage could be studied through conventional method by a study of parents, Fl and F2's (or back cross progenies). If at a locus, the genotypes of two parents are designated as A and B, and that of Fl is designated as H (codominant), the distribution of genotypes at different loci in an F, or backcross segregating population can be used through a computer device to obtain the genetic map. This has been done in Drosophila and mouse. (ii) Recombinant inbred (RI) strains have been produced in mouse, through continuous sib-mating in segregating populations, so that the RI strains which differ at two or more loci can be used for two-or three point crosses to generate information on recombination frequencies. The recombinant inbreds can also be used just like F, population along with the parents from which the population of RIs has been derived. (iii) In the above two approaches, sometimes difficulty is encountered in identifying allelic differences. To overcome this problem, in mouse interspecific crosses (Mus musculus x M. spretus) were used. Mus spretus is the most distantly related species and gives fertile Fl hybrids with laboratory mouse. The Fl hybrids can be backcrossed in either direction and the backcross progenies may be used for generating multilocus linkage maps of mouse genome. (iv) In human beings, crosses between desired parents can not be attempted. Similarly, RI lines can not be produced and interspecific crosses are not possible. In view of this, a panel of known families are used for ordering and spacing (recombination frequencies) of molecular markers. LOD scores are calculated and used for linkage analysis (for LOD scores see later in this chapter). (v) Thousands of genes and other molecular markers in Drosohhila have also been mapped through in situ hybridization onto polytene chromosomes.

## Linkage and Recombination between Molecular and Phenotypic Markers

A difference between two individuals for a molecular marker can be used as a genetic marker in much the same way as any other phenotypic marker. The polymorphism at the molecular level may be due to presence or absence of a restriction site or due to a RAPD or due to difference in length of a microsatellite. Although, there may be linkage between a molecular marker and a phenotypic trait, the change in molecular map (restriction map) may not affect the phenotype. In order to study linkage relations between a molecular marker and a phenotypic marker, we may examine the restriction fragment patterns and the phenotypes of two parents that differ morphologically as well as for molecular markers. By making crosses between two such parents, we can estimate the recombination frequencies between molecular markers and phenotypic markers in F, generation. (For details, consult Chapter, Fig.). A tight linkage of such a molecular marker with a disease may also be used for diagnosis of the disease. These molecular markers are also recommended as tools, (i) for identification of cultivers in many of our crop plants, (ii) for selection of desirable plants during plant breeding, and (iii) for isolation of genes, so that these will be increasingly utilized in future (see Chapter for some details).

## Computer software for construction of linkage maps

Several computer programmes are currently available for the construction of multipoint linkage maps. These programmes are, used extensively for a variety of animals and plants. (i) In 1985, LINKAGE was used for preparation of several maps. (ii) In 1987, MAPMAKER and CRI-MAP became available and were found to be more useful than LINKAGE and were used for mapping several chromosomes in humans , mouse and several plant materials. CRI-MAP takes advantage of more genotypic data, inferring genotypes, wherever possible.

Regardless of computer programme, LOD scores are used for quickly finding the best genetic order for large number of loci. (For LOD scores, see later in this chapter). It has been shown that the various computer programmes produce maps with comparable genetic orders.

**Molecular Genetic Maps in Humans (homo sapiens)**

The concept of RFLPs as genetic markers in humans was proposed as early as 1980 (or even earlier), with an objective to facilitate early diagnosis of diseases. Both genomic DNA (gDNA) and cDNA clones have been used, although it was suggested that cDNA sequences being conserved would not reveal much polymorphism. Random genomic clones arc also sometimes unusable due to the occurrence of repetitive DNA sequences in human genome. In view of this, DNA clones suitable for RFLP mapping are first collected using the criteria outlined above.

As discussed earlier, RFLPs are inherited as simple Mendelian codominant markers. Linkage relationships among RFLPs in humans can be established using pedigree analysis involving several generations. Restriction fragment patterns in individuals can be obtained easily using small volumes of peripheral blood, which should give sufficient lymphocyte DNA for analysis. The data for restriction patterns from each probe/enzyme combination are recorded for each individual and analysed for linkage using standard LOD score analysis (explained below).

Although in mouse, recombinant inbred lines (RILs) are produced by inbreeding (sib mating) and have been used successfully for mapping a large number of RFLPs, in humans inbreeding can not be practised and time required for raising another generation is long. Therefore, alternative approaches have been used. DNA markers are first ordered and their spacings (recombination frequencies) determined by analysis of segregation patterns of the markers within a panel of reference families, which represent finite and limited sample sues. Once ordering and spacing are determined between different RFLPs, models are fitted by maximising the likelihood of the parameters. The likelihoods of different models are compared by calculating the likelihood ratios. When the likelihood ratio is 1000 : 1 (or alteast 100 : 1) then the odds of one model are 1000 or at least 100 fold greater than the odds of another, and therefore, the first model is accepted. The $\log_{10}$ likelihood ratio is called LOD score (e.g. for 1000 : 1 likelihood ratio, $\log_{10}$ = 3, LOD score = 3; for 100 : 1 likelihood ratio, LOD score = 2), and is used for linkage analysis. A LOD score of 1 0to 2 is interesting, 2 to 3 is suggestive and > 3 is proof of linkage. It has been shown that no adjacent pair of markers is separated by more than 32% recombination.

Primary genetic linkage maps with markers spaced approximately 15 cM apart are now (1992) available for most human chromosomes. These primary genetic maps are being used for the construction of high resolution linkage maps with markers spaced only 1-2 cM apart. Such high resolution maps have already been prepared for some specific regions containing specific disease genes, but eventually they will be obtained for the entire chromosomes. They will also be used for the preparation of a physical map of the human genome.

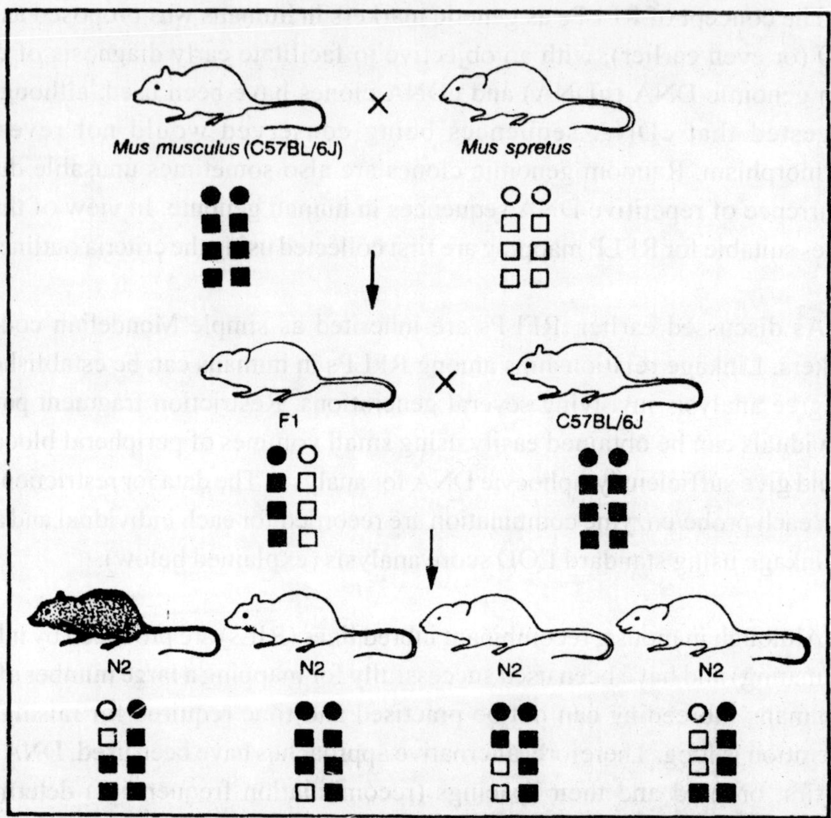


**Fig.** Breeding scheme used to generate the interspecific backross mapping panel; for comparison only one pair of chromosomes is shown: M. Spretus alleles are shown by open boxes, M. musculus (C57BL/6J) alleles are shown by closed black boxes; note that the recombination events of $F_1$ can be scored in $N_2$ backross progeny

For the construction of the above primary genetic linkage maps, a panel of 59 families (27 from Utah , USA; 29 from France; 3 from Venezuela) containing 1212 informative meioses was used. To provide a permanent source of DNA, cell lines have been established from these 59 families. Centre for Study of Human Polymorphism or CEPH (Centre d'Etude du Polymorphism Humain) in Pairs has undertaken to distribute DNA from these cell lines to laboratories collaborating to map the human genome. (CEPH was created by Jean Dausset using money for his Nobel Prize and a large legacy.). The decision to use the same material allowed markers studied in different laboratories to be combined in a single linkage map, and by the end of the year 1993, resolution of 1 cM may prove feasible. A complete 1 cM resolution map of human genome may contain as may as 7000-8000 loci, of which about 3000 loci were already mapped in 1991.

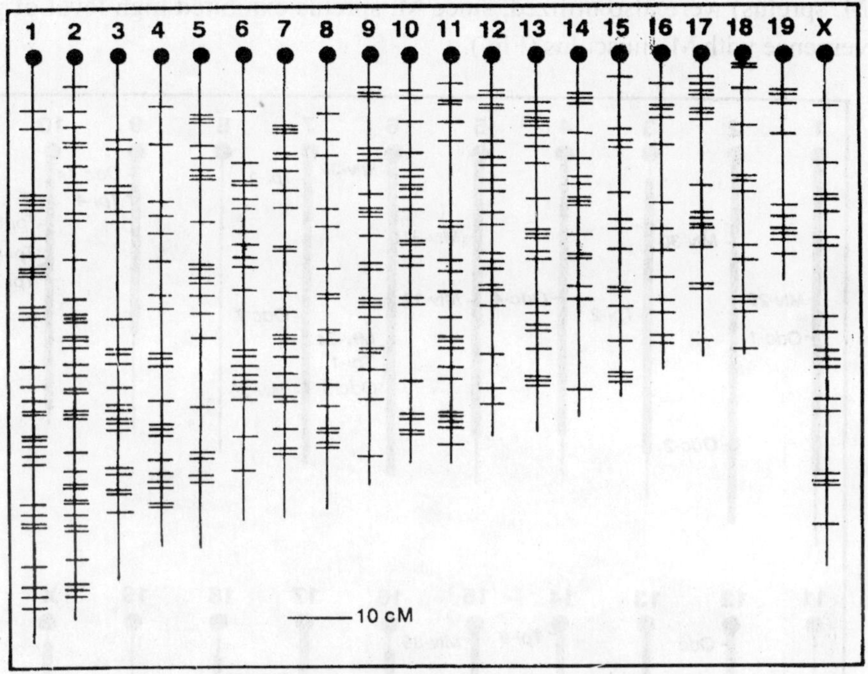


**Fig.** Diagrammatic representation of interspecific backross linkage map of mouse; mouse chromosomes are acrocentrics (centromeres are indicated by filled circles); the distances between centromers/telomeres and markers are not known)

The primary linkage map can be utilized for locating genes for a specific disease on a specific region of a chromosome. This is done through the study of cosegregation of disease with a molecular marker in a reference family. Once linkage of disease with a marker is known through a study of cosegregation, if the marker is already mapped, the chromosome region carrying the disease gene is also automatically known and can be subjected to the development of high resolution map for this region. From a closely linked marker, one could walk or jump to the disease gene using the technique of chromosome walking/jumping to isolate the gene eventually. This technique has been used for isolation of gene for cystic fibrosis (CF), a recessive disorder of some importance in humans.

## Molecular Genetic Maps in Mouse

The genetic maps in mouse have been prepared essentially through the use of the following material : (i) non-identical laboratory strains were used for crosses leading to the production of F, and backcross segregating populations; (ii) recombinant inbred (RI) lines were used either for crosses or like F-, individuals with the parents which differed for specific marker loci; (iii) inter-specific backcrosses (Mus musculus x M. spretus) were also utilized, since M. spretus exhibited high level of genetic divergence with M. musculus (Fig.).

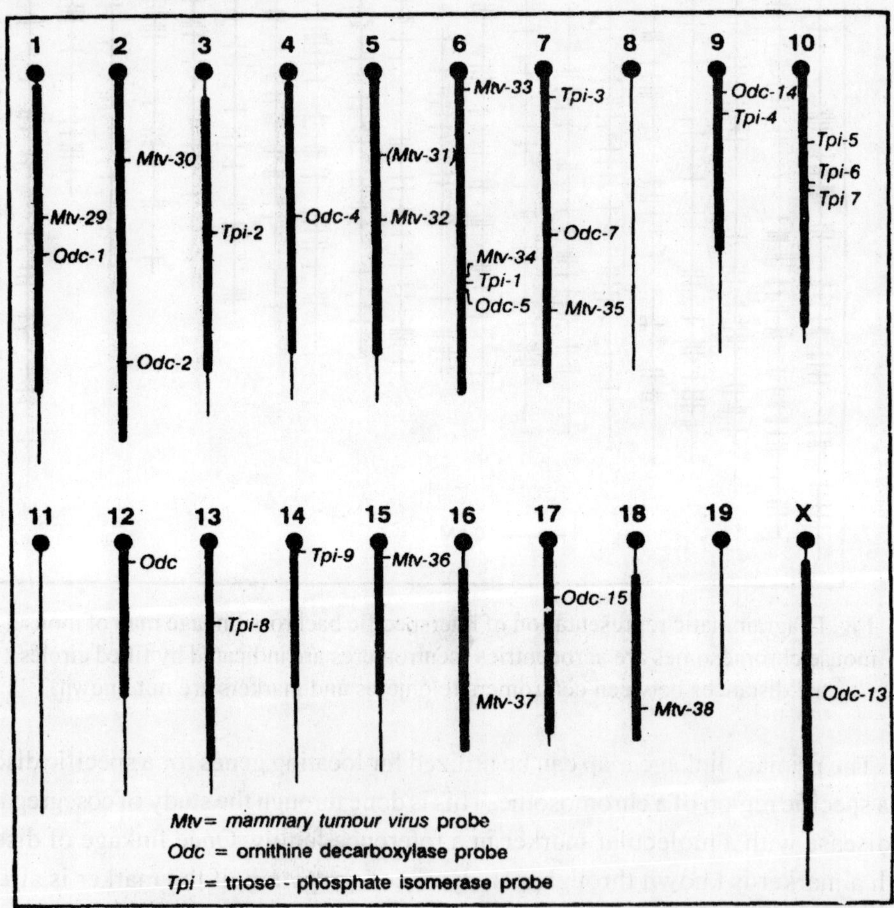


**Fig.** A mouse linkage map showing approximate locations of molecular probes Mtv (mammary tumour virus). Odc (ornithinc decarboxylase) and Tpi (triose-phosphate isomerase) loci on mouse chromosomes)

The mouse has a genome, which is as big as that of humans. RFLPs, RAPDs, and SSRs (microsatellites) were used as molecular markers. In 1989, a linkage mar with 965 loci was presented. This map included phenotypic markers, biochemical markers, cloned genes, and anonymous DNA markers. Subsequently in 1991, 600 loci were positioned on the map using interspecific crosses, which gave a resolution of 3cM (Fig 13 - 2, 13 - 3). More than 300 SSR loci were also mapped in 1992. It is expected that by 1995, a mouse linkage map with a resolution under 1 cM. will be available. Most of this work will be done in USA, by MUGO (Mus musculus Genome Organization).

**Molecular genetic maps in other animal species**

Molecular genetic maps are also being prepared in other farm animals for commercial and medical reasons. These maps of farm animals will enable genetic engineering for breeding programmes to increase yield, growth rate and other commercially important traits. It has also been shown that microsatellites are conserved in related animals. For instance, 50% of cattle microsatellites are amplified by primers from sheep and vice verse. Therefore, heterologous primers are being used for development of SSR or microsatellite maps of related species.

Mouse and rat models can be used for elucidating the genetics and biochemistry of human diseases. Therefore, development of molecular genetic maps in these rodents will be a great help in human and livestock health care.

## Cytogenetic Maps

The first step in mapping of chromosomes at the molecular level is to locate the markers on specific chromosomes. If linkage groups corresponding to individual chromosomes are available, then the molecular markers may be assigned to these chromosomes through study of linkage of available markers with molecular markers (RFLPs) already mapped. In other cases, hybrid cell lines of known cytogenetic composition may be used. These hybrid cell lines often have full set of chromosomes from one rodent (like mouse) and only one known specific human chromosome, which thus enables us to assign a marker or a protein encoding gene to a specific human chromosome. This has allowed the construction of meiotic linkage maps for each of the 23 human chromosomes. These maps have proved useful for locating genes for specific human diseases on specific linkage sites of individual chromosomes. In some cases, these maps also helped in the isolation of specific genes. Despite these advances, on the available maps, 1cM is equivalent to 1 megabase ($Mb = 10^6$bp) and in situ hybridization can not resolve less than several million base pairs. Pulsed field gel electrophoresis (PFGE) is also being used but due to technical limitations, DNA sequences more than a few hundred kilobase pairs can not be ordered. These techniques and some of the results achieved through these techniques will be discussed in this section.

**Cytogenetic Maps Using Chromosome Specific Markers**

Use of somatic cell hybrids of known chromosome constitution. Chromosome specific markers could be isolated from libraries prepared from somatic cell hybrids like mousehuman or hamster-human hybrids, which contain only one human chromosome of known identity (e.g. chromosome 10, 16, 20, 21, 22, etc.). The libraries prepared from such somatic cell hybrids were screened for those clones which contained human DNA inserts (using whole human genomic DNA as a probe), so that these selected clones will carry DNA sequences from a specific human chromosome. For instance, in an experiment, mouse-human hybrid cell line CHI-16, containing human chromosome 16 as the only human component, was used to construct a cosmid library with DNA fragments of approximately 40 kilobase pairs in length. From this library, cosmids containing chromosome 16 specific human DNA inserts could be selected through hybridization with 3'P- labeled human genomic DNA. These cosmid probes when used for mapping would give a cytogenelic map for chromosome 16 and so on.

Use of flow sorted and microdissected chromosomes. In mouse and human, individual chromosomes can be separated from mouse- human hybrid cell lines using the technique of flow sorting. Both normal and translocated human chromosomes could be obtained through flow sorting. In mouse, the chromosomes being similar in size, only five peaks were obtained in flow sorting and individual chromosomes could not be easily separated except when all chromosomes except chromosomes 19 and X were fused in a metacentric arrangement. This technique allowed the construction of chromosome specific libraries in A (lambda) vector. X-chromosome specific cDNA clones were also identified by probing λ-phage libraries prepared from flow sorted X chromosome with $^{32}P$ labeled cDNA probes derived from hamster-human cell hybrid cell lines containing the X chromosome as the only human chromosome. Only X-specific cDNA clones will hybridize, which will represent the transcribed sequences on X- chromosome and will thus allow the construction of a map of Xchromosome, involving functional genes only. A phage library was also prepared using flow sorted chromosome material enriched for chromosome 16.

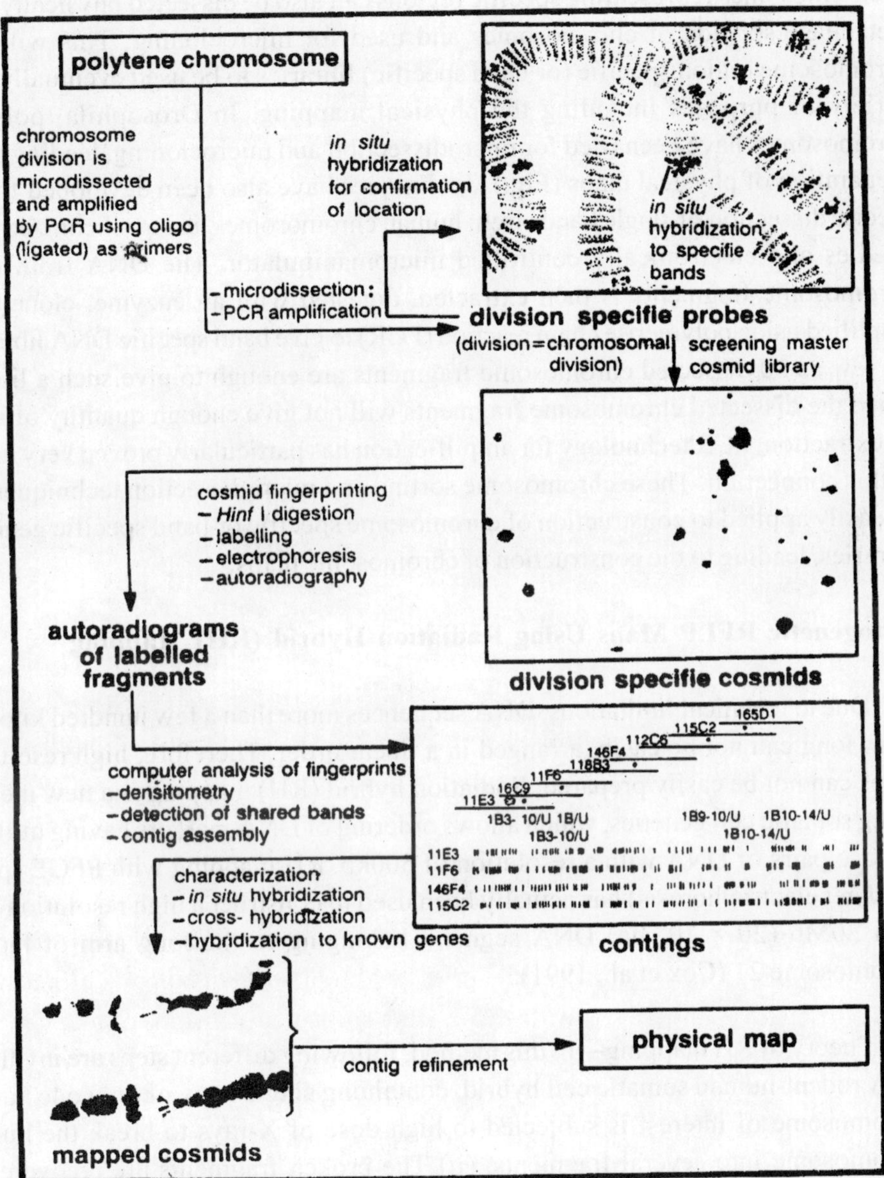


**Fig.** The strategy for constructing a Drosophila cosmid map; the pathway from polytene chromosomes to physical map is outlined and examples of data are presented

In recent years, micromanipulators have also been used for isolating specific chromosomes (where specific chromosomes can be identified), both in animals and plants. Once these specific individual chromosomes are separated in large number, they can be used for microcloning to give chromosome specific clones. Fragments of chromosomes representing specific regions can also be dissected physically from metaphase spreads of chromosomes and used for microcloning. This will give chromosome region specific (or band specific) libraries to be used eventually for a variety of purposes including the physical mapping. In Drosophila, polytene chromosomes have been used for microdissection and microcloning, leading to the preparation of physical maps (Fig.). Techniques have also been developed for the precise dissection of single bands from human chromosomes, using siliconized glass needles and electronically controlled micromanipulator. The DNA from these chromosome fragments is then extracted, digested with an enzyme, cloned and amplified using polymerise chain reaction (PCR) to give band specific DNA libraries. As few as 20 dissected chromosome fragments are enough to give such a library. Since the dissected chromosome fragments will not give enough quantity of DNA on extraction, PCR technology for amplification has particularly proved very useful in this connection. These chromosome sorting and microdissection techniques can be easily applied to construction of chromosome specific or band specific genomic libraries, leading to the construction of chromosome maps.

## Cytogenetic RFLP Maps Using Radiation Hybrid (RH) Mapping

Due to technical limitations, DNA sequences more than a few hundred kilobase pairs long can not be easily arranged in a linear order. Therefore, high resolution maps can not be easily prepared. Radiation hybrid (RH) mapping is a new method using somatic cell genetics, which allows ordering of DNA markers having millions of base pairs of DNA with a resolution of 500kb. RH mapping with PFGE (pulse field gel electrophoresis) has actually been used to construct a high resolution map of a 20Mb ($20 \times 10^6$ bp) DNA segment belonging to the long arm of human chromosome 21 (Cox et al., 1991).

Theory of RH mapping—In this method, following different steps are involved: (i) A rodent-human somatic cell hybrid, containing single copy of only one human chromosome of interest is subjected to high dose of X-rays to break the human chromosome into several fragments. (ii) The broken fragments are recovered in rodent cells and about one hundred such rodent-human hybrid clones are analysed for the presence or absence of specific human DNA markers. (iii) When two markers are found on two fragments, we assume that break has occurred between loci of these two markers. Frequencies of clones showing breakage between different individual pairs of markers are recorded, which will be an index of distance between the two specific loci. Utilizing this information , it is possible to arrange the loci of molecular markers in a linear order on the map, which is analogous to meiotic mapping.

An example of RH mapping—In a recent experiment, Cox et al. (1991) used a Chinese hamster-human somatic cell hybrid (CHG3) containing a single copy of human chromosome 21. The cell line was exposed to 8000 rad of X-rays, fragmenting human chromosome 21 giving a mean of five pieces of this chromosome per cell. Often these pieces were available as translocations with Chinese hamster fragments, clue to breakage and fusion. A dose of 8000 rad being lethal, irradiated cells were rescued by fusion with non-irradiated hamster recipient cells and using a medium that selects the hybrid cells. These hybrid clones were assayed for retention of 14 markers of chromosome 21. Since markers were used for Southern hybridization, retention of more than one fragments in a hybrid created difficulties. For instance, if two markers hybridize in a hybrid, one does not know whether there are two fragments or only one fragment with two markers. However, assuming that retention of one fragment is independent of the retention of the other, following relationship was used for estimation of breakage frequency 'Q' (analogous to recombination frequency in meiosis):

$$Q = \{(A^+ B^-) + (A^- B^+)\}/\{T (R_A + R_B - 2R_A R_B)\}$$

where $(A^+B^-)$ is observed number of hybrid clones retaining marker A, but not B; $(A^- B^+)$ is the observed number of hybrid clones retaining marker B, but not A; T, is the total number of hybrids analysed; $R_A$ is the fraction of hybrid clones analysed for A and retain A; $R_B$ is the fraction of hybrid clones analysed for B and retained B. A 'Q' value of zero means two markers are never broken and Q = 1 means two markers are always broken and are thus not linked. Necessary tests for significant linkage are also performed. A mapping function D = log (1 – Q) is used to calculate distance in centiRAys (cR) analogous to centiMorgans, where 1 $cR_{8000}$ means 1% frequency of breakage between markers nn exposure to 8000 rad of X-rays.

Although for individual pairs of markers, Q and D can be calculated manually as above, but for a large number of pairwise combinations, computer programmes are used.

Confirmation of RH map by PFGE. RH maps could be confirmed when, DNA from a cell line (e.g. CHG3 with chinese hamster nucleus plus human chromosome 21) was cleaved with rare cutter restriction enzymes like NotI, EagI, SalI, BssHI and MluI, which will produce large DNA fragments. In each case, digested DNA was separated on agarose gel by the technique of pulsed field gel electrophoresis (PFGE) and hybridized using Southern blot hybridisation technique. The markers close to each other on RH map are expected to hybridize with same DNA fragment on the Southern blots. A comparison of RH map for human chromosome 21 with its PFGE map, demonstrated that the order of the markers and the distances between them are similar. These PF(JE maps are physical maps, because actual lengths of DNA fragments are being estimated.

RH mapping is specially useful, because, single copy of a human chromossome is analysed , while meiotic mapping requires two homologous chromosomes showing DNA polymorphism. Therefore even non-polymorphic DNA markers, which can not be used for meiotic mapping, can be used for RH mapping.

## Cytogenetic RFLP Maps Using PCR

Assignment of protein encoding genes to specific chromosomes is a major goal of human cytogenetics. This can be routinely achieved by hybridizing a cloned gene with Southern blots or dot blots prepared from DNA samples extracted from hybrid cell lines of known cytogenetic composition. The Southern blots in this case allow rapid identification of a chromosome specific restriction fragment of the gene of interest. However, in this exercise, pseudogenes and genes related with the probe or sequence, in a highly conserved gene may give a false positive signal, so that not all restriction fragments so identified may represent the gene of interest. Therefore, a gene may not be assigned to a specific chromosome with complete certainty. In view of these difficulties in chromosome mapping, an alternative appro ich was recently utilized to map the gene p68 in humans. For this gene, the conserved region was already known and was used for the synthesis of primers to be used in PCR amplification. Species specific differences could be detected in PCR products, the difference (15bp difference) being due to intron. lengths. For p68 genes, it was shown that PCR products using two specific primers differed in mouse and humans in band position. This property was used in screening a number of hybrid cell line of known cytogenetic composition and the results from a number of cell lines could be compared to find out the chromosome on which the gene may be located. The hybrid cell line PCTBA. 1 . H contained only one human chromosome i.e. human chromosome 17 and DNA from this line gave a human specific restriction fragment in PCR product thus confirming that p68 gene was present on chromosome 17. Later, hybrid cell lines having either only one or more fragments of human chromosome 17 were derived. The use of these cell lines enabled the mapping of p68 gene to the distal part of the long arm of chromosome 17 in the region between the bands 17q23 and 17q125.

## Physical Maps

The physical maps of DNA fragments, chromosome fragments or of whole individual chromosomes are based on distances between genes or distances of genes from centromeres or from specific molecular markers measured either in terms of base pairs (megabase pairs = 106bp) or in terms of relative physical lengths. For instance, the distance of a gene from centromere may be expressed as a fraction of the length of the whole arm taken as unity, so that the distance of a gene from confromtere will be indicated as OM, if it is located at a distance of 1/10th of the whole arm. Sometimes, physical maps of individual loci on small chromosome fragments are also prepared, where the DNA sequences on either side of the locus are shown in the form of restriction sites for a number of restriction enzymes. This may be extended to the preparation of restriction map of a whole chromosome. Therefore, one should realize that in physical maps, distances are never expressed in centiMorgan units which are proportionate to recombination frequenceies. The distances in physical maps expressed in megabase pairs are sometimes not proportionate to recombination frequencies and, therefore, the physical maps are considered more important than the genetic or linkage maps. Some of the techniques used for the preparation of these physical maps will be described in this section.

### Physical Maps Using Chromosomal Aberrations (deletions)

Physical mapping of specific DNA sequences available in the form probes can be achieved using chromosomal aberrations such as translations, duplications and deletions. For instance, the disease choroidermia or tapeto-choroidal dystrophy (TCD) has been described in various patients having cytogenetically visible or submicroscopic deletions spanning the $X_q$21 band. This means that TCD locus must fall within the region these overlapping deletions. TCD is an X-chromosome linked disorder retina and choroid, causing progressive right blindness and central blindness during the age of 20-40 years. In order to determine the location of TCD locus more precisely, several of these deletions were characterized with the help of several anonymous cloned DNA sequences from the proximal region of $X_q$. This led to the division of $X_q$21 band into seven different intervals and TCD was assigned to interval 3 spanning three problem (DXS95, DXS165, DXS233). It was also found that several TCD patient had deletions which included the locus DXS 165 but not the other two loci. This suggested that TCD may be close to DXS 165. Further mapping was achieved by chromosome jumping and walking and TCD was assigned to a DNA segment of 15-20 kilobases.

**Physical Maps Using in Situ Hybridization**

Mapping of single copy DNA sequences by in situ hybridization (ISH) using nonradioactive label (e.g. biotin or digoxigenin) has also been successfully achieved in some cases, particularly for human chromosomes.

**Physical Maps Using Yeast Artificial Chromosomes (VAC) and ISH**

The isolation and characterization of the following three elements essential for replication made it possible to form completely new artificial chromosomes (i) origin of DNA replication, (ii) centromeres and (iii) telomeres. Single copies of each of these elements can be linked to specific DNA segments to form artificial chromosomes, which can be inserted in yeast cells and tested for their functions

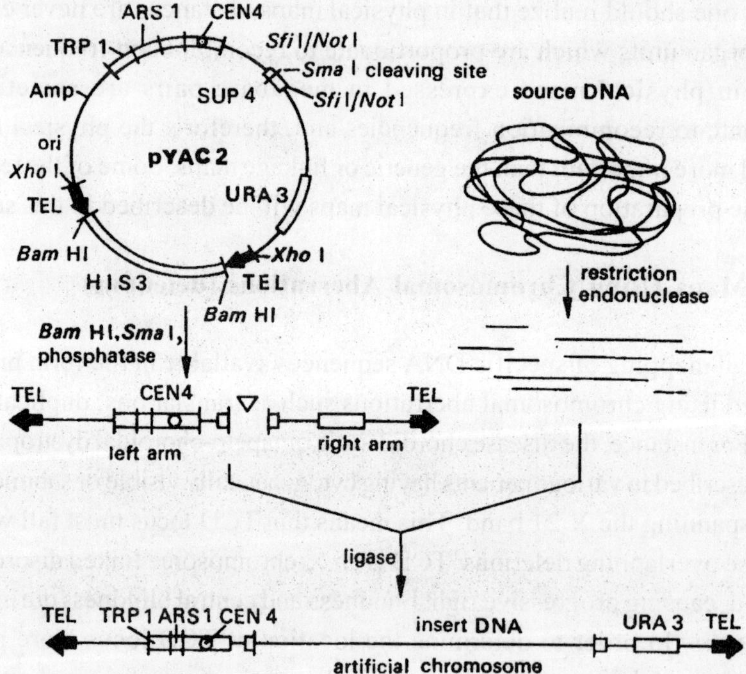


Fig. An artificial yeast chromosome vector, which may carry a long DNA segment to be cloned in yeast cells

(Fig.). Using these artificial chromosomes, molecular maps in Drosophila have recently been prepared. Actually, for large scale physical mapping of a eukaryotic genome, Drosophila is considered a model organism (though not as attractive as yeast or Aspergillus) for the following reasons (i) small genome with 165,000 kilobase pairs (kbp) (ii) the availability of large number of mutations and chromosome rearrangements and (iii) the use of transposon 'P' for germline transformation. Although a number of localized regions in Drosophila genome have been analyzed by chromosome walking, it is often limited to only few hundred kilobase pairs and requires the use of overlapping lambda (A) or cosmid clones (consult next section for chromosome walking) of the size below 40 kbp. It has been estimated that at least 10,000 cosmid clones will have to be analysed to include all sequences of the genome, so that the task will be formidable. Since DNA fragments, several hundred kilobase pairs long, can be cloned in YAC vectors, a YAC library of 1500 clones (instead of 10,000 with cosmids) is sufficient for mapping the Drosophila genome. During the years 1989-92, as many as more than 1000 Drosophila YAC clones (called DY clones) were cytologeally located.

The DY clones were prepared, using the following steps : (i) high molecular weight (HMW) DNA was isolated from D. melanogaster DNA, using caesium chloride gradient centrifugation; (ii) molecules larger than 120 kbp were separated by size fractionation in sucrose gradients; (iii) the large fragments (120kbp) separated as above were ligated onto YAC vector arms; (iv) the selected DY clones were used for transformation of yeast cells.

When DY clones were to be used for mapping, they were separated from yeast chromosomes by FIDE (field inversion gel electrophoresis) and later in each case, Drosophila DNA fragment was isolated from DY clone by enzyme cleavage. This DNA is then used for in situ hybridization (ISH) on Drosophila salivary bland chromosomes. This enabled mapping of the entire range of salivary gland chromosomes, although gaps between them do exist. The fragments used as probes can be used for further detailed restriction mapping or sequencing according to requirements. In humans also, a detailed YAC contigs was published in 1992.

The YAC clones may also be independently used for preparing physical maps of the adjoining regions of a cloned gene in the form of restriction maps. For instance, an effort was made to prepare restriction maps (involving rare cutting restriction enzymes = RE) of some loci in a region of human chromosome 18. As another example, using cDNA of PAI-3 as a probe (plasminogen activator inhibitor type 2, a gene responsible for hears attacks), three clones were selected from 13000 clones of a YAC library. These clones contained, complete or a part of PAI- 3 locus, as shown by colony hybridization. The YAC clones were digested with rare cutting REs and the resulting fragments were size fractionated, and restriction maps were prepared.

## Physical Maps Using Chromosome Walking and Jumping; (or hopping)

The techniques of chromosome walking and jumping were earlier described in Chapter. The centromeric sequences on a yeast chromosome could be isolated using chromosome walk from two cloned genes CDC10 and LEU 2 on either side of the centromere. Walks have also been used to establish the structure of some very large Drosophila genes that are much too long to be cloned even in cosmids. Unfortunately, chromosome walks are difficult in higher eukaryotes, due to the presence of repetitive sequences flanking majority of the genes, which allow cross hybridization. Therefore, the fragments of DNA can not be arranged along the length of chromosomes in higher eukaryotes.

Chromosome jumping technique has been used to cross large distances to approach closer to a gene of interest (consult Chapter for details). Size selected DNA fragments are generated by partial Mbol digestion and a jumping library is prepared. Jumping has the additional advantage of allowing one to jump over otherwise unclonable sequences. In humans, following jumping libraries, besides others, have been prepared.

(1) 100 kb jumping library to move towards Cystic Fibrosis (CF) locus (2) 200 kb jumping library to move toward Huntington Disease (HD).

In these cases, clones are derived which jump from a given locus (probe) towards the gene of interest. The jumping library (with $2 \times 10^6$ clones) is screened using closely linked probes that were labeled. Positive phage clones are subcloned and used for further chromosome jumping. The gene responsible for Huntingtion Disease (HD) in humans, is one such gene that is identified through its close linkage to the DNA marker DHS10, which, by in situ hybridization and somatic cell genetics, has been mapped near the lip of the short arm of chromosome 4 (4h). Chromosome jumping was used to move from DHS10 to HD. Similarly, cystic fibrosis gene was also mapped using chromosome jumping.

## Physical Maps of Large DNA Fragments by Chromosome Fragmentation

Linear DNA molecules of length > 10 megabases could be separated using alternating field gel electrophoresis, and in some cases, represented chromosome sized DNA molecules. Through hybridization with probes, these cloned sequences can be assigned to chromosomes. This may be followed by mapping of DNA segments along each chromosome. This is really useful particularly when a chromosome can not be mapped following classical methods of genetic analysis.

Fragmentation of a chromosome—This involves breaking of a chromosome at the position of a cloned DNA segment. A technique for breaking the chromosome at the site of a cloned segment is available in yeast, where a diploid yeast cell is transformed twice. (i) First it is transformed with a small linear DNA molecule, one end of it having DNA segment to be mapped and the other end having the telomere. Due to recombination of this molecule with complete chromosome, a shortened stable DNA molecule is obtained. (ii) The diploid yeast cells are again transformed with another linear DNA molecule, one end of which has DNA segment to be mapped and the other end has a telomere and a centromere on the opposite end (relative to linear molecule used in the first transformation). Recombination in this case will give another shortened chromosome with stable ends. The stable chromosome fragments (obtained from two separate transformations) and the cloned DNA segment can be positioned relative to the chromosome ends.

Using the above technique, three genes were mapped (MELT. G.ALI. LAY2). For this purpose, a number of different yeast chromosome fragmentation vectors (YCFl, YCF2, YCF3, YCF4) were utilized (Fig.).

Positioning of a cloned sequence on the chromosome—The position on chromosome can also be located using the following steps : (i) The two linear molecules are designed as YCF3 and YCF4 vectors. Since the centromere/telomere orientation of the segment is not known, four molecules must be constructed with both orientations of segment inserted into each vector. This can be done by cueing the vector at a specific site in polylinker and introducing the segment with compatible ends. The vector carrying the segment is linearized, leaving the telomere at one end and the DNA segment at the other. (ii) The four linear molecules are constructed and used to transform the strain YNN290, having useful markers (iii) Transformants are streaked onto rich media plates and grown for three days to allow the colony colour to develop. (iv) Intact chromosome DNA is prepared from 5-10 pink colonies for each, YCF3 and YCF4. (v) DNA is separated by alternating CHEF gel electrophoresis and transferred to nylon membrane. (vi) The blots are hybridized with mapping sequence. I (vii) Two bands were identified to represent two chromosome segments and were sized using phage DNA fragments or the chromosomes of YPH149 (a strain of yeast). (viii) The mapping sequence is positioned and confirmed by rehybridizing the membrane with one or more probes earlier mapped. The advantage of the present method also lies in the fact that a cloned sequence can be mapped without preparing detailed restriction map as done in chromosome walking.

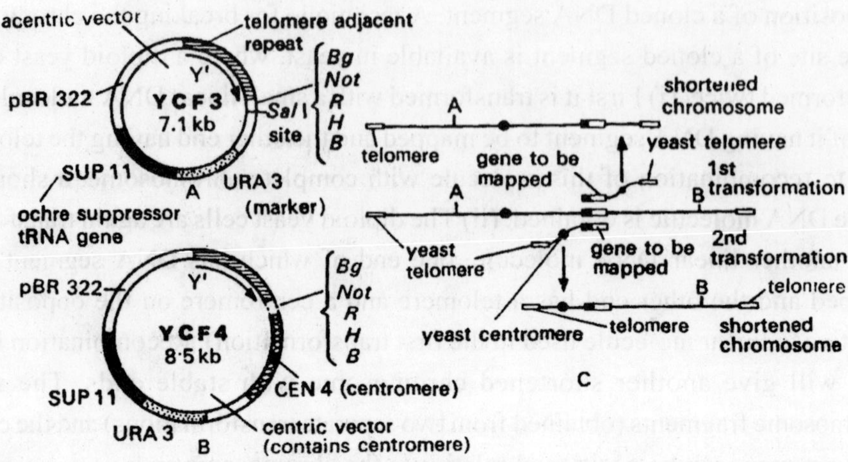


**Fig.** Two yeast chromosome fragmentation (YCF) vectors (A, B) showing details; use of these vectors for fragmentation of chromosomers, leading to physical mapping is shown in C.

Physical mapping of chromosome ends using telomeric DNA fragments A universal deficiency in genetic maps of higher organisms is the absence of markers that identify the ends of chromosomes. Unique, physically defined genetic markers associated with individual telomeres would provide the boundaries for genetic maps. In view of this, availability of a molecular marker representing the telomeric sequence, will be a great help in preparing genetic and physical maps in eukaryotes. This telomeric sequence has been found to be a six nucleotide sequence that is repeated in the telomeric region over and over again without mistake. The sequence has been isolated in Ambidopsis, Homo sapiens, Tetrahymena, Paramecium, Oxytricha, Sncchnrontyces, etc. and is G-and-C rich with a general formula (T or A) $_m$(G)$_n$, which is (TTAGGG)n in humans. When this telomeric sequence was used for in .situ hybridization, the hybridization took place in a wide variety of vertebrates, showing that the sequence is conserved and variation though exists, but should be only minor. Biotinylated synthetic oligomers, (TTTAGGG)$_6$ and (CCCTAAA) homologous to the consensus sequence of the telomere in Arabidopsis thaliana, were also used for in situ hybridization to root tip chromosomes of barley and rye. Their presence only at the ends of chromosomes was detected (Schwarzacher and Hes- 1991). Chromosome specific subtelomeric sequences for human chromosomes have also been cloned, which can be used to find the ends of specific chromosomes (Riethman et al., 1989).

**Physical Maps for Mammalian Chromosomes (including humans)**

Top down (restriction mapping) vs. bottom up (Contig mapping) approaches—One of the early goals of what is now called "Human Genome Project" is to make physical maps of each of the 22 autosomes and two sex chromosomes (X, Y) in humans. Two conceptually different types of strategies are being tried : (i) The first strategy involves preparation of restriction maps of large DNA fragments (macro restriction analysis), generated by rare enzymes that cut the genome infrequently. It is achieved through analysis by pulse-field gel electrophoresis (PFGE). This approach is reductionist or top down approach, because we start with isolation of individual chromosomes or their large segments using hybrid cell lines of known chromosome composition. Then from the existing genetic maps of these isolated individual chromosomes (or segments), we use probes that serve as anchor points on the physical map. The gaps between these anchor points are then filled by extensive Southern analysis. This technique has already yielded restriction maps for yeast (Saccharomyces pombe) chromosomes. (ii) The second strategy involves preparation for complete ordered genomic library. This is a bottom up approach, because we start with individual clones, which are fingerprinted and through detection of overlapping of DNA segments in clones, the latter are arranged in a large number of contiguous sets of overlapping clones (called contigs), which coalesce into a map of the whole genome.

Above two approaches are complementary, but have their limitations and benefits. For instance, the top down approach of restricition mapping of large DNA fragments, provides information about distantly spaced genomic regions. However, it does not provide purified DNA in an immortal amplifiable form. In the second i.e. honour up approach of ordered complete library, on the other hand, it is easy to start with individual clones, but it is difficult to complete the library, there being many gaps. However, it provides a high resolution information and a permanent source of purified DNA. It is believed that both types of maps are useful and should be constructed in parallel.

Utilizing the above two approaches involving restriction mapping and contiguous mapping, a hybrid map technology has also been suggested. In hybrid maps, we make use of restriction maps, linkage maps (RFLP maps), cytogenetics and somatic cell genetics, in order to orient and align a series of contigs. In such a map, the discontinuity of contig maps is largely overcome and clones will be available that will cover most of the mapped region.

**Sequence Tagged Sites (STS) and Expressed Sequence Tags (EST)**

Due to the availability of polymerase chain reaction (PCR) technology, in recent years strategies for mapping human genome have regularly been reviewed and scrutinized. Consequently, sequence tagged sites (STSs) suggested by Olson et al. (1988), are becoming standard markers for the physical mapping of human genome. In this technique a short tract of DNA from the clone that has been physically mapped is sequenced (300,500bp) and the site is now defined by this sequence and called an STS A PCR assay for an STS could be implemented simply by synthesizing two short (20 nucleotides) obligodeoxynucleotides, chosen to be complementary to opposite strands at the opposite ends of the STS sequence. A DNA sample can then be tested for the presence of the STS sequence through its capacity to amplify using the above primers in PCR.

While STSs represent unique sequences, which may or may not be transcribed, recently analogous sequences called expressed sequence tags (ESTs) have been utilized, which would have an additional feature of representing only expressed genes. In humans, cDNA libraries derived from brain mRNA have been utilized for this purpose. An EST is simply a segment of sequence from a cDNA clone that corresponds to a mRNA. It is shown that ESTs longer than ISObp (150-4006p) are useful even for search of similarity and for mapping by workers in different laboratories. Using nucleotide sequence databank (GenBank) and protein sequence database (Protein Information Resource = PIR), the ESTs can also be matched with existing sequences and assigned to specific genes with the help of computer programmes. Using EST sequences for preparing specific primers, and using hybrid cell lines of known chromosome constitution as template in PCR, an EST can also be assigned to a specific chromosome by the mere presence of amplified DNA. A large number of ESTs could be assigned to specific chromosomes using this technique. Thus, ESTs will extend the physical mapping of expressed genes in the human gnome at a much faster rate than hitherto conceived.

## Integrated Genomic Maps

In the past, genetic maps were prepared by different groups separately, particularly, using RFLPs. These 'private maps like private armies induce fights' (E. Robson). Therefore, the data from different sources are now being used and integrated composite maps are being presented. These maps would have information on linkage distances, chromosome location and physical distances in term of number of base pairs, etc.

During October 1992, two comprehensive ways of the entire human genome became available. (i) In the first of these maps prepared jointly by hundreds of scientists from National Institute of Health (NIH) and CEPH collaborative group, 1416 loci spread over 90% of the genome were mapped. This map contains mostly RFLP markers, but also included 279 genes (and expressed sequences) and 339 microsatellites. Only 351 markers (including 205 microsatellites) had heterozygosity of 70% or more (heterozygosity is a measure of variability; maximum heterozygosity for a two allele marker is only 50%). (ii) Another map, published by about 20 workers at Genethon (a factory for human genetics in France), was based on 792 (C-A) repeat microsatellite markers (600 of them are highly informative). The above two maps may be combined, since both used CEPH families. A large yeast-artificial chromosome (YAC) Contig of human chromosomes was also published in 1992 and it is hoped that the genetic map will soon be integrated with cytogenetic and physical maps (contigs). It is hoped that by the end of 1993, a map with an average marker spacing of less than 1 cM will become available.

Integrated genome maps are also being prepared in other animals particularly mouse and Drosophila. In mouse, a microsatellite based map was published in 1992 and will be integrated with RFLP and RAPD markers earlier mapped. In Drosophila also, integrated molecular maps are being prepared. In 1992, more than 1000 YAC clones were placed on a physical map through in-situ hybridization. In another approach, specific chromosome regions were microdissected and used for preparing overlapping cosmid clones (contig). An integrated genetic map for Drosophile using linkage data, cytogenetic data and other information has been published in 1992 in a special issue of Drosophila Information Service'.

## Sequencing of Genomes

In recent years, considerable effort is being directed towards sequencing of whole genomes in a variety of micro-organisms and eukaryotes. The largest genone sequenced completely so far belongs to cytomegalovirus (2,00,000 base pairs). The genome of E. coli (4 - 8 x 106bp) has also been sequenced to 76% level. The progress on sequencing of whole genome in a variety of organisms is shown

**Table**

Progress toward complete genome sequencing
(as on ,tune .1, 1992 from Biotechnology-July 1992).

| | *Organism* | *Genome size (Mb)* | *Base pairs deposited* | *% genome sequenced* |
|---|---|---|---|---|
| 1. | E. coli | 4.8 | $3.4 \times 10^6$ | 76.0 |
| 2. | S. cerevisiae | 14.1 | $4.0 \times 10^6$ | 27.0 |
| 3. | C. elegans | 100.0 | $1 \times 10^6$ | 1.1 |
| 4. | D. melanogaster | 165.0 | $3.0 \times 10^6$ | 1.8 |
| 5. | M. musculus | 2300.0 | $8.2 \times 10^6$ | 0.3 |
| 6. | H. sapiens | 2800.0 | $1.8 \times 10^7$ | 0.6 |

in Table. It is expected in very near futur complete genomes of human beings as well as several other organisms will be sequenced.

# NINE

# Fertilization and Transfer

The techniques utilized for culturing cells, tissues and organs in animals have achieved significance in recent years due to their extensive use in manipulation of genetic material using recombinant DNA technology. Now you will learn that plant cells are totipotent and can develop into whole plants irrespective of their origin from any part of the plant. This is not true about animal cells (including human cells). In case of animals, therefore, only the gametes (oocytes and spermatocytes) can be manipulated and used for the development of an animal. In view of this, it was important that the techniques are developed which will allow (i) recovery of oocytes in large number; (ii) fertilization of these oocytes in vitro (outside the body of the animal); (iii) transfer of the fertilized ovum to an appropriate place in the body of the animal followed by proper development of the embryo. During the last one decade, great effort has been directed in refining these techniques and sufficient progress in this direction has actually been made not only in animals (cattle, goats, sheep, swine, chicken, etc.), but also in human beings. It is hoped that these developments will lead to genetic improvement of farm animals leading to high growth rates and increased milk production. In human beings, after successful recovery of viable oocytes, in vitro fertilization (IVF) and embryo transfer (ET) followed by normal development of the fetus, the birth of a baby named Louise Joy Brown (on July 25, 1978), is considered to be a sensational event of great scientific importance and human health care. The development of these techniques has already led to the treatment of a variety of infertilities in human beings. The above techniques involving collection of oocytes, in vitro fertilization, sexing of embryos and embryo transfer (particularly in human beings, but to some extent also in animals like cows, goats, pigs, sheep, swine, etc) will be discussed in this chapter. In the next chapter, we will discuss the techniques used for the transfection of eggs or embryos in animals using recombinant DNA technology, leading to the development of transgenic animals.

## In Vitro Fertilization (IVF) in Humans

The techniques of recovery of human oocytes, their in vitro fertilization and embryo culture have been under intensive investigation since late 1960's. These development will be briefly described in this section.

### Types and causes of infertility

In vitro fertilization (IVF) is usually needed for a couple, who may have infertility due to any one of a variety of reasons. Following are some examples of infertility.

Female infertility—(a) Tubal infertility—Prior to the availability of IVF as a clinical procedure, often the damaged or non-functional fallopian tubes (fallopian tubes ate the tubes which receive oocytes from the follicles of ovaries to allow fertilization) were repaired through surgery with low success rate. Other approaches to cure tubal infertility include ovarian implantation, tubal transplantation or replacement of tubes by grafting. The IVF technique is gradually replacing the above approaches of surgery for the treatment of tubal infertility, because once fertilization has taken place in vitro, no fallopian tubes, but only functional uterus is required for embryo transfer.

(b) Nonfunctional or inaccessible (or absent) ovaries—When the ovaries of the female partner are nonfunctional or inaccessible, oocyte may be obtained from a donor, which can be inseminated by the husband's spermatozoa in vitro. The developing embryo may then be transferred to the infertile recipient. This would also require exogenous supply of steroid hormones if ovaries are non-functional.

(c) Nonfunctional (or absent) uterus—When uterus is absent or nonfunctional, the oocytes from the female partner may be subjected to IVF and the fertilized embryo may be transferred to a surrogate mother for the period of the pregnancy.

(d) Idiopathic infertility—Another class of infertility is called idiopathic infertility, since no explanation is available for this class of infertility. In most cases, this may involve failure of fertilization or abnormal fertilization. IVF and ET provide a useful therapeutic method for the cure of this idiopathic infertility also.

Male infertility—For normal fertility the sperm concentration in human semen should be 15-20 million per ml. However, for IVF as little as 10,000-50,000 motile spermatozoa per ml are adequate. A reduced level of sperms in a male is often described as oligospermia, while the absence or very low concentration of motile sperms (although non-motile sperms may be found) is described as azoospermia. In view of the above, it is possible that a patient suffering with oligospermia may contain sufficient motile spermatoza for IVF but not for the fertilization in vivo.

## Assessment of patients

The female partner included for IVF programme should fulfil the following criteria: (i) She should be physically fit to carry a resultant pregnancy and evidence should be available to indicate that surgical procedures involved in IVF and ET will not create any problem. It has been observed that obesity and previous abdominal operations make oocyte collection more difficult and hazardous. (ii) Her ovaries should be accessible for oocytes. If the ovaries are obscured due to adhesions, preliminary laparotomy and ovariolysis may need to be performed. (iii) Her uterus should be capable of accepting and sustaining pregnancy for full term of 9 months. (iv) She should have easily negotiable cerivcal canal (cervix) for ET, failing which an alternative route is used.

The male partner is examined for only an appropriate concentration of motile sperms, failing which alternative male with appropriate semen may have to be used.

## Treatment of patients for IVF

The female patients for IVF are required to record the commencement of each of their menstrual cycles for atleast six months, so that decision may be taken for (i) admitting the patient in the hospital, (ii) date of urine sampling for luteinising hormone (LH) surge and (iii) date of administration of human chorionic gonadotrophin (hCG) to control the final stages of follicular and oocyte formation.

Nature of the cycle for follicular development—The nature of the treatment prescribed for the female partner of the couple will depend upon which of the following three cycles is utilized for the oocyte recovery for IVF:

(1) In the natural cycle, spontaneous luteinising hormone or LH surge (under whose influence all fully developed follicles of both ovaries burst, ovulate and ova are collected by the fallopian tubes) is detected by urine or plasma sampling at 3-6h intervals. The spontaneous cycle has the advantages of early term pregnancy and of natural hormonal environment for embryo transfer (ET). The disadvantages include the availability and recovery of only one oocyte leading to low success rate of IVF and ET.

(2) In the stimulated cycle, multiple follicular development is stimulated by clomiphene, and/or human menopausal gonadotrophin (HMG), but LH surge is spontaneous and detected as above. Human chorionic gonadotrophin (HCG) is also administered to check inhibition of LH surge caused by hMG. The advantage of stimulated cycle is the availability of multiple follicles and oocytes so that several embroys are available for ET leading to higher success rate. The disadvantages include difficulty in interpreting the variable hormonal levels, necessity of ultrasonic scanning to assess the number of maturing follicles, recovery of occytes at different development stages (some of these occytes are abnormal), abnormal environment and increased risk of abnormal pregnancy.

In the controlled cycle, the development of the follicles is arrested at the optimum stage of maturation by the administration of hCG. The advantage of this cycle is that oocyte recovery can be performed through laparoscopy (see later) at a time convenient to the doctor and the patient so that the patient's hospital stay and anxiety are reduced. The difficulty, however, is in predicting optimum time for hCG administration, which whould also vary for different preovulatory follicles.

Of the above three cycles, the stimulated and controlled cycles are known to have given higher success rates in oocyte recovery, embryo transfer and pregnancies achieved.

Ovarian stimulation—Once the decision has been taken that the stimulatory cycle needs to be used, as mentioned above, the patient needs to be given doses of clomiphene and hMG. Clomiphene is given at the optimum rate of 150 mg per day from the 5th day to 9th day of the cycle. Some patients do not respond to five day course so that prolonged administration for 10 days may be required. This prolonged treatment will, however, also prolong the anti-oestrogen effect so that uterus receptivity for embryo implantation may be reduced leading to increased spontaneous abortion rate. However, the abortion rate is low if hMG is used instead of clomiphene. The only disadvantage of using hMG is the increased cost; so that a combination of clomiphene and hMG is often used. The use of hMG also leads to hyperstimulation which may not support pregnancy and instead lead to a variety of complications including abdominal discomfort, ascites, pleural effusion and thromboembolic phenomena, sometimes leading to deaths.

(a) Monitoring ovarian stimulation—The following parameters are used as indicators for ovarian stimulation to help collection of oocytes at the optimum stage of development: (i) statistical prediction of the day of LH (luteinising hormone) surge on the basis of data on lengths and dates of menstrual cycles (although this prediction is fairly accurate in normal cycle, its accuracy in stimulatory cycles is yet to be established); (ii) temperature chart ( the temperature rise at preovulatory and post-ovulatory stages); (iii) changes in cervical mucus score indicating follicular development; it begins to rise 5 days before LH surge, reaches a maximum level a day before and falls sharply on the day of LH surge; (iv) plasma or urinary oestrogen levels, which can be estimated at 24h urine collections M, and are used for timing the LH surge and hCG injection; established method of Brown et al. (1968) is used, but more recently radioimmunoassay M has been successfully utilized; (v) ultrasonic determination of follicular size; (vi) detection of LH surge and progesterone determination.

(b) Spontaneous LH surge—For accurate prediction of ovulation time, M, LH peak is most useful, since if laparoscopy is performed after ovulation, no oocyte is recovered and if performed more than 6h earlier, fertilization may not occur (however, due to refinement of techniques, oocytes can now mature in culture making the time of oocyte collection less critical). LH can be assayed in urine (3 hourly intervals) either by haemagglutination or by radioimmunoassay to allow correct predicition of time for laparoscopy.

Plasma progesterone concentrations are also measured since they begin to increase 24h prior to LH surge in stimulated cycles. Therefore, an increase in progesterone means that spontaneous LH surge will occur within the next 24h. Indirectly, this will also confirm suspected ovulation. LH determination is also needed for the administration of hCG, because no rise in LH should be observed before administering hCG which is used to control the time of ovulation. If LH level has already risen prior to the administration of hCG, then no hCG is administered and laparoscopy 1 is timed 21-28h after the midpoint of the first sample which showed elevated levels of LH.

(c) Administration of hCG for controlled ovulation—As discussed earlier, , hCG is administered so that ovulation is controlled. The decision to administer hCG is made on the basis of (i) the day of the menstrual cycle, (ii) absolute levels and changes in the levels of ocstrodial 17 j (iii) size and number of follicles estimated through ultrasonic examinations and (iv) cervical mucus score. The ultrasonic scan should confirm if there are one or more follicles of atleast 1.9 cm diameter. If the follicles are smaller, hCG is delayed until the follicles attain preovulatory sire. Further, the urinary LH level should not rise before hCG is administered as discussed above.

In patients with no indication of spontaneous LH surge, laparoscopy is timed 28-36h after hCG injection.

**Oocyte recovery (laparoscopy)**

Earlier methods—Several methods for oocyte recovery were tried in the past and included the following: (i) the follicular fluid is totally aspirated and oocytes identified in these follicular aspirates; if an oocyte is identified in the aspirate, further aspiration is not necessary; (ii) the excised ovary is minced for oocyte recovery; (iii) intact follicles are dissected prior to puncture to study the developmental stage; (iv) follicles are dissected under the microscope, and oocytes recovered.

Equipment and technique for laparoscopy—Since the earlier methods listed above caused extensive damage to ovary, their application was restricted. The development of laparoscopy, therefore, was a major advancement in the techniques of oocyte recovery. In this technique, while looking at the follicles in the ovary using a microscopic equipment, one aims at the ovulating follicles, which are aspirated with the aspirating apparatus, consisting of (i) stainless steel needle, 23 cm long and having an external diameter of 2.1 mm and internal diameter of 1.6 mm; the needle is lined by tiflon reducing the internal diameter to 1.0 mm; (ii) a tubing, 52 cm long, and continuous from the sharp bevel, along the needle and through a silicone rubber hung into a 5 ml or 10 ml tissue culture tube. The total fluid space between needle and the tubing is 1.2 ml. An 18 gauge needle perforates the rubber hung and connected to a suction pump operated by foot. The suction should be of the order of 80-100 mm of mercury. For the insufflation of the peritoneal cavity, carbon dioxide is often used, although nitrogen mixtures (90% N., 5% O,, 5% COQ) are also recommended.

The presence of omentum or adhesions overlying the ovary may create difficulties and should be dealt with laparoscopic scissors. The ovarian ligament is grasped gently and follicles carefully inspected. Ovulating follicles are selected for aspiration. Follicles that may have ova capable of undergoing fertilization are smaller than 1.5 cm, although if the follicle is within the ovarian tissue, its size can not be accurately assessed. The suction apparatus should be tested using culture medium to avoid spillage of fluid and loss of oocytes due to failure of the pump.

The follicular fluid (4 ml-12 ml) is aspirated completely from ovulating follicle as described above. It should be straw coloured, although it may sometimes be contaminated with blood. As soon as blood appears in the aspirating tube, heparinised culture medium should be added to the aspirate to avoid clotting, because if blood clot appears in aspirate, the ovum may not be easily identified, isolated and fertilized.

As the fluid is aspirated, walls of the follicle collapse. The aspirate may be examined and oocyte identified (in which case, further aspiration is not required), failing which repeated flushing of the follicle with heparinised culture medium is performed to gel the oucytes through reaspiration of distended follicle. This may be repeated as often as necessary until the oocyte is recoverd.

For recovery of fully mature oocytes as above, follicular monitors and the timing between LH surge and hCG administration should be carefully interpreted., since one would like to culture the oocyte in vitro for a minimum period before fertilization. This will increase the chances of fertilization and subsequent embryo development.

**Oocyte culture and IVF**

Culture of oocytes—The naked oocyte is identified under the microscope and is handled with great care. The oocytes are incubated for 5-10 hours after oocyte recovery depending upon the estimated maturity of the follicles and oocytes. The following four culture media have been tried for oocyte culture and IVF: (i) modified Ham's F11 medium; (ii) Earl's solution; (iii) modified Whitten's culture medium; (iv) Whittingham's T6 medium (for details consult Trounson & Counti, 1982). The culture medium is prepared weekly and tested by culturing mouse embryos from 2-cell stage to blastocyst stage. The use of the same medium for fertilization and embryo culture has actually simplified the fertilization and culture techniques.

Preparation of semen—Semen is collected at site by masturbation, 60-90 minutes before insemination. After liquefaction, the semen is centrifuged and the resultant sperm pellet is resuspended in culture medium. After recentrifugation, the washed spermatozoa arc incubated in culture medium at 37°C for 30-60 minutes and a sample is taken from the surface, so that it contains the most active spermatozoa.

In vitro fertilization—A total of 10,000-50,000 motile spermatozoa, prepared as above, are added to 10-11 ml of culture medium containing the oocyte. After insemination, the oocyte is left for 12-13 hours and then inspected in a petri dish containing culture medium. The oocyte should contain two pronuclei and two polar bodies. Any abnormality like presence of multiple pronuclei, granulation or vesciculature of the cytoplasm or abnormal shape are noted and such embroys are not used for transfer. ) The first cleavage takes place 24-30 hours after insemination and each subsequent division should occur within 10- 12 hours thereafter. If cleavage does not occur within 24 hours, this suggests abnormal oocyte, which should not be used for transfer.

Besides the study of cleavage, best embryos may he selected through fluorescein diacetate test, although in humans its utility is doubtful, since oocytes and emborys both may appear similar morphologically. Another approach is to remove one cell of the embryo and to examine it under electron microscope. This test may also be evasive, since the cell may not be representative of the embryo (degenerating cells may also be found in embryo) and the appearance of the cell may not be indicative of the biological function.

## Embryo Transfer (ET) in Humans

### Time of embryo transfer

Success has been achieved from the transfer of 1-16 celled embryos to the uterus (embryos in vivo stay in fallopian tubes upto 8-16 cell stage and then move to uterus), but not from the more advanced embryos. Prolonged culture of an embryo in vitro reduces viability and thus reduces pregnancy rate. The best stage is perhaps 2-4 cell, although these early embryos may not survive in uterus. Multiple embryos may also be transferred to increase the chance of subsequent pregnancy, but this may lead to multiple pregnancy, which the patient may not like.

### Technique of embryo transfer

The patient is given 10mg diazepam (valium) orally before embryo transfer, which is performed in the operating theatre. No anaesthesia or analgesia are needed and ambryo transfer is negotiated through cervical canal. Following steps are involved in embryo transfer: (i) the patient is placed in the knee chest or litholomy position with enough head down tilt to ensure that the fundus is lower than the cervical canal; (ii) a sterile bivalve vaginal speculum is gently inserted to visualize the cervix ; (iii) the uterus is manipulated gently to align the cervical canal and uterine cavity; (iv) when the cervix is exposed and ready for transfer, the embryo is drawn into a teflon catheter (external diameter 1.27 mm, internal diameter 1.0 mm) in 10 E.el of tissue culture medium; teflon is used, since due to its low adhensiveness, it reduces the chance of carrying cervical secretions into the uterine cavity; (v) the teflon catheter is then passed down on outer teflon sheath which protects the inner catheter from vaginal contamination; (vi) the catheter is inserted into the uterine cavity just short of the fundus, using the information about the; distance from the fundus to external cervical canal (this information is earlier collected through ultrasound measurements); (vii) the embryo is gently injected in culture medium and the catheter and cannula are gently withdrawn; (viii) catheter is examined under the microscope to enure that the embryo has been expelled.

The above steps should be performed with extreme care, since incorrect placement of the catheter, or the use of excessive quantity of fluid for transfer may lead to the expulsion of the embryo from the uterine lumen or to possible tubal pregnancy. Such complications in pregnancy following embryo transfer have actually been observed.

After embryo transfer, the patient is advised (i) rest for 2-7h and (ii) to abstain from intercourse for atleast a week. No hormonal support is advised after embryo transfer.

**IVF, and Embryo Transfer in Livestock in Farm**

In recent years, a new industry has also developed for getting genetically improved farm animals in good number. The technique involves the following two steps: (a) collection of fertilized eggs from genetically superior females, which either ovulate spontaneously or are induced to produce a number of eggs (superovulation); (b) the embroys are transferred into genetically less desirable females thus achieving the goal of getting many mop progeny from superior females. The embryos can also be cultured artificial media for future use. The sexing of embryos is also done permitting more female progeny if required. Another technique involves splitting embryos into segments, each segment developing into an offspring using the above technology, so that 20-30 calves per year can be obtained from one valuable donor.

It is also possible to replace the nucleus of an egg of a genetically less desirable female animal with that of a somatic cell of a superior female so that superior progeny may be obtained. This will overcome the problem of non-availability of superior eggs in large number. Methods are also being developed to transform the egg or embryo before embryo transfer is affected. Transgenic animals in this manner have been produced, in mice, cows, sheep, goats, chicken, pigs, etc. This aspect will be discussed in some detail in Chapter.

In farm animals, more efficient methods are being developed for superovulation, in vitro fertilization, embryo sexing, embryo culture, tract formation of embryo, and embryo transfer. The nutritional requirements for culturing embryos of farm animals have generally been extrapolated from work with laboratory animals like mice and rabbits. However, embryos of most mammals can be maintained outside the bodies of animals for half a day or more with no irreversible damages. If transgenic animals produced by the above methods are genetically superior and can be multiplied using the methods outlined above, the expense involved will be rewarded by producing many superior animals through super- ovulation, and embryo transfer in surrogate mothers. The genetic potent of an embryo can be multiplied by dividing the embryo into group cells to obtain indentical twins, triplets or quadruplets. The technology for freezing semen and embryos has also been developed, and will provide enormous flexibility in using these techniques for genetic improvement farm animals.

## Embryo Transfer in Cattle

Some 40 years ago, artificial insemination (where semen from superior male could be used to inseminate a large number of females) emerged as a new force in animal breeding and led to significant improvement in our livestock. This work was facilitated due to our ability to store semen for prolonged periods and to transport it to distances, whenever required.

### Technique for embryo transfer in cattle

Embryo transfer (or transplantation) in cattle is a more recent development, where newly formed embryos, prior to their implantation (after fertilization), are removed from a superior 'donor' female animal and transferred into the reproductive tract of other relatively inferior females (often described as surrogate mothers), where they develop to term. Techniques have also been developed where embryo can be divided into two or more equal parts, each part developing into an offspring giving twins, triplets, etc. after embryo transfer. Therefore, embryo transfer technique is used for faster multiplication of genetically superior high yielding cattle. The offspring in such cases will derive its genes partly from the donor female and partly from males to which the donor were bred, but none from the surrogate mothers, who work only as incubators for the development of embryos from the donor. Following steps are involved in the embryo transfer in cattles: (i) A genetically superior high milking cow is selected as donor and low milk yielding cow is selected to be used as recipient (surrogate mother). The oestrus (heat) of both is synchronized using prostaglandins. (ii) Follicle stimulating hormone (FSH) is injected in the donor to induce superovulation to get increased number of ovules. (iii) The donor cow is subjected to artificial insemination using forzen semen from genetically superior top pedigree bull. (iv) Seven days after the artificial insemination, the fertilized eggs (embryos) are collected (flushed) from the uterus of the donor cow, using special nutrient medium (in a year, six flushings of embryos can be obtained from a donor cow, each flushing giving six good embryos so that 36 embryos in a year can be obtained from one donor cow). (v) The embryos are isolated from the flushings and examined under stereo- microscope; good embryos are selected. (vi) Selected embryos are maintained in special medium at 37°C in an incubator. (vii) One good embryo (or a part, where embryo is split) is transferred to each surrogate mother, which is low milk producing cow (on an average 50% success is achieved, so that 18 superior calves can be produced from one donor cow in a year, while in the natural course, only one calf is produced by a cow in a year).

**Technique for freezing embryos in cattle**

In order to freeze embryos for long term storage, the embryos are dehydrated with the help of a 'cryoprotectant' like glycerol. After the equilibration is achieved, embryos are cooled very slowly (in a controlled manner ) from 25°C to-38°C. The cooling is actually done in a programmable controlled rate freezer. The embryos at -38°C are plunged into liquid nitrogen (-196°C) and stored for posterity and used whenever required.

**Advantages of embryo transfer in cattle**

Following are the advantages of the technique of embryo transfer:

(1) Embryo transfer increases the reproductive capacity of a superior heifer or a cow (sometimes their number is limited, since they are imported), since semen from one male can be used for a large number of females, and because a single superior female can be made to give a large number of ovules through the technique of superovulation.

(2) Old superior cows, that are unable to maintain pregnancy can still donate ovules for embryo transfer.

(3) We may not need to import cows in only the frozen embryos (frozen embryos can be stored for 10 years or more without losing viability), and grow them in surrogate mothers. It is much cheaper than importing cows. The newly developing offspring also develops an immune system to deal with pathogens in the new environment.

(4) Embryos can be frozen and preserved in an embryo bank and used whenever required.

(5) Embryo transfer can be used to expedite the programme of bull testing.

## Transgenic Animals

Although, there is a variety of applications of animal cell and tissue culture, one of its important application also involves transfer of foreign genes into our livestock. In organisms like bacteria and other microbes, or even in higher plants, the uptake of genes by cells is often described by the term 'transformation'. However in animals, this term has been replaced by the term 'transfection', because the term 'transformation' in animal cell culture is used to descirbe phenotypic alteration of cells (see Chapter). The usage of the term 'transformation' for 'cell alteration' has been unfortunate and discontinuation of its usage for this purpose is suggested. Transfection or gene transfer in animals may be carried out at the cellular level to get transfected cells, which may be used for a variety of purposes such as production of chemicals and pharmaceutical drugs. It may also be undertaken for basic studies involving study of structure and function of genes. Although many mammalian cell lines have been regularly utilized for these purposes, transfection has also been achieved successfully for the production of transgenic animals. The improvement in livestock through transgenesis has already led to the following encouraging results: (i) increased milk production in cattle; (ii) increased growth rate of livestock and fish (iii) large scale production of valuable proteins in milk, urine and blood of livestock, enabling the use of transgenic animals as 'bioreactors' for 'molecular farming', (iv) improvement of wool production through production of transgenic sheep. These encouraging results have made transfection and production of transgenic animals a fascinating thrust area of research. In this chapter, different methods of transfection and other aspects of trans genesis in animals will be discussed. The available examples of successful production of transgenic animals will also be discussed.

### Gene Transfer or Transfection

Fusion can be exercised between isolated nuclei, or between a nucleus and an enucleated whole cell. Similardly, the transfer of genetic material to whole cells can be exercised at the level of (i) individual chromosomes or fragments or (ii) isolated genes or gene fragments.

### Transfection of fertilized eggs or embryos

The production of transgenic animals requires transfection of specialized cells or embryos, since only eggs or embryos can develop into whole animals. Contrary to this, in plants any cell or protoplast may be transformed, since undifferentiated plant cells, being totipotent, can develop into whole plants through differentiation and growth. The eggs may be modified by transfer of whole nuclei, whole chromosomes or parts thereof or DNA fragments.

Transfer of whole nuclei (or split embryos)—Transfer of whole nucleus from a somatic cell of the superior donor to the enucleated egg can be achieved using the following steps (Fig.) : (i) enucleation of unfertilized egg is achieved by centrifuging cytochalasin-B-treated cells, such that the nuclei detatch from the eggs and pellet at the bottom of the tube, leaving enucleated eggs in the supernatant; (ii) karyoplasts (nuclei with only some residual plasma membrane) are similarly obtained from the blastula stage of the developing embryos of the donor; (iii) karyoplasts derived from the donor are incubated with enucleated eggs in the presence of PEG (polyethylene glycol), and fusion is achieved; (iv) the manipulated egg is transferred to the uterus of a surrogate mother for development.

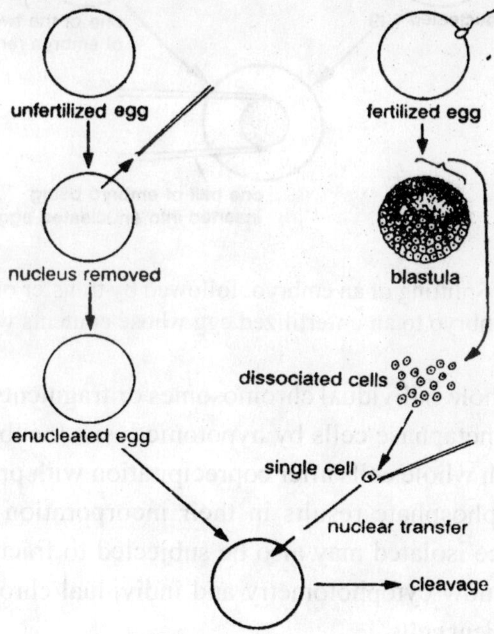


**Fig.** Fusion of karyoplasts with enucleated cells, in the presence of PEG (polythylene glycol)

Techniques have also been developed, where a superior developing embryo may be bisected into two parts using a surgical blade. Each bisected part of the embryo may be separately transferred to an enucleated unfertilized egg (Fig.). Such a manipulated egg may then be transferred to the uterus of a surrogate mother for further development.

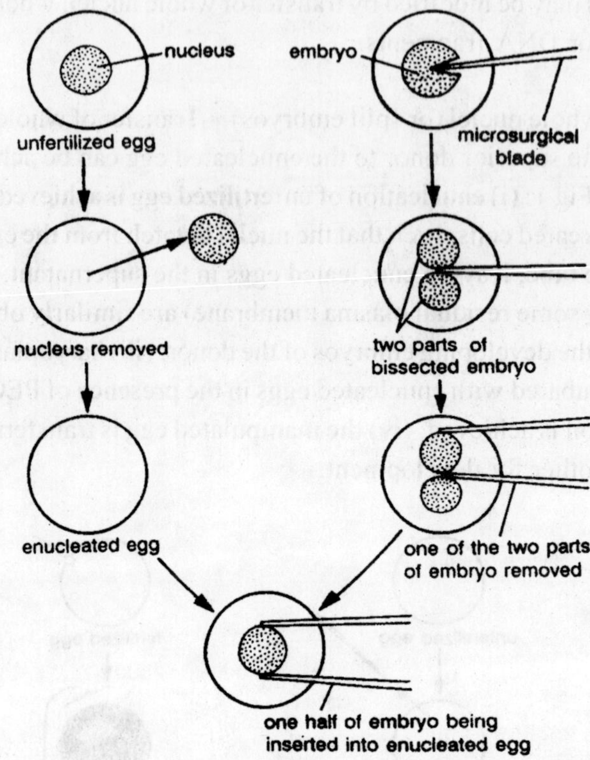


**Fig.** Splitting of an embryo, followed by transfer of one half of the embryo to an unfertilized egg whose contents were removed

Transfer of whole individual chromosomes or fragments—Chromosomes may be isolated from metaphase cells by hypotonic lysis. Incubation of these isolated chromosomes with whole cells after coprecipitation with production of transgenic animal, calcium phosphate results in their incorporation into the nuclei . The chromosomes once isolated may also be subjected to fractionation using density centrifugation or now cytophotometry and individual chromosome pairs may be inserted into recipient cells.

DNA microinjection into the egg—Transfer of individual cloned genes is generally achieved by inserting into a fertilized egg, a large number of copies of recombinant plasmid carrying the gene of interest, using the technique of DNA microinjection. (Fig.). The DNA is microinjected into a fertilized egg before the fusion of male and female nuclei. The egg is first immobilized by applying mild suction to the large, blunt holding pipette and DNA is then injected through the sharp end of a narrow glass microneedle. Through recombination at the DNA level, the inserted gene may get integrated into the host genome and inherited in a Mendelian manner. Transgenic mice have been produced in many laboratories, utilizing this technique. The method has also been successfully utilized for production of transgenic chicken, cows, fish, pigs, rabbits and sheep.

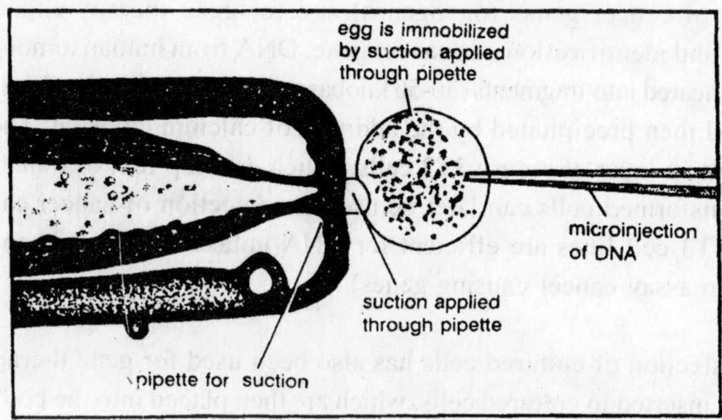


**Fig.** Microinjection of foreign DNA into a fertilized egg for production of transgenic animals

In Drosophila, the P elements (transposons) are used as vectors, so that a recombinant P element carrying the foreign gene is used for microinjection. Although the microinjected flies may not carry the inserted gene in all cells, some of their progeny may carry the gene in all the cells.

Virus mediated gene transfer to embryo—Retrovirus infection of embryos has also been used for the production of transgenic mice. This virus has been found to be an efficient vector system for animals (see later for details). The virus carrying the gene of interest transfers it into the genome of embryonic cells leading to its integration and production of transgenic animals.

**Transfection of cultured mammalian cells**

Many types of cells, from bacterial to mammalian cells, can be transfected through transfer of genes, either using viruses as vectors or by physical or chemical methods. The viruses that have been found suitable as vectors include retrovintses, vaccinia virus, adeno-associated virus (AAV), herpes viruses and bovine papiloma virus. In mammalian cells, transfection frequency also increases dramatically, if DNA is co-precipitated with calcium phosphate in the recipient cells. In all these methods, some of the transferred DNA may integrate with the chromosomes of the cultured cells, thus resulting into stable transfection. If the donor DNA encodes a genetic trail, the recipient cells will express this trait.

Transfection of cultured mammalian cells has been utilized extensively for detection of cancer genes (oncogenes) and for gene therapy experiments. For detection and identification of a cancer gene, DNA from human tumour cell lines is purified, sheared into fragments (30-50 kilobase pairs in length), dissolved in phosphate buffer and then precipitated by the addition of calcium chloride. The solution is poured unto a layer of mouse 3T3 cells, which develop foci of transformed cells. These transformed cells can later be used for detection of cancer causing genes. (Mouse 3T3 cell lines are efficient for DNA uptake and have been decades for decades to assay cancer causing genes).

Transfection of cultured cells has also been used for gene therapy where the genes are inserted in cultured cells, which are then placed into the body of a patient for gene therapy. A variety of virus vectors are used for this purpose. Retroviruses were used in most of the early clinical trials (in 1990-91) including those on the two young girls suffering with adenosine deaminase (ADA) deficiency. A modified adenovirus and adeno- associated virus are also being used. 'Naked DNA' is also used, for gene therapy, using 'particle gun', 'liposomes' or 'electroporation device'. Future research will identify the safest and most efficient delivery methods for gene therapy.

## Targeted gene transfer

In most of the methods for gene transfer described above, the integration of the foreign gene occurs at apparently random chromosomal positions. Although these 'misplaced' genes are subject to correct regulation under the control of regulatory sequences, random integration does not lead to precise replacement of defective gene at a homologous site. The gene transfer at homologous sites in the host genome is popularly described as targeted gene transfer. Initially, such targeted gene transfer by homologous recombination for replacement of a resident wild or mutant gene was possible only in bacteria and yeast. Later, for the first time in 1998, such targeted gene transfer was successfully achieved in human cells by Oliver Smithies at Madison (USA). He transferred a human P globin gene into the /3 globin gene of recipient cells by recombination. The targeted gene transfer was posssible due to the presence of homologous DNA sequences at the targeted site and in the vector carrying the foreign gene. Marker genes are also used for selecting cells transfected with transferred gene at the targeted site. Such a selection has been possible by several methods: (i) by assaying the activity of an enzyme, as in hypoxanthine phosphoribosyl transferase (HPRT) enzyme; (ii) by using a marker gene for antibiotic resistance that would enable cells to grow in a medium containing the antibiotic that would otherwise kill the cells; (iii) by polymerase chain reaction (PCR), if sequence of the gene is available for making PCR primers (see Chapter). Chimeric mice have actually been prepared by targeted gene transfer, where HPRT gene or homoeo box containing genes have been modified in some embryonic cells. If these cells contribute to the production of egg or sperm, the transgenic mice will be produced. Success in targeted gene transfer has actually been achieved in many other cases also.

Gene targeting using embryonic stem (ES) cells—In this technique, embryo derived stem cells are cultured (for stem cells, see Chpater) . These stem cells are transfected with a vector containing the desirable gene. This allows targeting of gene to a particular site by homologous recombination. The rare targeted cells are isolated, multiplied and used for injection of expanded blastocysts, which are then implanted into surrogate mothers to allow the development to be completed. The developed animals can be checked for transgenesis. The different steps of this technique are illustrated in Figure.

## Transgenic Animals

### Transgenic mice and other animals

One of the first reports of transgenic animals published in December, 1982, involved transfer of growth hormone (GH) gene (from rat) fused to the promoter for the mouse metallothionein 1 (MT) gene. Since then, many transgenic animals, including those in cattle, sheep, goats, pigs, rabbits, chickens and fish have been produced (Table and will be utilized in future for a variety of purposes including (i) their efficiency in utilizing feed; (ii) ability to give learner meat, (iii) their ability to grow to marketable size sooner and (iv) their resistance to certain diseases. More than this, recently efforts are being made to use transgenic animals as living bioreactors. Transgenic animals produced for this purpose will secrete valuable recombinant proteins and pharmaceuticals into their milk, blood and urine which can be used for extraction of these drugs. This new possibility of manufacturing drugs through transgenic animals is often described as 'molecular farming' or 'molecular pharming'.

**Table**

A list of transgenic animals produced and the promoter, enhancer and structural transgenes involved in each case.

| *Animal* | *Genes transferred (promoter or enhancer/structural genes)* |
|---|---|
| 1. Mouse | mMT/rGH, mMT/bGH, mMT/oGH, mMT/hGH, mMT/hGRF, mMT/hFIX |
| 2. Chicken | ALV, REV |
| 3. Cow | BPV, lactoferrin |
| 4. Fish | mGH, mMT/hGH, mMT/bGal, cd-Crystallin SV/hygro, AFP |
| 5. Pig | mMT/hGH, mMT/bGH, hMT/pGH, MLV/rGH, bPRL/bGH |
| 6. Rabbit | mMT/hGH, hMT/hGH, rbEu/rb, c-myc |
| 7. Sheep | mMT/hGH, mMT/TK, mMT/bGH, mMT/hGRF, oBLG/hFIX, oBLG/a.1 SY, oMT/oGH |
| 8. Goats | A variant of tPA gene (LAtPA) |

Note: The list of genes; promoter/or enhancer is separated from structural gene by a slash (/). The source of the gene is given by a lower preceding the gene : b = bovine(ox): c = chicken ; h = human ; m = mouse ; o = ovine: p = porcine ; r = rat; rb = rabbit.
Gene abbreviatons-ALV = avian leukosis virus : $\alpha$1AT = $\alpha$ 1 anti-trypsin ; BPV = bovine papilloma virus; Eu = immunoglobulin heavy chain; FIX = factor IX. GH = growth hormone; GRF = growth releasing factor: $\beta$Gal = $\beta$ gulctosidase: hygro = hygromycin; BLG = $\beta$-lactoglobulin: MT = metallothionein : MVL = moloney murine leukemia virus; c-myc = a cellular proto-oncogene; REV = reticuloendotheliosis; PRL = prolactin ; SV = SV 40'; TK = thymidine kinase: AFP = antifreeze protein: *t*PA = human tissue-type plasminogen activator.

Although, initially many experiments leading to the production of transgenic animals did not give commercially attractive results, success has been obtained in some recent cases. Some of these cases will be discussed.

## Transgenic sheep

The rate of transgenesis in sheep is very low (0.1 to 0.20% ). This can be improved, if only transgenic viable embryos (after necessary checking) are transferred to surrogate ewes (female sheep). Embryos at 8-1 > cell stage can be split into two parts, one for continued culture and the other for detection of integrated genes using polymerase chain reaction (PCR). Although microinjection is the most common method for DNA delivery, gene targeting may be increasingly used in future. In this approach, embryonic stem (ES) cells in culture are transfected with a vector which targets the gene to a particular site by homologous recombination (as discussed above). This technique, though successfully used in mice, has yet to be applied to sheep, where ES cells will have to be isolated first.

The first reports of transgenic sheep were published by J.P. Simons (1988) of Edinburgh. Two transgenic ewes were produced, cacti carrying about 10 copies of human anti- hemophilic factor IX gene (cDNA) fused with the 10.5 kb BLG gene (BLG = f3lactoglobulin) BLG gene was used, because it is necessary for specific expression of gene in mammary glands. Consequently, the gene had a tissue specific expression and ewes secreted human factor IX (or alpha-1 antitrypsin) into their milk, this human factor IX is active, even though the expression of transgene is k); I he transgenic ewes were born in early summer of 1980 and were successfully mated same year in December. In 1987, each ewe gave birth to a single lamb. Each lamb inherited BLG-F IX transgene and secreted factor IX in the milk. This programme of the production of transgenic animals by J.P. Simons at Edinburgh was funded by Pharmaceutical Proteins Ltd(Cambridge, UK), due to its commercal appeal.

In another report (published in 1991), also from Edinburgh, five transgenic sheep were produced (Alan Colman and colleagues). In all these cases, transgene involved fusion of the ovine $\beta$ lactoglobulin gene promoter fused to the human $\alpha_1$ antitrypsin (h$\alpha_1$AT) gene. Four of these animals were female and one male. In one female the protein h$\alpha_1$AT reached a level of 35 grams per litre of milk. The protein purified from milk had a biological activity indistinguishable from human plasma derived antitrypsin. The deficiency for h$\alpha_1$AT leads to a lethal disease emphysema, which is a common hereditary disorder among Caucasian males of European descent. Therefore any strategy giving high yield of this protein economically will be most welcome. In view of this, transgenic sheep with h$\alpha_1$AT gene will prove very useful as a bioreactor.

Recombinant DNA technique can also be used to increase the ability of sheep for wool growth. For this purpose, genes essential for synthesis of some important amino acids found in keratin proteins of wool have been cloned and introduced in embryos to produce transgenic sheep. For instantce, genes (cysE and cysts) for two enzymes (serine acetyltransferase = SAT and O-acetylserinc sulphydrylase = OAS), involved in systeine biosynthesis, were isolated from bacteria and cloned in a vector. These genes were introduced in sheep cells, ultimately leading to the production of transgenic sheep, where these genes are expressed. Growth hormone (GH) genes have also been introduced and can be used to promote body weight. Other genes involved in wool production have also been cloned and will be used for transgenesis, thus increasing the potential of wool production through genetic engineering.

### Transgenic pigs

The efficiency of the production of transgenic pigs is still very low compared to that of the production of transgenic mice. In mice, 2.5% to 6% of the microinjected eggs developed into transgenic mice, but in pegs this frequency was as low as 0.6% even when as many as 7000 eggs were injected. Despite this low frequency, transgenic pigs carrying growth hormone (GH) gene from bovine (of 'ox' origin) or human, and sheep globin gene have been produced (by V.G. Purse) at Agriculture Research Service, Beltsville, USA. The pigs carrying hGH gene showed different levels of expression and only 66% of these animals showed detectable levels of hGH and hGH in their plasma. The animals grew a little faster but did not become large; similarly pigs with sheep globin gene did not show any expression of the transgene for unknown reasons. In these transgenic pigs, however, a modest increase of 10-15% in daily weight and 16-18% in feed efficiency was observed, which is though lower to those in mice, but are comparable to those obtained due to daily injection with pig growth hormone. It was also observed that there was a marked reduction in the subcutaneous fat in some of these transgenic pigs suggesting the possibility of producing cleaner meat with lower fat content. These results may have a significant impact on the 9.5 billion dollar annual pig industry in U.S.A.

It is also reported that a long term elevation of growth hormone was generally detrimental to health. The pigs had incidence of gastric ulcers, arthritis and several other diseases. Therefore, techniques will have to be developed to manipulate better the transgene expression by a variety of methods (e.g. changing genetic background or modifying the husbandry regiment).

**Transgenic goats**

Transgenic goats were successfully produced recently by groups headed by John McPherson and Karl Ebert both in USA. These transgenic goats expressed a heterologous protein (a variant of human tissue-type plasminogen activator = LAtPA) in their milk. This protein is used for dissolving blood clots i.e. for treatment of coronary thrombosis. A cDNA representing LAtPA was linked with either the murine whey acid promoter (WAP) or a P casein promoter in an expression vector and used for injecting early embryos obtained surgically from the oviducts of superovulated dairy goats. These injected embryos were either immediately transferred to the oviducts of recipient females (surrogate mothers) or cultured for 72 hours (blocked at 8-16 cell stage) before transfer to the uterus of recipient females. Of 29 offspring from 36 recipients, one male and one female contained the transgene. The transgenic female delivered five offsprings, one of which was transgenic showing expression of LAtPA at a low level of few milligrams per litre of milk. In another case of a transgenic goat, few grams of LAtPA per litre of milk could be obtained. At this conce.ıtration, the dairy goat may become an economically viable bioreactor for human pharmaceuticals.

**Transgenic cows**

In most earlier attempts (in Canada) for the production of transgenic cows, embryos or fertilized oocytes produced in vivo were utilized. Fertilized oocytes or proembryos were surgically retrieved from superovulata and artificially inseminated cows. Microinjectcd zygotes were then transferred by surgery either directly into the oviduct of recipient cows or into temporary hosts like sheep or rabbits. In view of two surgical operations, this method is labour intensive and more expensive.

In 'The Netherland', recently (1991) a technique has been developed for in vitro embryo production. In this new procedure, oocytes obtained from the ovaries of slaughter house cows, were matured and fertilized in vitro. Their pronuclei were microinjected with a construct containing a bovine alpha -S1-casein promoter (bovine =ox) driving a cDNA encoding the antibacterial human iron binding protein, 'lactoferrin'. The embryos were cultured to morula/blastula stage and then non-surgically transferred to recipient females. Two of the 19 calves born from 103 transferred zygotes were transgenic (one male and other female). This procedure may facilitate the use of cows as bioreactors at the commercial level.

**Transgenic fish**

Attempts to produce transgenic fish started in 1985 and some encouraging results have been obtained. The genes that have been introduced by microinjection in fish include the following: (i) human or rat gene for growth hormone, (ii) chicken clone for delta-crystalline protein, (iii) E.coli gene for P-galactosidase, (iv) E. coli gene for neomycin resistance, (v) winter flounder gene for antifreeze protein (flounder=flat fish), (vi) rainbow trout gene for growth hormone.

The technique of microinjection has been successfully used to generate transgenic fish in many species such as common carp, catfish, goldfish, loach, mcdaka, salmon, Tilapia, rainbow trout and zebrafish. In other animals (e.g. mice, cows, pigs, sheep and rabbits), usually direct microinjection of cloned DNA into male pronuclei of fertilized eggs has proved very successful, but in most fish species studied so far, pronuclei can not be easily visualized (except in mcdaka), so that the DNA needs to be injected into the cytoplasm. Eggs and sperms from mature indiviudals are collected and placed into a separate dry container. Fertilization is initiated by adding water and sperm to eggs, with gentle stirring to facilitate the fertilization process. Egg shells are hardened in water. About 106 to 10 8 molecules of linearized DNA in a volume of 211 ml or less are microinjected into each egg (1-4 cells stage) within the first few hours after fertilization. Following microinjection, eggs are incubated in appropriate hatching trays and dead embryos are removed daily.

Since in fish, fertilization is external, in vitro culturing of embryos and their subsequent transfer into foster mothers (required in mammalian systems) is not required. Further, the injection into the cytoplasm is not as harmful as that into the nucleus, so that the survival rate in fish is much higher (35% to H()%).

Human growth hormone gene transferred to transgenic fish allowed growth that was twice the size of their corresponding non- transgenic fish (goldfish, rainbow trout, salmon). Similarly, antifreeze protein (AFP) gene was transferred in several cases and its expression was studied in transgenic salmon. It was shown that the level of AFP gene expression is still too low to provide protection against freeze. There is also a report (in 1991) of the production of transgenic zebrafish from an Indian laboratory (Madurai Kamraj University). In this attempt, a plasmid containing rat growth-hormone gene was microinjected into fertilized zebrafish eggs. Presence of the same in adult fish was later confirmed.

# TEN

# Immunotechnology

The defense mechanisms of animals against parasites so that these animals are normally capable of resisting the infection by most pathogens are dealt with in Immunology. However, the biochemical and molecular basis of such resistance has been understood only during the last few decades. The pathogens are always recognized by the host cells through chemical interactions; a specific antigen or immunogen, present in the parasite's body, elicits in the host the production of a specific glycoprotein complex (an antibody) or an immunoglobulin (Ig). The antibody or immunoglobulin is always a protein, covalently bound with an oligosaccharide. In contrast, although most often, antigen is a protein, it may also be a polysaccharide or nucleic acid or any other substance. It is also possible that a foreign protein (not necessarily belonging to a pathogen) may act as an antigen so that when injected, it may induce antibody formation. Certain low molecular weight molecules called haptens, may though bind to antibodies, do not individually stimulate antibody production. However, they may become antigenic and stimulate antibody production, if they are tightly bound to certain macromolecules such as proteins, polysaccharides and nucleic acids. It should also be recognized that the entire surface of an antigen molecule is not necessary for its antigenicity; instead specific group of atoms called antigenic determinant or epitope, consisting of 5-8 amino acids, is needed for immune response and antibody production. The different areas of immunotechnology involving both basic and applied aspects, will be briefly discussed in this chapter. In the next chapter, genetic engineering and uses of monoclonal antibodies will be discussed.

## Natural and Acquired Immunity

### 'Natural immunity'

'Natural immunity' of an animal is also known as innate immunity, native immunity or inherited immunity. This relates to a general or non-specific type of resistance which prevents infection by different kinds of pathogens. The extent of this natural immunity differs in different organisms. For example, man can easily get mumps while cats and dogs are immune to this disease. This level of natural immunity may vary not only between species but also between races or strains, and sexes. This may also he controlled by nutrition, hormones and many other factors.

**Acquired immunity**

'Acquired immunity develops during the life time of an individual and refers to the immunity, which in specific individual displays against a specific pathogen. This is frequently related to the presence of antibodies or interferon in the blood. Acquired immunity based on antibodies is the most efficient type of acquired immunity and it may be either (i) actively acquired or (ii) passively acquired.

Actively acquired immunity—Actively acquired immunity may be either natural or artificial. Actively acquired natural immunity results from infection from which a person recovers. During the infection, antibody production for that specific pathogen is stimulated so that when there is a subsequent infection by either the same or antigenically related pathogon, the antibodies assist in the body's defense.

Actively acquired artificial immunity is the most common method of immunization or vaccination. The immunogens are injected in the body in controlled quantity to stimulate the production of immunoglobulins. Killed and attenuated strains of bacteria and viruses are now used widely for immunization against many diseases (cg., typhoid, small pox, poliomyelitis, yellow fever, measles, etc.). Attenuated organisms produce mild infection and induce natural immunity. Vaccines are also being developed now by a variety of other methods involving recombinant DNA technology.

Passively acquired immunity—Passive immunity may also be acquired either by natural or artificial means. While passive immunity acquired in natural means, involves the transfer of antibodies from mother to her unborn child, through the placenta during the later part of pregnancy, passive immunity of the artificial type refers to the original production of antibodies in some other individual (human or lower mammal) followed by injection of these antibodies with the help of a needle or syringe. Several drug manufacturing companies are involved in the large scale production of antibodies in horses and cows by active immunization.

**Immune Responses**

The injection of an antigen into an animal may initiate several important changes in the animal. The first response is to eliminate the antigen through extensive phagocytosis in the liver, lung and spleen, so that nearly 90% of the antigen is removed from the circulation in about 10 to 20 minutes of its introduction. In the second phase, which lasts for about 4 to 7 days, the antigen is eliminated by enzymic hydrolysis and digestion. However, the most important phase in antigen elimination is the immune response involving formation of antibodies, which provide resistance to the animal body against infection.

**Lymphoid system (B cells and T cells)**

The immune system of the body is monitored with the help of white blood cells called lymphocytes, which circulate between blood and the lymph system as illustrated in Figure (lymph is a colourless fluid and to circulated like blood with the help of vessels that make a lymphatic system just like blood circulation system). The lymphocytes are of two type. :

B cells and T cells—The B cells synthesize antibodies causing agglutination of unwanted antigens, while T cells carry out cellular immune responses where cells having unwanted antigens on their surface are attacked without any mediation by antibodies. The former response is called humoral immunity and the latter the cellular immunity (see Figure for details). The T cells interact with antigen through cell surface proteins called 'I' cell receptors, which are specially attached to the cell membrane. B cells and T cells, both originate in bone marrow, which is the primary source of all lymphocytes. In mammals, B cells mature within bone marrow, but T cells migrate out and mature in special organs like thymus. When they mature, 13 cells leave bone marrow and T cells leave the organ like thymus and travel to secondary lymphoid organs like lymph nodes, tonsils and spleen. In these organs they can be activated to produce antibodies.

These cells then pass to blood system and back again to the lymph. T cells can be (i) cytotoxic T lymphocytes (CTLs), which destroy infected cells of the host; (ii) helper T cells that help B cells and the cytotoxic T cells to mature and perform their immune response related function (for details, see Fig.); (iii) suppressor T cells, which block B and T cell responses, whenever required. On the surface of cylotoxic T cells are found specific cell surface proteins called T cell receptors, which are analogous to antibodies in structure, having C (constant) and V (variable) regions with enormous variability for recognition of a large variety of antigens. The T cell receptors, like antibodies, have been identified to be glycoproteins (mw 80-90 kd), each being a dimer consisting of alpha (cx) and beta (/3) chains (each 40-50 kd), which are linked with the help of a disulphide bond and other non-covalent bonds.

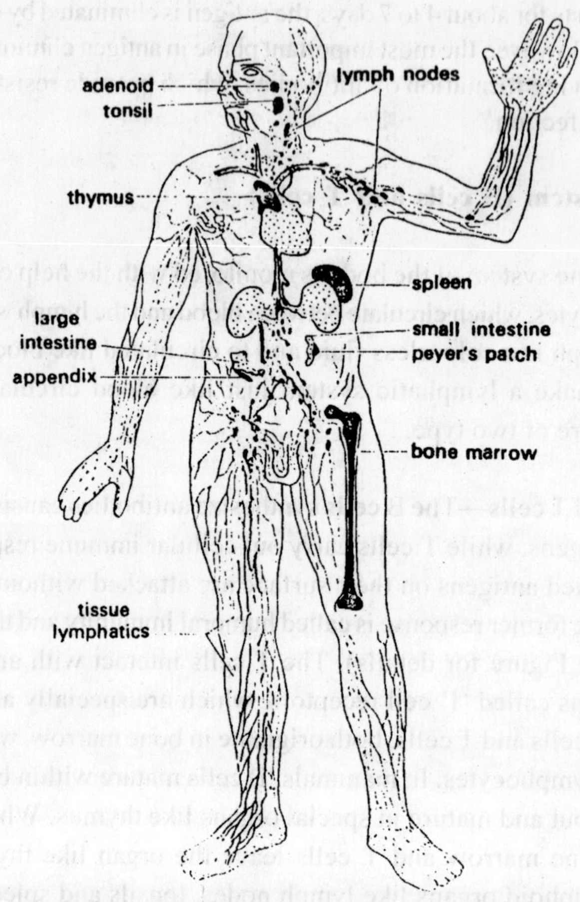


**Fig.** An outline of lymphoid system in a human being

While B cells can be stimulated by the free antigens in the circulating system of the body, T cells respond to only those antigens which are bound to cell surface, and that too only in the presence of another determinant protein called major histocompatibility complex or MHC (Fig.). This helps the T cells in recognising only the virus infected cells of the host to which the T cells belong, and does not allow the recognition of the cells which are either normal uninfected cell of the same organism or the infected cells of another organism to which the T cells do not belong. This dual specificity of T cell receptors is dependent on large variety of receptors available just like antibodies. Since they resemble antibodies in structure and variability, their genes though distinct are similar to the Ig genes meant for the synthesis of immunoglobulins or antibodies.

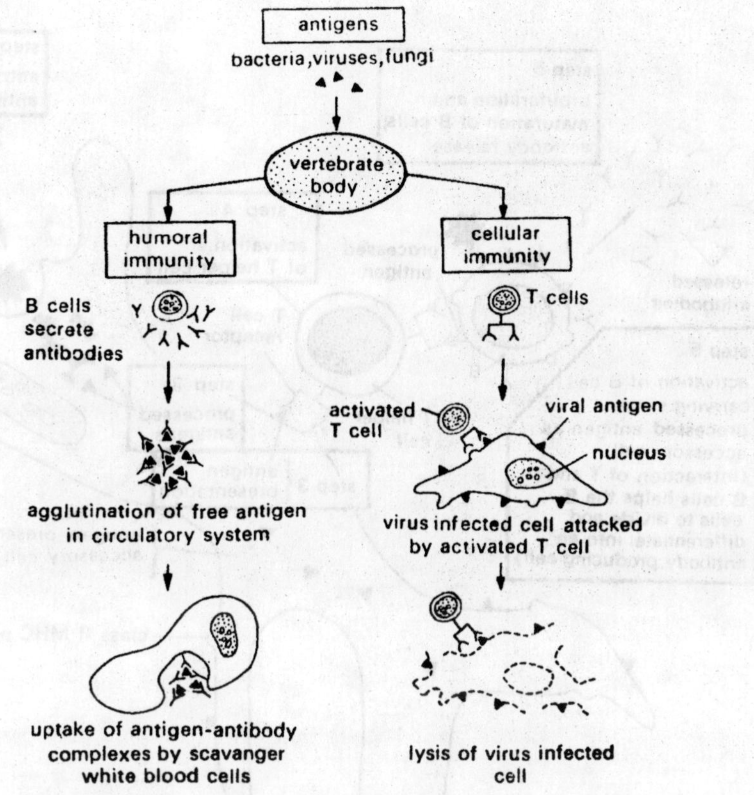


**Fig.** Immune response to an antigen, involving humoral immunity through B cells and cellular immunity by T cells

## Clonal Selection Theory

Various theories have been proposed regarding the formation of antibodies, although only the clonal selection theory seems to be acceptable in most situations. According to this theory, lymphocyte cells are able to make a specific antibody against a specific antigen. The cell may not be making the antibody continuously but may get transformed into an antibody synthesizing cell when the antigen contacts the proper lymphocyte cell. B cells, while not engaged in active synthesis of antibodies, carry specific antibodies on their surfaces. Due to these surface bound antibodies, a specific antigen may get bound with specific B cells and may thus trigger the proliferation and maturation of that particular B cell. The cell responds with repeated cell divisions until a clone of antibody synthesizing cells called plasma cells is formed. These cell then form large number of same antibodies (Fig.), at a rate of 10,000 molecules per second per cell.

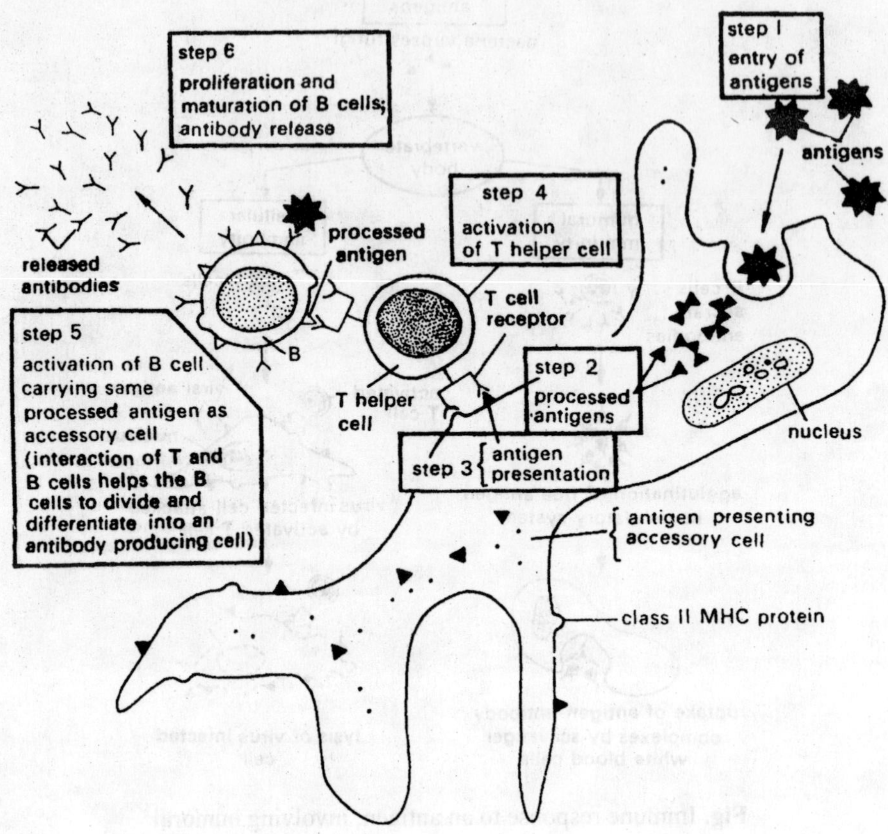


**Fig.** Different steps showing the role of helper T cells in activation and maturation of B cells into plasma cells for production of antibodies

The antibodies are produced by B-lymphocytes, and there is a specific antibody for each specific antigen. Antibodies are also produced by other cells in the body, such as 1-lymphocytes, macrophages, granulocytes and must cells, which, like B-lymphocytes also differentiate from stem cells in bone marrow. A concerted effort of all these cells is required to produce the immune response.

When an animal first encounters an antigen, it produces a low primary response, since only few B cells or T cells can respond to antigen. However, the next encounter with the same antigen leads to a far more rapid and intense response called secondary response, since many cells are now' available for response (Fig.). Due to first exposure to antigen, not only mature antibody producing B cells are produced, but immature B and T memory cells are also produced, which can respond on a second exposure to the antigen (Fig.).

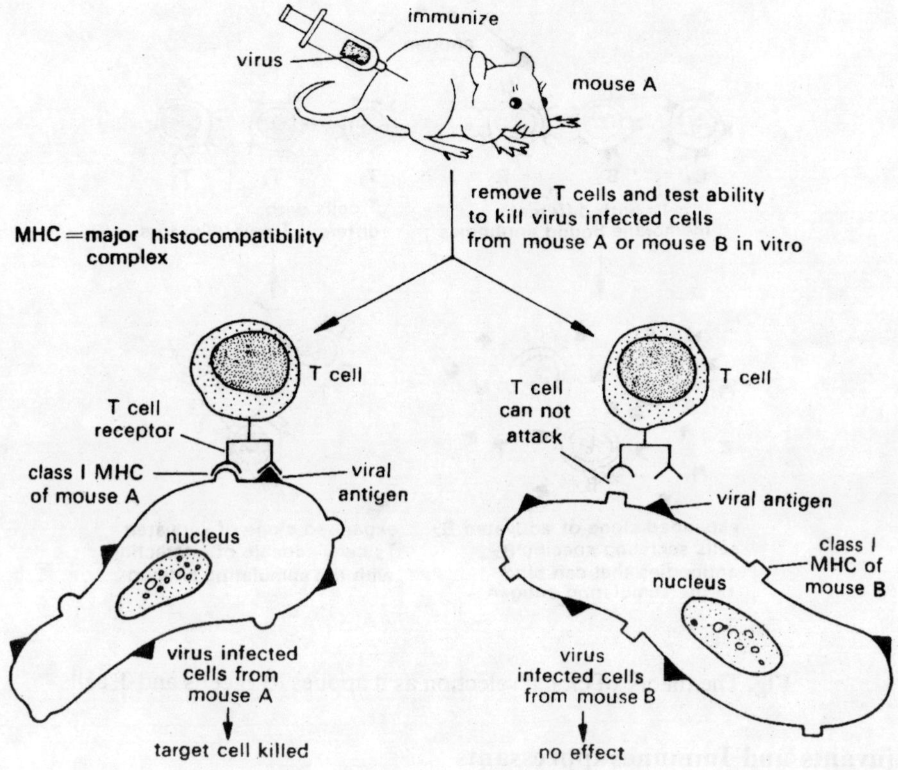


**Fig.** An outline of the experiment demonstrating that T cells recognize antigen only in combination with a normal cell surface protein, MHC (major histocompatibility complex)

About 1,0,000 different antibodies are known to exist in a human body at a given time (there are 1012 lymphocytes with 106 — 108 specificities). Obviously, thousands of plasma cells throughout the circulatory System are in operation for the production of antibodies at any given time. About 90-95% of the protein produced in plasma cells give rise to antibodies, which are sometimes stored in large quantities in these cells, and can be visible as crystals within the cell. With the introduction of an antigen inside the body, the antibody producing plasma cells increase in number, their RNA content increases and the entire protein synthesizing machinery is activated for use in antibody production.

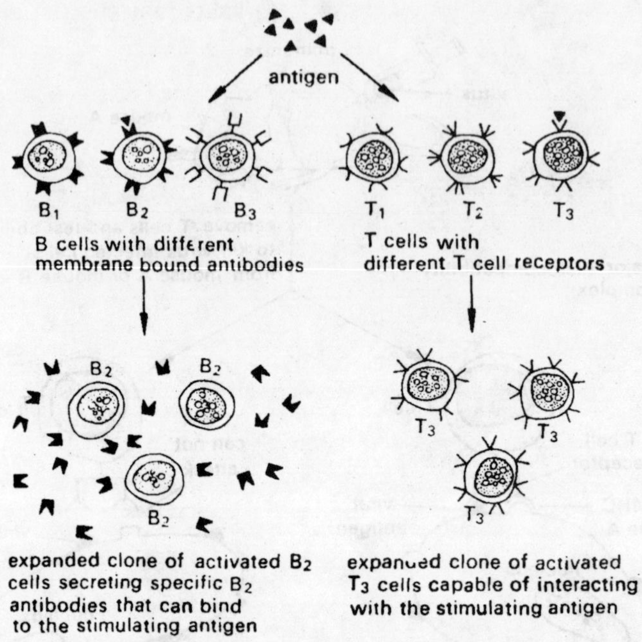


**Fig.** The theory of clonal selection as it applies to B cells and T cells

## Adjuvants and Immunosuppressants

Several types of treatments modify the immune response in animals. Adjuvants, which may or may not be antigenic, increase the level of circulating antibodies, and thus improve immune responses consistently. Examples of antigenic adjuvants are gram negative bacteria ,such as Bordorellla pertussis endotoxins of gram negative bacteria, acid last micro-organisms, phytohemagglutinins and antibodies. Non-antigenic adjuvants include potassium aluminінum tartarate, calcium phosphate. minerals mil. lanolin, surface active agents, calcium alginate, collodian or acrylic particles and polynucleotides. Often a mixture of several such substances is arced as a complete adjuvant. Freunds complete adjuvant is such a mixture which contains killed mycobacteria, an oil and an emulsifier.

In contrast to adjuvants, there are immunosuppressants, which include several physical agents (such as X-rays or gamma rays), chemical reagents and biological agents that suppress the immune response in animals. Some corticosteroids, purine analogs, folic acid antagonists and alkylating agents are potent immunosuppressants. The enzymes ribonuclease and asparaginase also suppress immune responses.

The suppression of antibody production can be brought about by any one of the following five general biological means: (i) There may be a deficiency in the ability

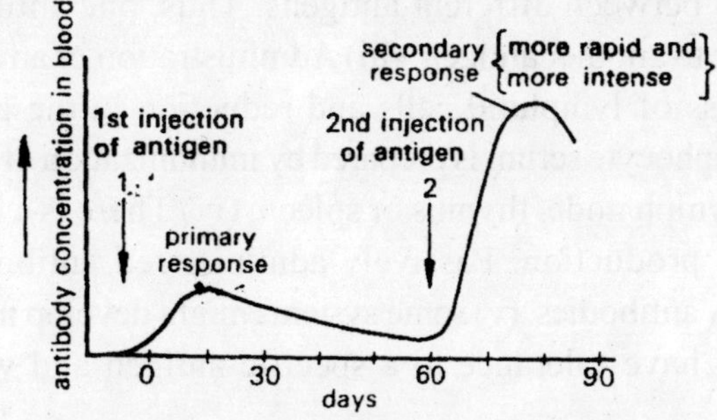


**Fig.** A graph showing the primary and secondary responses of antibody production in a rabbit.

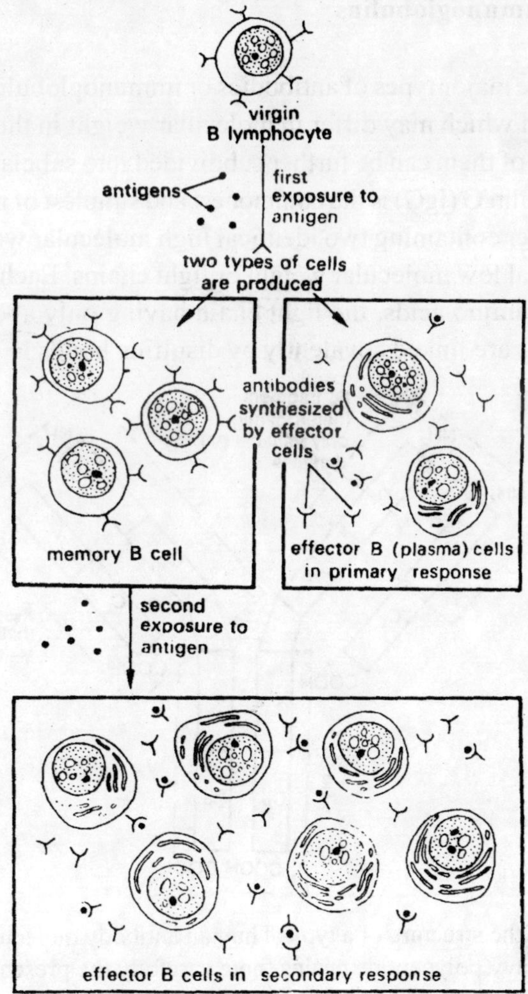


**Fig.** Cellular basis for greater speed and intensity of a secondary antibody response

of cells to synthesize gamma globulins (components of immunoglobulins). This deficiency is called hypogammoglobulinemia or agammoglobulinemia. (ii) There may be a competition between different antigens. Thus, one antigen may suppress the immune response to another antigen. (iii) Administration of an antilymphocyte serum (ALS) causes loss of lymphoid cells and reduction in the antibody response tea antigens. Antilymphocyte serum is prepared by immunization of a heterologous species with cells from lymph node, thymus or spleen. (iv) There is a feedback inhibition of immunoglobulin production. Passively administered antibody often inhibits the production of new antibodies. (v) some systems might develop immunologic tolerance. The system may have tolerance to a specific antigen and would not produce its antibodies.

## Structure of Immunoglobulins

There are five major types of antibodies or immunoglobulins, namely IgA, IgD, IgE, IgG and IgM which may differ in molecular weight in the range of 150,000 to 720,000. Several of them can be further subdivided into subclasses like IgG1, IgG2, etc. Immunoglobulin G (IgG) is the commonest and simplest of the immunoglobulins, which is a tetramer containing two identical high molecular weight or heavy chains and two identifical low molecular weight or light chains. Each heavy chain is made up of about 500 amino acids, the light chain having only about half this number. These four chains are linked covalently by disulfide bonds in a Y-shaped structure (Fig.).

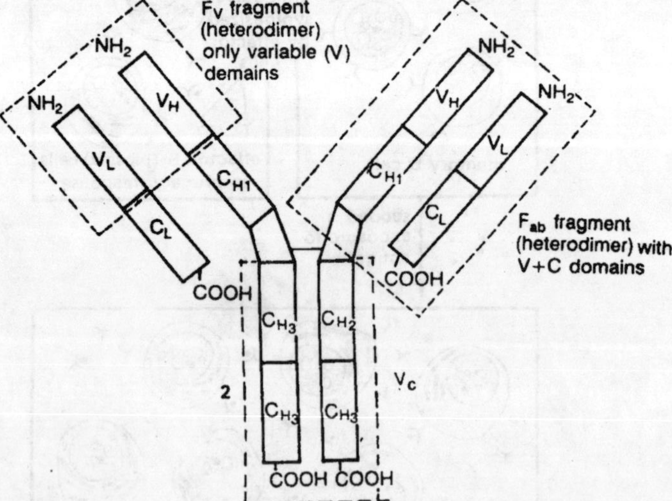


**Fig.** A model of the structure of a typical human antibody molecule, composed of two light and two heavy polypeptide chains (note carefully the presence of disulphide S-S bonds and carbohydrate moieties; also note constant regions i.e. $C_L$ and $C_H$ and the variable regions, $V_L$ and $V_H$)

The Y shaped structure of antibody can be divided into following functional domains (Fig): (i) antigen binding; domains-H'v and Fab fragments and (ii) effector domain-Fe fragment, which helps in elimination of antigen. Fv fragment is a heterodimer, consisting of the variable domains of the heavy chain and the light chain ($V_H$, and $V_L$). Fab fragment is also a heterodimer and consists of Fv fragment + $C_{H1}$ and $C_L$. In other words, Fab consists of the entire light chain and the $V_H$, and $C_{H1}$ domains of heavy chain. $F_c$ fragment consists of a homodimer with $C_{H2}$ and $C_{H3}$ domains of both heavy chains. The entire antibody can be enzymatically cleaved at the hinge region into Fab and Fe fragments. The identification of these fragments of antibody has allowed the application of protein engineering and genetic manipulation for the production of novel antibodies without the use of hybridomas or animal cells (see next chapter for details).

There are five types of heavy chains, designated as α (alpha), β (beta), s (epsilon), γ (gamma) and μ (mu) in IgA, IgD, IgE gig and IgM respectively. However, there are only two types of light chains, designated as ω (kappa) and λ (lambda). All five types of heavy chains occur in alternate forms containing either h or .1 chians, thus resulting in ten different combinations.

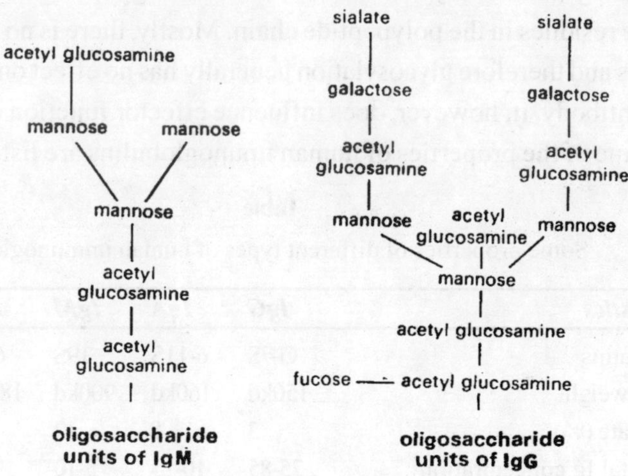


**Fig.** Details of the structure of carbohydrate moieties (oligosaccharide units) found in the antibodies $I_gM$ and $I_gG$

Each Ig chain consists of two distincts parts: (a) C-terminal (carboxyl-terminal) constant part or C-region and (b) N-terminal (amino terminal) variable part or V-region. For each C-region ( $C_a$, $C_\mu$, $C_g$, etc.), there can be several V sequences. Separate sets of V sequences arc known for heavy chains ($V_H$) and light chains

($V_k$ = for kappa and $V_l$ = for lambda). The different V sequences have fair degree of homology only in some variable portions. There are distinct genes for the synthesis of C and V regions of polypeptides.

Immunoglobulin A (IgA) is the major antibody found in secretions (tears, mucus and saliva), which is the first defense against invading bacteria and viruses. The functions of IgD and IgE are not known, although it is known that IgE differs from other immunoglobulins in being thermolabile, other immunoglobulins being stable at 50-60°C for half an hour to 4 hours. IgM, which appears early in response to many invading pathogens, is bigger than IgG (which replaces IgM later) and constitutes about 5-10% of the total immunoglobulins in the adult human, making about 125 ± 45 mg per 100 ml of serum.

Antibodies are also glycosylated, mostly carrying carbohydrate moieties in $C_{H_2}$ region. The carbohydrate component of the immunoglobin is an oligosaccharide with variable mono-saccharide units. It may include simple mannose rich units as in 1gM or complex N- acetyllactosamine rich units as in IgG (Fig.). In 1gG the carbohydrate portion is about 2.5% of the total molecule and it is equally divided between the tin, gamma heavy chains. It is linked through amino acids, aspartic acid and threonine residues in the polypeptide chain. Mostly, there is no glycosylation m Fab fragments and therefore glycosylation generally has no effect on antigen binding property of antibody. It, however, does influence effector function controlled by Fc fragment. Some of the properties of human imunoglobulins are listed in Table.

**Table**

Some properties of different types of human immunoglobin

| *Characterstics* | *IgG* | *IgA* | *IgM* | *lgD* | *IgE* |
|---|---|---|---|---|---|
| Svedberg units | G-7S | 6-11S | 19S | 6.1S | 8.2S |
| Molecular weight | 150kd | 160kd | 900kd | 185kd | 200kd |
| Carbohydrate (%) | 3 | 8 | 12 | 13 | 12 |
| Per cent total Ig concentration | 75-85 | 10-15 | 5-10 | 1 | 0.002 |
| Concentration (mg/100 ml of serum) | 1275 | 225 | 125 | 3 | 1 |

## Genetics and Molecular Biology of the Production of Antibodies or Immunoglobulins

Antibodies or immunoglobulins are coded by genes, which undergo DIN' rearrangements during cell differentiation, thus suggesting that although the genome of an organism remains constant in different cells and in ill germline, it does undergo a change in B and T cells for the immunoglobin genes in mammals. As discussed earlier in this chapter, it has also been shown that each heavy or light chain consist of ow or more constant regions (coded by C genes) and a variable region (coded by ;a V I-l unc) so that each polypeptidc is derived from two genes. This has led to the formulation of two genes one polypeptide hypothesis in contrast to the popular one gene one polypeptide hypothesis applicable like most gene systems.

The immmunoglobin gene system is unique in the sense that thousands of different antibodies are synthesized in different cells, which are supposed to have the same DNA consent. Surprisingly, we (it) now, have as many genes as the variants of antibodies in an organism, but instead sequences coding for different types of antibodies are actually derived in different cells through rearrangement of the master DNA sequence available in the germline. This has been achieved by a remarkable mechanism which has been discovered only recently. In view of the significance of this discovery, the 1987 Nobel Prize for Physiology and Medicine was awarded to S. Tonegawa of MIT (USA) for his valuable work on this subject. This work will be briefly discussed in this section.

**Ig gene in germline and differentiated cells**

By using cDNA probes for different regions of an immunoglobulin, it was possible to demonstrate that in the antibody producing cells, the DNA sequences coding for C and V regions of an Ig gene are very close to each other, though not contiguous. On the other hand, in the germline these sequences are separated from each other, sometimes by long distances. This was expected because in earlier studies the same C region was found combined with different V regions in the same individual. Later experiments showed that C-C splicing at the DNA Level is responsible for generating the coding genes for both light and heavy chains. It was also shown that although the genes for an immunoglobulin molecule may be found on more than one chromosome, but cutting and rejoining of DNA sequences takes place only within a chromosome and not across chromosomes. In addition to C and V sequences, both light and heavy chain coding sequences have another sequence called J (junction or joining sequences). In heavy chain sequence, in addition to C, J and V sequences, another shore sequence D (diversity) is also found.

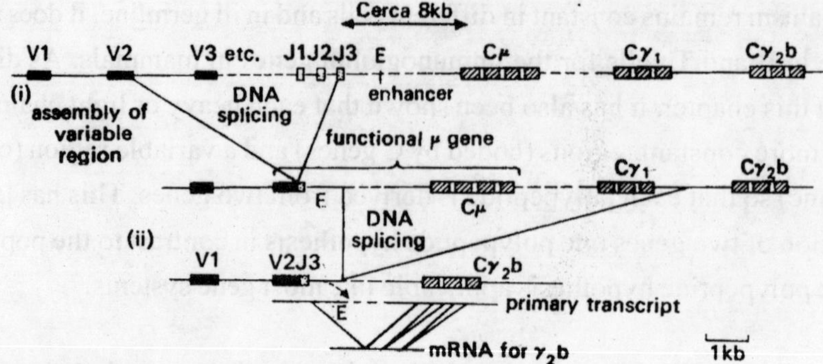


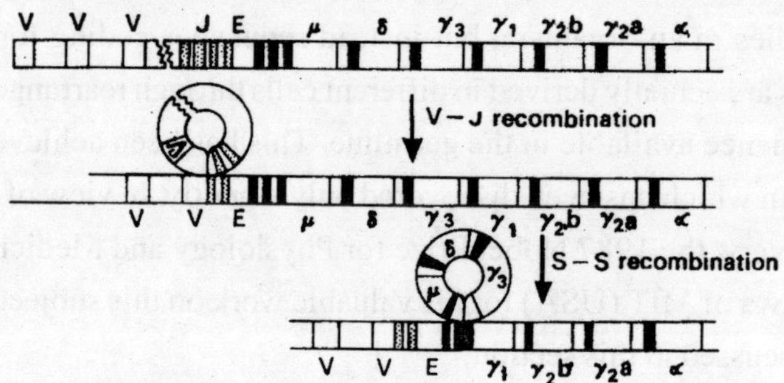


Fig. (A) Structure of immunoglobulin gene and the steps involved in DNA rearrangements in different regions. (B) A model explaining DNA rearrangements in heavy chain switching.

## V-regions through DNA splicing

Splicing and rearrangement of DNA sequences take place in antibody producing cells to create a large variety of Ig genes. As an example, we will discuss DNA splicing only for the heavy chain. There are two rearrangements in this case: (a) the first generating the DNA sequence coding for $V_H$ and (b) the second leading to association of a specific $V_H$ with a definite $C_H$ sequence. Using molecular probes in different kinds of cells, steps involved in rearrangements leading to association of different V regions could be worked out as shown in Figure. In IgM sequence, it is shown that in embryo or germline cells, $V_H$ and Cat regions have a small region of J sequence. Between V and J sequences, presence of D sequence was also shown. Although there is diversity in V, J and D sequences (C region is constant), but variation is also provided by the manner in which DNA splicing and rearrangement is brought about.

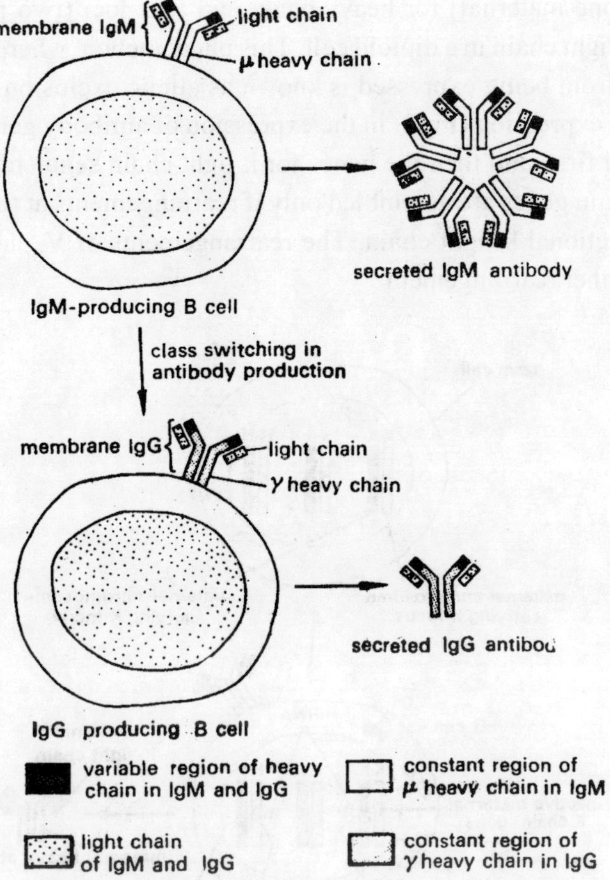


**Fig.** During antibody production, attachment of the same $V_H$ region to a different $C_H$ region for $I_g$G class switching

## V-C association through DNA splicing

As indicated above, the second phase of DNA rearrangement involves association of a V segment wish a C segment. It has been shown that the same V segment can attatch with different C regions, thus providing for H-chain class switching (Fig.). This is achieved by successive deletions of DNA, leading to the production of IgM in primary response and IgG and IgA as a secondary immune response. IgM has all heavy chains, while in IgG$_3$, segment C is missing and in IgG$_1$, $C_\mu$ and $C_{4\gamma}^{3}$ are missing. Further deletions may occur in other cases as shown in Figure.

## Allelic Exclusion

When a B cell differentiates and matures into a plasma cell, it synthesizes solitary types of light and heavy chains. This happens despite the presence of two loci (one paternal and one maternal) for heavy chain and four loci (two paternal and two maternal) for light chain in a diploid cell. This phenomenon, where some of the loci are excluded from being expressed is known as allelic exclusion and is unique to antibody gene expression (Fig.). In the expression of antibody genes, $V_H$ segments are rearranged first, and then the genes for k light chain segments are assembled. The λ light chain genes are assembled only if rearrangements at the $V_k$ locus fail to produce a functional k light chain. The rearrangements of $V_H$ and $V_k$ genes thus inhibit any further rearrangement.

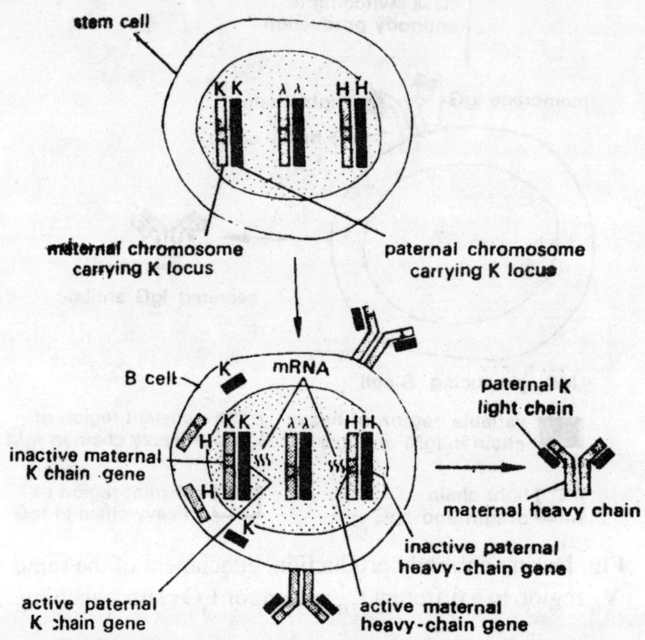


**Fig.** Allelic exclusion at the antibody loci, where a particular B cell synthesizes only one type of heavy chain and one type of light chain (once a functional gene is assembled, further DNA rearrangements are inhibited)

## Interferons

Interferons, which were discovered for the first time in 1957 by two British Scientists (Alick Isaacs and Jean Lindermann), are proteinaceous substances (mw 18,000 to 100,000) produced inside the body for defense against virus infection. They are also produced in response to many nonantigenic chemicals including polysaccharides, endotoxins, double stranded DNA, RNA, etc. Unlike immunoglobulins, interferons are stable for several hours at room temperature at pH 2.0.

Two mechanisms for the control of virus infection by interferons have been suggested: (i) The interferons affect the cellular membranes, so that the virus particles are not disseminated to the other cells; (ii) interferons control the multiplication of virus particles by inhibiting their protein synthesis. It is believed that the interferons set off a number of enzymes in the cell which can stop the virus using the cells protein synthesizing machinery for its own protein synthesis. It has also been observed that ribosomes from interferon infected mammalian cells do not form polysomes with viral RNA, but when they are exposed to host (mammalian) RNA, they readily form polysomes and protean synthesis is initiated. It shows that ribosomes in the presence of interferon can recognise specific mRNA and regulate the overall protein synthesis.

Because of their inhibitory nature, the interferons are being tested for their effectiveness in controlling cancer, though with only limited success. In some cases, interferons inhibit tumour cell growth and, stimulate immune cells to destroy cancerous cells. With some exceptions, the tumour that respond to interferon treatment are also responsive to chemotherapeutic agents. Many companies in U.S.A., Japan, U.K., W. Germany, France and Switzerland are working towards large scale production of interferons (also see Chapters).

## Vaccine Development and Immunization

### History of vaccine development

Vaccine production and immunization or vaccination is the most important application of immunotechnology, as far as human welfare is concerned. The first experiment in immunization was performed by Louis Pasteur, a French chemist turned biologist, on July 6, 1885, when he treated a young boy against rabies. He extracted fluid from the spinal cord of a rabid dog and injected it in small amount to the boy, who was bitten several times by a rabid dog. Apparently, the extract from the spinal cord had stimulated the production of antibodies against the rabies virus. The science and application of the vaccine technology, however, has made significant progress in recent decades, so that the period 1950-70 is considered to be the golden age for vaccine development. During this period, vaccines for poliomyelitis, measles, mumps and rebella were developed. In majority of the vaccine recipients, there is no significant ill effects and mortality rate among vaccinees varies from about one per 200,000 to one in 3,000,000. Thus the vaccines are quite safe and cost effective. The protection given by measles vaccination is documented in Table.

**Table**

Benefits of vaccination against measles (data liken from A.K. Galazka and others, Bull WHO 1984, 62: 357).

| *Adverse reaction* | *Incidence (no. per 100,000 cases) following* | | *Ratio* |
|---|---|---|---|
| | *Infection* | *Vaccination* | *infection/vaccination* |
| Pneumonia | 5000 | 1 | 5000 |
| Convulsions | 720 | 60 | 12 |
| Brain damage | 280 | 0.1 | 2800 |
| Early death | 10 | 0.02 | 500 |
| Chronic infection, | 2 | 0.075 | 26 |

However, production of vaccines against certain important infectious agents have been difficult. The problems involved in the production of vaccine in such cases include the following: (i) Antigenic variation or presence of many serotypes as in influenza, rhinovirus and in human immunodeficiency virus (HIV). (ii) Integration of viral DNA/cDNA in the host cell genome as in hepatitis B and retrovirus. (iii) Large animal reservoirs as in influenza and human HIV. (iv) Infection transmitted by cells which may or may not express viral antigen as in HIV. (v) Crucial cells of the immune system are infected as in HIV. If a pathogen has most or all of these characteristics, it is difficults to produce an effective vaccine. The HIV causing AIDS at present seems to fit this description.

## Nature of vaccines

In broader sense, the term vaccine is used for any biological material injected in the body to stimulate the active immunity. Immunization of the body through vaccination is generally done for the prevention of disease. The vaccine is administered in advance so as to give the body time to set active immunity, before invasion by the pathogen occurs. The objective of vaccination is to stimulate the production of antibodies without the person actually having to develop the disease. The vaccines may be of following types:

Living organisms (with attenuated virulence) as vaccines—Some vaccines involve use of organisms, whose virulence is greatly attenuated or diminished. These organisms are unable to produce disease but induce immunity when injected in the host. There are at least five different ways of producing—attenuated strains of pathogens: (i) Many successive passages of pathogen in some animals other than the usual host of the pathogen or in cultured cells. (ii) Selection of mutants of low virulence. (iii) Treatment of pathogens with chemicals. (iv) Cultivation of pathogens under unfavourable conditions such as high temperature. (v) Deletion or insertion of genes leading to attenuation of pathogens as tried for vaccinia virus.

Some examples of vaccines prepared by the attenuation of virulence are given in Table.

**Table**

Vaccines prepared by attenuation of virulence

| ***Disease/pathogen*** | ***Method of attenuation*** |
|---|---|
| 1. Rabies | Cultivation in chick or duck embryo |
| 2. Yellow fever | Cultivation in chick embryo |
| 3. Smallpox | Animal passage (calf); genetic engineering |
| 4. Polio | Mutant strain (cabin oral polio vaccine) |
| 5. B.C.G. (Bacillus of Calmette Gucrien) | Cultivation of organism on media containing bile |
| 6. Bacillus autlrracis (Anthrax) | Cultivation at 40-50°C |

The concept of recombinant 'vaccinia' virus as alive vaccine gained momentum in 1990's due to its earlier effective use as a smallpox vaccine. Its cost range makes it suitable for livestock also. Following attributes of Vaccines (used as an expression vector) made it attractive for vaccine development: (i) stimulation of humoral and cell mediated immunity, (ii) low cost, (iii) heat stability, (iv) simple method of administration, (v) visible proof of successful vaccination and (vi) long duration of effectiveness after single dose.

Genetically engineered viruses as vaccines—Although in the past, attenuated strains of vaccinia virus were obtained by serial virus passage, genetic engineering methods have been used for this purpose in 1980's and 1990's. Deletion of several genes (thymidine kinase, a growth factor, hemagglutinin, 13 . 8 kD secreted protein, and ribonucleotide reductase) decreases its virulence. Similarly, insertion of some lymphokine genes (e.g interleukin-2 or IL-2) also lead to decreased virulence. While reducing virulence or pathogenicity, immune response of the recombinant protein (coded by gene inserted in the virus for developing specific vaccine) must be increased.

More than one genes can also be stably integrated into attenuated vaccinia virus for developing recombinant virus to be used as a vaccine against multiple pathogens. A glyeoprotein gene of rabies virus has been integrated and the resulting rabies virus vaccine has been field tested in Europe and U.S.A. giving promising results. Other vaccines against HIV are being developed using the gene for HIV-I envelope glycoprotein. In future, the widespread use of Vaccinia as a live vaccine will depend on improving safety, while achieving higher response to the recombinant protein.

Dead organisms as vaccines—In some cases, the dead bodies of the pathogen, when injected into a host, stimulate the production of antibodies without causing any infection. This method has been used in immunizing against typhoid fever, whooping cough, blackleg in cattle and a few other diseases.

Bacterial toxins and toxoids as vaccines—Exotoxins of most pathogenic bacteria are highly antigenic; they stimulate the production of antibodies (antitoxins) in animal or human body. These exotoxins are highly unstable and they lose toxicity during storage and are converted gradually into toxoids. Toxoids retain the antigenic power but they are not toxic. Therefore, they can be used in producing immunity against corresponding toxins. The transformation of toxins into toxoids is accelerated by heat, formaldehyde and other chemicals. Alum precipitated toxoids are highly antigenic. Toxoids have been used to develop immunity against several diseases including diphtheria, tetanus, gengrene, etc.

Immunizing sera as vaccines—An immunizing serum (plural 'sera') is the actual blood serum of a person or animal and contains an immunizing material or antibody. It is generally obtained by injecting a toxoid into an animal (usually a horse), permitting establishment of active immunity in the animal, and then withdrawing some of its blood. The corpuscles are removed from serum, which is then standardized as to the strength of antibodies and stored in sterile tubes or bottles.

Immunizing sera are used mostly for cure rather than prevention. They may be given for prevention after exposure has taken place, but their immunizing effects are too short lived to give any benefit if administered much in advance of the disease. As a curative agent, the action of immunizing serum is very rapid. Immunizing serum has been used in the cure of diphtheria and tetanus.

Vaccines of defined chemical nature—Sine the chemical nature of many antigens is now known, it has been possible to produce these antigens either by biological or by chemical means. Perhaps the first defined molecular species to be used as a vaccine were the polysaccharide vaccine for Neisseria meningitides (meningitis) and Streptococcus pneumoniae (pneumonia). The pure polysaccharide preparations from the bacterial capsule could be used as an antigen in adults or children of over two years of age. Addition of some other chemicals or biological principles often increases the potency of the polysaccharide antigen. For example, Haeniophilus influenzoe (influenza) B polysaccharide is very poor in inducing antibodies in the susceptible host on its own. However, its antigenie property is greatly enhanced if it is mixed up with diphtheria tuxoid Proteins and nucleic acids have also been identified as immunogenic molecules. An outer membrane protein-polysaccharide complex from meningeoocci bacteria in association with aluminium phosphate or hydroxide elicits immunogenic responses and hence the mixture can be used as a vaccine. Antigens produced using cloned genes, synthetic peptides and proteins along with certain adjuvants are also often used as vaccine. Recently, D.C. Wraith and others (1987) have proposed that the nucleoprotein of influenza virus can be used as a universal vaccine for influenza.

The advent of recombinant DNA and monoclonal technologies coupled with the recent advances in immunology, has opened a new era into the production of vaccines. It seems that isolated immunogens can now be produced in unlimited quantities and vaccine development can proceed without hindrance. Several vaccines based on the new technologies have been licensed for commercial production. Some of these include the following: (i) Yeast produced vaccine for hepatitis B virus based on cloning the DNA sequence for the antigen; (ii) Escherichia call vaccines for pigs with extra plasmids encoding the antigen for K88 strain of E. coli; (iii) Monoclonal antibody against K99 strain of E. coli for protection of calves.

You already know that the immune system, provides defense mechanisms in animals. In this connection, antigen-antibody interactions and the detailed structure of antibodies were also described. The role of B cells and T cells in immune responses and the rearrangement of DNA sequences for providing diversity in antibodies were also describe. For human health care, the subject of the production and use of antibodies has become a very important area of research, not only for academic purposes but also for its relevance to industrial growth for production of vaccines and drugs for diagnosis and therapeutic uses.

For purposes of research and use in medicine, antibodies are often isolated by sacrificing animals after hyper-immunizing them with antigens. From such hyperimmunized animals, the blood serum may be taken and antibodies may be isolated from this serum. However, the antibodies, whenever separated from serum after induction due to an antigen, are usually heterogeneous, because the cells keep on producing a variety of antibodies. Consequently, the antibody preparation made from serum will not have the desired specificity and , therefore, can not be utilized for diagnosis or screening. Monoclonal antibodies (Mabs), on the other hand, are homogeneous immunological reagents of defined specificity, so that these can be utilized for diagnosis and screening with case and certainty. The discovery, production, genetic manipulation and uses of these monoclonal antibodies will be described in this chapter.

## Hybridoma Technology and the Production of Monoclonal Antibodies

### Basic Hybridoma Technology

Monoclonal antibodies can be produced in specialized cells through a technique now popularly known as hybridoma technology. This technology was discovered in 1975 by two scientists, Georges Kohler of West Germany and Cesal Milstein of Argentina (now working in U.K.), who jointly with Niels Jerne of Denmark (now working in Germany) were awarded the 1984 Nobel Prize for Physiology and Medicine. The term hybridoma is applied to fused cells resulting due to fusion of following two types of cells : (i) an antibody producing lymphocyte cell (e.g. a spleen cell of mouse immunized with red blood cells from sheep), and (ii) a single myeloma cell (bone marrow tumour cell) which is capable of multiplying indefinitely. These fused hybrid cells or hybridoma have the antibody producing capability inherited from lymphocytes and have the ability to grow continuously (immortal) like malignant cancer cells. Following steps are involved in the production of monoclonal antibodies using hybridoma technology (Fig.) : (i) Immunize a rabbit through repeated injection of a specific antigen for the production of specific antibody, facilitated due to proliferation of the desired B cells. (ii) Produce tumours in a mouse or a rabbit. (iii) From the above two types of animals, culture separately spleen cells (spleen cells are rich in B cells and T cells) that produce specific antibodies, and myeloma cells that produce tumours (the myeloma cell line used, is unusual in two ways ; it has stopped synthesizing antibodies and it is a mutant called HGPRT that can not synthesize the enzyme hypoxanthine-guanine phosphorihosyl transferase or HGPRT). (iv) Induce fusion of spleen cells to myeloma cells, using polyethylene glycol (PEG) to produce hybridoma ; the hybrid cells are grown in selective hypoxanthine aminopterin thymidine (HAT) medium. HAT medium contains a drug aminopterin, which blocks one pathway for nucleotide synthesis, making the cells dependent on another pathway that needs HGPRT enzyme absent in myeloma cells. Therefore, myeloma cells that do not fuse with B cells will die since they are $HGPRT^-$. B cells that do not fuse will also die because they lack tumorigenic property of immortal growth. Therefore, HAT medium allows selection of hybridoma cells, which inherit HGPRT gene from B cells and tumorigenic property from myeloma cells. (v) Select the desired hybridoma for cloning and antibody production; this is facilitated by preparing single cell colonies that will grow and can be used for screening of antibody producing hybridomas ; only one in several hundred cell hybrids will produce antibodies of the desired specificity (vi) Culture selected hybridoma cells for the production of monoclonal antibodies in large quantity ; these hybridoma cells may be frozen for future use and may also be injected in the body of an animal so that antibodies will be produced in the body and can be recovered later from the body fluid (Fig.).

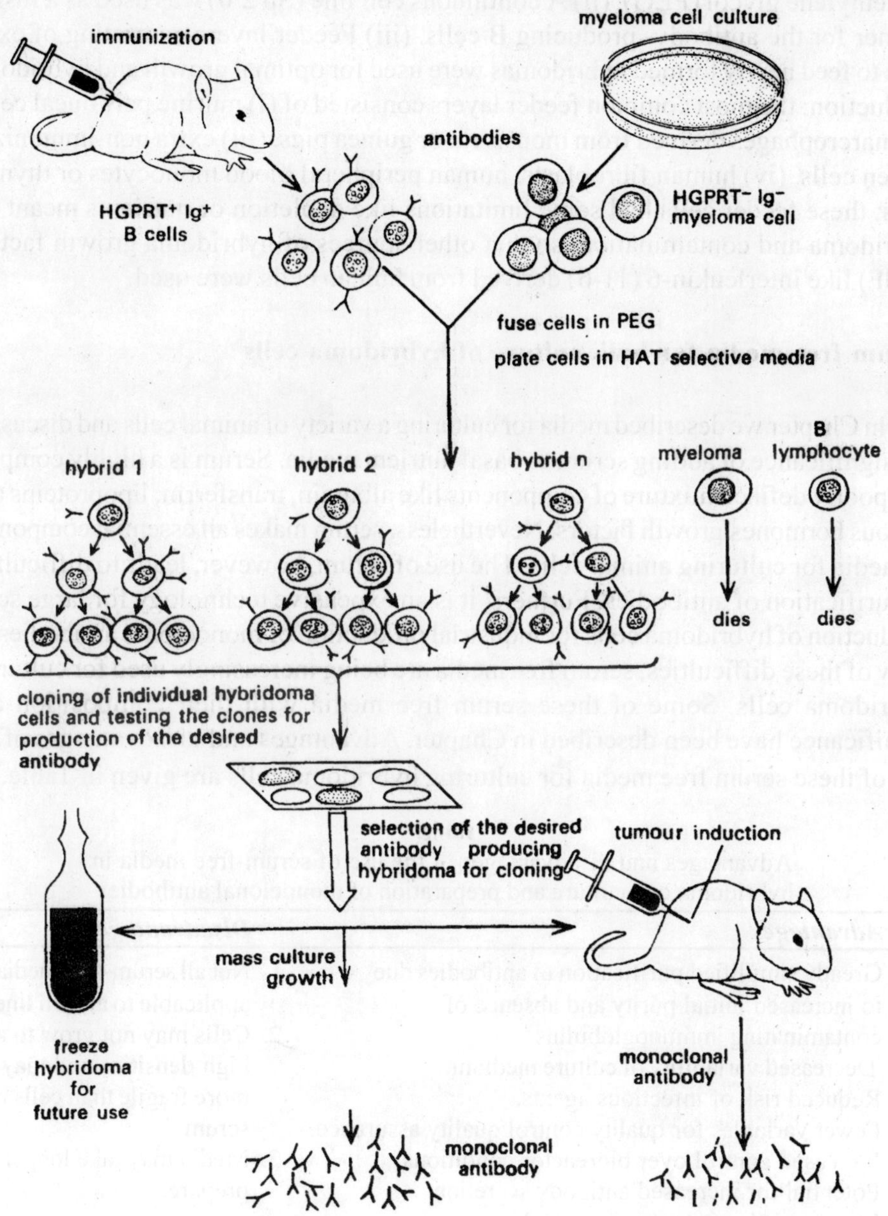


**Fig.** Procedure for making and selecting hybridomas for the production of monoclonal antibodies

## Improvements in hybridoma technology

Considerable efforts during the last 10-15 years have been made to improve the yield of monoclonal antibodies using hybridoma technology. These efforts included the following : (i) As indicated above, cell fusions were facilitated through the use of polyethylene glycol (PEG). (ii) A continuous cell line (Sp 2/0) was used as a fusion partner for the antibody- producing B cells. (iii) Feeder layers consisting of extra cells to feed newly formed hybridomas were used for optimal growth and hybridoma production; the most common feeder layers consisted of (i) murine peritoneal cells, (ii) marcrophages derived from mouse, rat or guinea pigs, (iii) extra non-immunized spleen cells, (iv) human fibroblasts, human peripheral blood monocytes or thymus cells; these feeder cells had some limitations like depletion of nutrients meant for hybridoma and contamination, so that other sources of hybridoma growth factors (HGF) like interleukin-6 (11-6) derived from human cells were used.

## Serum free media for bulk culture of hybridoma cells

In Chapter we described media for culturing a variety of animal cells and discusseu the significance of adding serum to basal nutrient media. Serum is a highly complex and poorly defined mixture of components like albumin, transferrin, lipoproteins and various hormones/growth factors. Nevertheless, serum makes an essential component of media for culturing animal cells. The use of serum, however, leads to difficulties in purification of antibodies. Furthers, it is an expensive technology for large scale production of hybridoma cells for industrial production of monoclonal antibodies. In view of these difficulties, serum free media are being increasingly used for culturing hybridoma cells. Some of these serum free media with their composition and significance have been described in Chapter. Advantages and disadvantages of the use of these serum free media for culturing hybridoma cells are given in Table.

**Table.**
Advantages and disadvantages of the use of serum-free media in hybridoma cell culture and preparation of monoclonal antibodies

| *Advantages* | *Disadvantages* |
|---|---|
| 1. Greatly simplified purification of antibodies due to increased initial purity and absence of contaminating immunoglobulins | 1. Not all serum-free media are applicable to all cell lines. |
| 2. Decreased variability of culture medium. | 2. Cells may not grow to as high densities and may be more fragile than cells in scrum. |
| 3. Reduced risk of infectious agents. | |
| 4. Fewer variables for quality control/quality assurance. | |
| 5. Increased control over bioreactor conditions. | 3. Media may take longer to prepare. |
| 6. Potential for increased antibody secretion. | |
| 7. Low or no dependence on animals. | |
| 8. Cost effective. | |
| 9. Overall enhanced efficiency. | |

## Purification of antibodies

Monoclonal antibodies may need to be purified before they are used for a variety of purposes. Before final purification, the cultures (whether cell cultures or bacterial cultures with cloned genes, as described later) may be subjected to cell fractionation for enrichment of the antibody protein. In *E. coli,* the antibodies may be secreted in the periplasm, which may be used for enrichment of antibody, so that further purification is simplified. Alternatively the antibodies may be purified from cell homogenate or cell debris obtained from the medium. Antibodies can be purified by any one of the following techniques (i) ion-exchange chromatography ; (ii) antigenaffinity chromatography.

## Antibody Engineering using Genetic Manipulations

Although hybridoma technology enables us to obtain monoclonal antibodies or homogeneous antibodies of predefined specificity, it can not allow production of novel antibodies of desired specificity. Such novel antibodies can be obtained by redesigning antibodies through the application of gene technology, an area of research which has revolutionized the potential of monoclonal antibodies. For application of gene technology, an initial but crucial problem was to find a convenient way for expression of the cloned antibody genes into proteins (antibodies). For this purpose expression could be obtained in (i) mammalian cells and (ii) bacterial cells, particularly *E. coli.* Initially, antibody genes were taken from hybridomas, cloned into plasmid vectors and expressed as complete antibodies in mammalian cells. However, later genes for different small fragments of antibodies (e.g. Fv, FAb, Fe or even single chain F v = scFv) could be cloned and expressed in bacteria. (Fig.). Expression in mammalian cells becomes important, when glycosylation of antibodies is desired. However, lack of glycosylation does not influence antigen binding capacity of Fv or Fab fragments or of complete antibody, although it does affect some of the effector functions attributed to Fe fragment. Therefore, for industrial production of antibodies, glycosylation is immaterial and *E. coli* can be used as a useful expression system. Following are some of the advantages of the use of *E. coli:* (i) fast growth; (ii) simple fermentation to allow large scale production in fermenters; (iii) fragments can be produced by Fv or Fab genes and not from complete antibody through proteolysis; these small fragments can be used for a variety of studies on structure (through NMR and crystallography) and function ; (iv) antibodies in the form of bifunctional molecules can be produced (e.g. antigen binding + toxin); (v) screening of binding activity or catalytic activity can be done on bacterial colonies or phage plaques, whithout isolating antibodies ; this screening procedure can be effectively used for screening antibody gene libraries (see later). Various strategies for antibody expression in E. coli are shown in Figure.

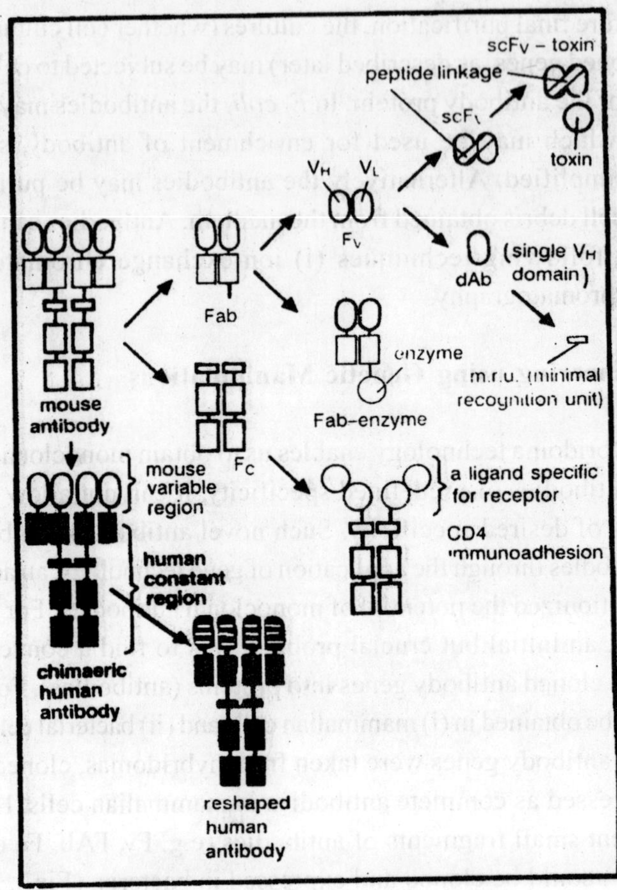


**Fig.** Engineered antibodies and fragments.
(See text for abbreviations used and for other details)

## Alternatives to hybridoma technology for monoclonal antibodies

In recent years, techniques have been developed, which allow production of monoclonal antibodies without hybridomas or even without the help of animals used for providing cells (lymphocytes and myeloma cells). For instance, antibody genes can be isolated from lymphocytes of immunized animals and then cloned and expressed in bacteria. The antibodies produced in bacteria under the control of cloned genes can be screened for binding to specific antigens. Thus, while hybridoma technology can immortalize antibody producing cells, gene technology immortalises antibody producing genes. Computer graphic techniques are also being used to built specific antigen binding sites in antibodies. Using this approach, some designer antibodies of practical value have already been produced. In this latter strategy, genes are not really cloned from lymphocytes, but are instead designed from a repertoire of antibody genes available in a collection. These approaches, where hybridoma technology and/or the use of lymphocytes can be dispensed with for the production of Mabs will be discussed in this section.

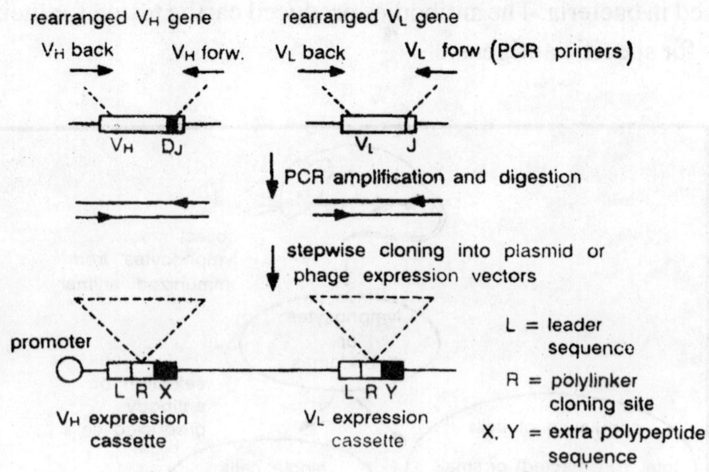


**Fig.** PCR cloning of rearranged $V_H$ and $V_L$ genes into expression vectors. (See text for details).

## Bypassing hybridomas and cloning of Mab genes

The $V_H$ and $V_L$ genes for antibodies can be amplified through polymerise chain reaction (PCR) using 'universal primers' (universal primers will carry conserved sequences for most antibodies). By building restriction sites in the above primers, the amplified $V_H$ and $V_L$ genes can also be cloned directly for expression in mammalian cells or bacteria (Fig.). The raw material for PCR may be hybridomas or B cells, which may be homogeneous (if derived from single cells) or heterogeneous. In the latter case, a variety of $V_H$ and $V_L$ genes will be amplified and will combine at random to produce as many as 106 clones for antibody genes (from 1000 different $V_H$ and 1000 different $V_L$ genes). These genes will be cloned in λ phage and their products (particularly Fab fragments) can be screened for antigen binding activities. From such a large number of combinations in a combatorial library, it is very difficult to recover the original pairs of V genes (e.g. $V_{Ha}$ . $V_{La}$ or $V_{Hx}$ .$V_{Lx}$ is an original pair: $V_{Ha}$. $V_{Ly}$ is a new combination). However, the complexity may be reduced by using antigen-selected B lymphocytes (filters coated with antigen can be used for screening). Different steps involved in cloning paired VH/VI genes are outlined in Figure.

## Designing and building of Mab genes

The antigen binding sites of antibodies have been studied in some detail in recent years. This led to modelling of entirely new antibodies, sometimes for their use as enzymes. This modelling through computer graphics can be used for alteration of antibody genes or for synthesis of entirely new genes. These genes can be cloned and expressed in bacteria. The antibodies produced can be tested for their specificity and affinity for specific antigens.

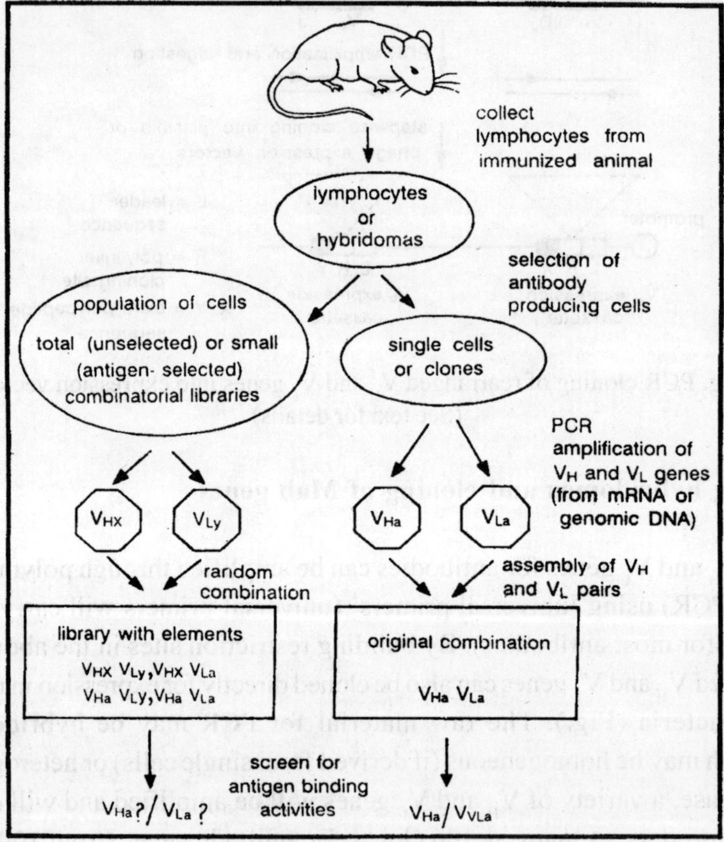


**Fig.** Strategies for cloning paired $P_H$ and $V_L$ genes from lymphocytes of an immunized animal

## Primary and secondary libraries for antibody genes

In this method, a repertoire of antibody genes can be prepared by using genes that can be obtained from a number of different sources including the following : (i)

rearranged V genes from animals obtained through the use of PCR (with universal primers); (ii) new V genes obtained through gene conversion, a process adopted in birds; (iii) rearranged genes obtained from mRNA through reverse transcription ; (iv) designing entirely new V genes or D segments. The next step is to allow the expression of library in bacteria and screen antibodies for antigen binding activities. The screening can be done on membrane filters coated with antigen. In future, the screening procedures may be replaced by methods of selection. In either case the selected VH and VL genes can be subjected to mutations to increase the affinity of an antibody for a specific antigen. A variety of methods for the above strategy are being developed , so that in future monoclonal antibodies will be produced without hybridomas and lymphocytes.

## Production of Human and Humanized Antibodies

Human antibodies are currently produced by the following methods: (i) fusion of mouse myeloma cells with human lymphocytes (blast cells in peripheral blood lymphocytes); (ii) immortalization of human cells by Epstein-Barr virus. Both the methods have limitations. Human-mouse hybrid cells have a tendency for preferential loss of human chromosomes, making them unstable. Similarly, Epstein-Barr virus does not allow preferential immortalization of blasts engaged in antibody response. In view of these difficulties, 'humanizing of rodent monoclonal antibodies' through genetic engineering is the most practical approach, which is being evolved and used for the production of mouse-human chimeric monoclonal antibodies to be used for tumour therapy or for manipulation of human immune system or against cell surface antigens. The humanized chimeric antibodies combine the rodent variable regions with the human constant (or constant + variable) framework regions. To further reduce the immunogenicity of rodent elements, humanized antibodies have been produced, which retain only the antigen-binding complementarity-determining regions (CDRs) from the rodent in association with human framework regions. In another approach, transgenic mice carrying human genes for V, D, J and C regions have been produced. These can be used for the production of human antibodies directly by hyperimmunization.

## Uses of Monoclonal Antibodies

### Monoclonal antibodies in diagnosis, screening and therapy

Monoclonal antibodies or specific antibodies are now an essential tool of much biomedical research and are of great commercial and medical value. For instance,

ABO blood groups could be earlier identified with the help of human sera carrying antibodies of known specificity. These human sera in U.K. have been replaced by monoclonal antibodies produced by hybridomas, for the identification of ABU blood groups. Thus, the diagnostic and screening value of the monoclonal antibodies through serological tests has been demonstrated. Besides the use of monoclonals in identification of blood groups in UK (UK blood typing), following three uses for monoclonals are described, although, only the first two of these make a definite market at present : (i) diagnosis (including ELISA test for detection of viruses and imaging), (ii) immunopurification; (iii) therapy.

In diagnosis, pregnancy can be detected by assaying of hormones with monoclonals. Similarly, pathogens can be detected in a few hours sparing several days of culturing of cells earlier needed. Immunopurification involves separation of one substance from a mixture of very similar molecules. For instance, individual interferons could be purified using monoclonal antibodies and could be used for inactivating T lymphocytes responsible for rejection of organ transplants. Removal of tumour cells from bone marrow is another therapeutic use of monoclonal antibodies.

For therapeutic uses, monoclonal antibodies are so designed that they will neutralize the reaction or response by one defined antigen, but still preserve the reaction of all other antigens. Several antigens of T cell receptor complex , including CD3, CD4 and CD8 have been the targets of specific antibodies for therapy. Most widely used monoclonal antibody is OKT3, which has been licensed for clinical use, particularly for the treatment of acute renal allograft rejections. Monoclonal antibodies have also been used for treatment of patients with (i) malignant leukemic cells, (ii) B cell lymphomas, and (iii) a variety of allograft rejections after transplantation. However, in several cases of therapy by monoclonal antibodies, there are undesirable side effects like high fever, vomiting, diarrhoea and respiratory distress; sometimes it also leads to broad immuno-suppression leading to increased incidence of infections. For many human-diseases as above, murine (of mouse) antibodies are utilized. Their effectiveness is, however, limited due to their short survival time in humans and relatively low cytocidal effect. For this reason, humanized antibodies are being produced by genetic manipulations and in future these will be, increasingly utilized for therapy. 'Monoclonal antibodies-cytotoxic agent = conjugates' called 'immunotoxins' have also been designed as carriers of cytotoxic substances to the target cells. These are described in greater detail in Chapter 24.

'Radiolabelled monoclonal antibodies' have also been developed for delivery of a cytotoxic effector to target cells and for 'radioimaging'. For therapy, one advantage of using radiolabelled antibodies is that they cankill cells from a distance and thus can also kill cells adjacent to antigen expressing cells. Further, the radiolabelled antibody need not be internalized to kill the tumour cells. For the purpose both β emitting (e.g. $^{731}$I, $^{90}$Y, $^{186}$Re, $^{188}$Re and $^{67}$Cu) and α emitting radionuclides ($^{211}$At, $^{212}$Pb, and $^{212}$Bi) have been used for patients with hepatoma, HTLV-1 (human T cell leukemia/lymphoma virus-1), ATL (adult T-cell leukemia), a variety of allografts etc.

Radioimmunodetection is also widely used. $(Fab')_2$ and Fab fragment_are preferred for imaging, because both targetting and blood clearance are rapid. Tumours as small as 0.5 cm, which are missed by other radiological methods, can be detected by radiolabelled antibodies or Fat fragments.

**Monoclonal antibodies in vaccine production**

Antibodies have also been used to immunize against certain diseases in humans and cattles. The most promising outcome in this area is the prospect of developing antimalarial vaccine in the near future. Monoclonal antibodies that inhibit the in vitro multiplication of Plasmodium, and the antigametocyte antibodies that inactivate male gametes have been developed. Monoclonal antibodies that destroy merozoite infected red blood cell have also been developed now. Such antibodies may prove useful as vaccines (see Chapter).

**Monoclonal antibodies as enzymes (abzymes)**

In recent years, considerable interest has been observed in the production of enzymes with specificities using techniques of enzyme engineering (see Chapter). Another approach is the use of antibodies as enzyme, that are described as enzymes. The antibodies may often bind specific ligands (haptens), but may not carry out chemical reactions. By modifying these ligands, antibodies may be generated that will catalyse specific reactions just like enzymes. Production of these abzymes is based on the following two principles : (i) enzymes work by binding the transition of a reactant better than the ground state ; (ii) antibodies which bind to specific small molecules can be produced by coupling this small molecule to a protein carrier and using this protein for immunizing experimental animals. If this small molecule is a transition state analogue (molecules that will mimic the shape and electronic configuration of transition state of reactant), then the antibodies that were produced to bind to this molecule will function as enzyme towards the substrate of this reaction. Considerable success in the production of enzymes catalyzing following reactions has been achieved : (i) acyl transfer reaction, (ii) carbon-carbon bond formation, (iii) carbon-carbon bond cleavage. A number of monoclonal antibodies using the above approach have been produced to be used as enzymes. Several of these antibodies were shown to accelerate the reactions, which they were supposed to catalyse.

**Monoclonal antibodies for purification and quantitation of other molecules**

Antibodies have been used for the purification and quantitation of certain molecules present in trace amounts, such as hormones, cyclic nucleotides, polypeptides, enzymes, antigens, etc. The assay is extremely sensitive and quantities as low as nano-or picomolar concentrations can be detected in small volumes (1 ml) of body fluids such as plasma, urine or cerebrospinal fluid. Following steps are involved in the purification and quantitation of enzymes and other proteinaceous molecules.

1. Raising the antibodies—Monoclonal antibodies for a specific enzyme or protein are raised using traditional hybridoma technology. The enzyme is first extracted and partially purified using most of the conventional steps, such as extraction, ammonium sulfate precipitation and gel/column filtration. The preparation is diluted and emulsified with 'Freunds complete adjuvant'. Then, it is injected intramuscularly into mice, to raise antibodies. Generally, several doses at intervals of a few days are injected to boost the antibody production. The spleen cells from the mice are collected and fused with myeloma (tumor) cell lines derived from a rat. The resultant hybridoma cells are screened for their capacity to produce antibodies and secrete them out (Fig.). The hybridoma cells can be thawed and recultured for the production of antibodies.

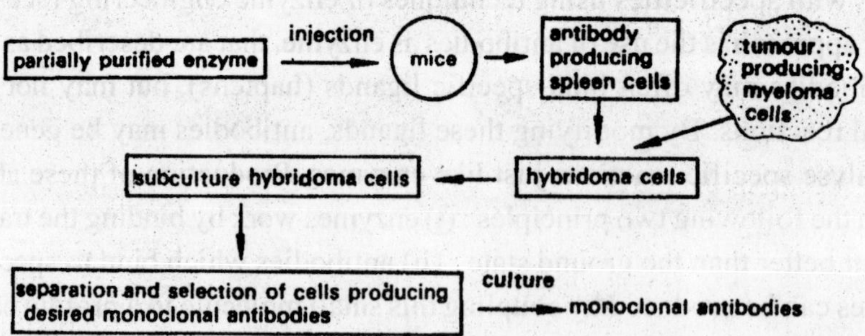


Fig. Different steps involved in the production of monoclonal antibodies

The amount of antibody production can be substantially increased by ascites fluid production method, instead of producing antibodies by hybridoma cell cultures. In this method, hybridoma is injected intraperitonealy into histocompatible mice to produce ascite cell tumours. One week to 10 days after injection the peritoneal cavities in the mice appear swollen. The ascites fluid from the cavity is taken out in heparin, using a needle, which is rich in antibody. Hybridomas in culture produce only upto 400, ug/ml of specific antibody, while in ascites fluid, the amount of antibody is about 100-200 times more. From one mouse about 10-15 ml of ascites fluid can be obtained.

2. Purification and quantitation—Either the supernatants from cultured hybridoma cells or the ascite fluid is used for detection and purification of an enzyme. Sometimes a specific antibody is purified before it is used for enzyme detection/purification. The most common methods for enzyme purification/detection are: affinity chromatography, radioimmunoassay and immunoprecipitation.

(a) Affinity chromatography—In affinity chromatography, the monoclonal antibody is generally immobilized on an insoluble matrix such as cross linked dextran or agarose beads, by a covalent linking agent (e.g. cyanogen bromide). The immobilized antibody is generally packed in a column, where its average concentration is 5-10mg antibody per ml of bed volume. Protein solution containing the specific enzyme is applied over these immobilized antibodies. The specific enzyme binds with the antibody while other components do not, and thus the specific enzyme is separated. The enzyme can be recovered later by using suitable reagents. Antigens can also be purified using affinity chromatography wish monoclonal antibodies. The antigen, which binds with the immobilized antibody, is eluted and recovered from the immoblized support by suitable washing procedures. Change in pH of the buffer often causes elution. Some important proteins such as alpha fetoprotein and leucocyte interferon are also purified using this procedure.

(b) Radioimmunoassay—Radioimmunoassay is an extremely sensitive method for the detection and quantitation of any substance that is antigenic and that can be labelled with a radioactive isotope. Basically the method depends upon the competition between labelled (known) and unlabelled (unknown) antigen for the same antibody. A known amount of labelled antigen (enzyme in this case), a known amount of specific antibody and an unknown amount of unlabelled antigen in the sample are allowed to reach equilibrium. By measuring the amount of labelled antigen bound to the antibody, a measure of the antigen in unknown sample can be obtained. By suitably adjusting the amounts of antibody and labelled antigen, even small amount of antigen in an unknown material can be detected.

In order to measure the amount of labelled antigen attached to antibody, it is necessary to separate the antigen-antibody complexes from the mixture. By measuring the radioactivity, the percentage of labelled antigen bound to the antibody can be calculated. The concentration of an unknown (unlabelled) antigen can be determined by reference to a standard curve constructed from data obtained by allowing varying amounts of unlabelled antigen to compete (Fig.).

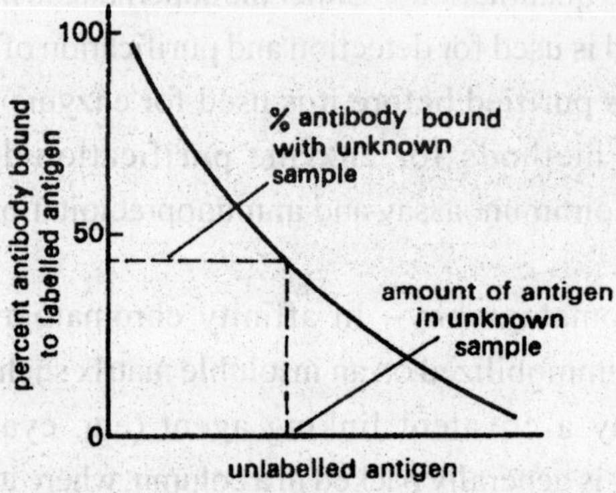


**Fig.** A standard curve used in radioimmunossay for quantitative estimation of antigen in an unknown sample

(c) Immunoprecipitation—When correct conditions are employed, antigen and antibody react to form a precipitate. This precipitation phenomenon has been used to separate and purify enzymes and other antigenic substances in a variety of ways.

One application of immunoprecipitation is immunoelectrophoresis. This is the most sensitive method for the defection of enzymes and antigens in a mixture. It involves combination of electrophoresis and gel diffusion. In this procedure, the mixture containing; the enzyme (or antigen) is placed in a small well, made in a layer of agar placed on a glass sheet, or a microscopic slide. It is then subjected to elect rophoresis by application of an electric current. The enzyme and other components of the mixture migrate through the agar at variable rates. Following electrophoresis, a trough is cut in the gel, parallel to the electrophoretic migration (Fig.). An antiserum (containing antibody of that specific enzyme) or monoclonal antibody preparation is then placed in the trough. The reactants diffuse towards one another and form separate arcs of precipitate for each antigen-antibody complex. If only one monoclonal antibody preparation is used, then only one precipitation arc is formed.

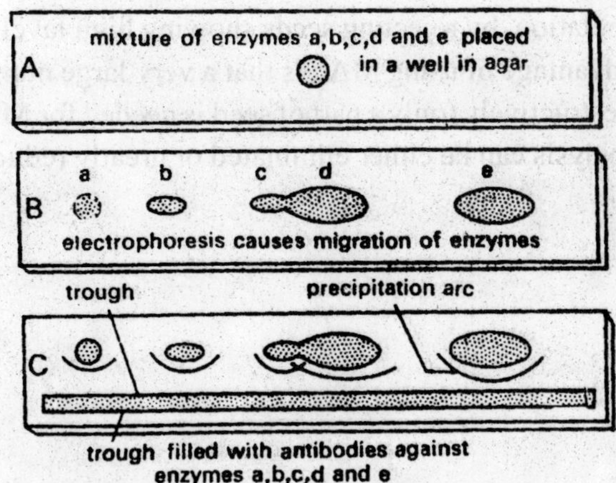


**Fig.** Technique of immunoprecipitation for detection and separation of enzymes in mixture

## Monoclonal antibodies for cytogenetic analysis

The technique of ELISA (enzyme linked immunosorbant assay), utilizing monoclonal antibodies has also been used for cytogenetic analysis in wheat. Monoclonal antibodies (MAbs) specific to chromosome, I B, OA and (D are already available and those for other chromosomes will become available in future, due to proteins coded by chromosomes of all the homoeologois groups (Table)

**Table**

Different proteins and the homoeologous groups having chromosomes with genes for these proteins.

| *Class of Proteins* | *Homoelogous groups having related chromosomes* |
|---|---|
| 1. Endosperm proteins | 1,6 |
| 2. Amylase inhibitors | 2, 3. 6 |
| 3. Albumins and globulins | 3, 4, 5, 6 and 7 |

Nullisomic and monosomic conditions for the chromosome 1B and the absence of 1BS in 1BL/IRS wheat-rye translocation (1BS carries gene for wheat gliadin protein) could be ascertained using the monoclonal antibody P24B, which is specific for IBS gliadins. Relative binding of P24B to 50% propanol extracts of half kernels could be used as a criterion for the identification of nullisomics (showing no binding or very low binding) and monosomics (intermediate binding), because disomics show high level of binding. Sometimes nullisomics and monosomics can not be distinguished from each other with complete certainty, so that the results need to be confirmed through chromosome analysis. But MAbs can be used to enrich the number of aneuploids in a population, by rejecting seeds showing high lavel of reaction with MAb. The major advantage of using MAb is that a very large number of seeds can be examined non-destructively (only a part of seed is needed for MAb reaction) and the chromosome analysis can be either eliminated or greatly reduced.

# SECTION - 4

# SIGNIFICANCE FOR PLANTS

# ELEVEN

# Cell and Tissue Culture

## Techniques of Culture Media and Cell Culture

Plant tissue culture first began in 1898, when a German botanist, G. Haberlandt successfully cultured fully differentiated individual plant cells, isolated from different tissues in several plant species. For about thirty five years (i.e. upto 1934), little further progress in cell culture research was made, although culture of embryos, roots and other tissues was achieved in this period. During 1934-1939, due to discovery of the importance of auxins and B- vitamins, the foundation of plant tissue culture was laid down by three scientistst (Gautheret, White and Nobecourt), even though only small pieces of tissue and not individual differentiated cells could be grown in cultures. During the next twenty years (1940-60), a variety of chemicals (hormones, vitamins, etc.) were identified for their effect on cell division, growth and differentiation. Thus, by early nineteen sixties, the methods of in vitro culture of plant cells, tissues and organs was reasonably well-developed. Research in this area was also initiated in early 1960's at the Botany Department, Delhi University in India. Consequently, media and culture techniques for a variety of plant materials became known and their uses suggested, which are now being extensively utilized due to the recognition of the significance of plant tissue culture in plant biotechnology. Plant tissue culture research has become a thrust area during the last decade due to the renewed emphasis it has received in all areas of crop improvement programmes. In several publications, including books, and in syllabi prescribed for biotechnology courses of our universities, plant tissue culture along with plant genetic engineering have become synonymous with plant biotechnology. Keeping this in view, this chapter and the next few chapters are being devoted to various aspects of plant tissue culture. This chapter is devoted to culture media and cell culture techniques.

## Laboratory Requirements

### Laboratory space

For a standard tissue culture laboratory, following minimum facilities are necessary: (i) a media room, where glassware and plasticware are washed, stored and used for preparation of culture media; this room should be equipped with working benches (including a laminar air flow cabinet), a deepfreeze, a refrigerator, weighing balance, a hot plate-cum - magnetic stirrer, a pH meter, a vacuum pump, and an autoclave; (ii) a culture room, where controlled temperature, diurnal illumination, and humidity are maintained and cultures are grown; the rooms should have racks for placing culture tubes, flasks, etc. and also a shaking machine for growing suspension cultures.

### Washing and sterilization

New glassware is always washed using detergents especially designed for the purpose to remove all traces of acids, etc. Finally, the glassware is rinsed in tap water and then in distilled water. Sometimes, the glassware may need to be autoclaved before washing to remove agar medium and to destroy microbial contaminants.

The culture medium (details of which arc given in the next section), and the glassware are autoclaved at 100°C for 20 min. If autoclave is not available, a pressure cooker may be used. Growth factors, such as GA3, zeatin, ABA, urea, etc. (which are thermolabile) should not be autoclaved, but sterilized separately by membrane filtration (for details consult Bhojwani and Razdan, 1983). Instruments, such as forceps, scalpels, needles, spatula are sterilized by dipping in 95% ethanol followed by flaming and cooling.

Plant material is surface sterilized using sodium or calcium hypochlorite solution (0.3 - 0.6%) for 15 -30 min. For ovule, embryo and endosperm culture, only the seeds are surface sterilized and the ovule or embryo or endosperm is dissected under aseptic conditions (usually in a laminar air flow cabinet). Similarly, for shoot apices, the buds are surface sterilized and the apices are dissected under aseptic conditions. After surface sterilization , plant material is rinsed 3-4 limes in sterile distilled water.

All operations including transfer of plant material are carried out under aseptic conditions, preferably under the hood of a laminar air flow cabinet, which is commercially available for tissue culture laboratories. These cabinets have now replaced UV-sterilisation wooden cabinets earlier utilised for this purpose.

## Culture Media

No single medium can be suggested for all types of plants and organs, so that the details of culture medium need to be worked out for each plant material separately. Several recipes have been developed during the last three decades and some of them are listed in Table. The most important of these media are MS media, which evolved due to a variety of modifications of the original MS medium (Murashige and Skoog, 1962). The pH of the medium is usually adjusted between 5.0 and 6.0.

**Table**

Composition (mg/1) of three media for plant cell suspension cultures.

| *Ingredient* | *Murashige and Skoog (MS) medium** | *Gamborg et al. (Bs) medium*** | *While medium*** |
|---|---|---|---|
| $NH_4NO_3$ | 1650 | - | - |
| $KNO_3$ | 1900 | 2500 | 80 |
| $Ca(NO_3)_2.4H_2O$ | - | - | 288 |
| $CaCl_2.2H_2O$ | 440 | 150 | 737 |
| $MgSO_4.7H_2O$ | 370 | 250 | - |
| $KH_2PO_4$ | 170 | - | - |
| $Na_2SO_4.10H_2O$ | - | - | 460 |
| $(NH_4)_2SO_4$ | - | 13.1 | - |
| $NaH_2PO_4.H_2O$ | - | 150 | 19 |
| KCl | - | hS | |
| KI | 0.83 | 0.75 | - |
| $H_3BO_3$ | 6.2 | 3.0 | - |
| $M_nSO_4.4H_2O$ | 22.3 | - | 1.5 |
| $MnSO_4H_2O$ | - | 10.0 | 0.75 |
| $Z_nSO_4.7H_2O$ | 8.6 | 2.0 | - |
| $Na_2MoO_4.2H_2O$ | 0.25 | 0.25 | - |
| $CuSO_4.5H_2O$ | 0.025 | 0.025 | 0.001 |
| $CoCl_2.6H_2O$ | 0.025 | 0.015 | - |
| $Na_2EDTA$ | 37.3 | 37.3 | 2.67 |
| $FeSO_4.7H_2O$ | 27.8 | 27.8 | 27.8 |
| Sucrose $MoO_3$ | 30,000 | 20,000 | 20.000 |
| Inositol | 100 | 100 | - |
| Nicotinic acid | 0.5 | 1.0 | 0.5 |
| Pyridoxine-HCl | 0.5 | 1.0 | 0. I |
| Thiamine-HCl | 0.1 | 10.0 | 0.1 |
| Glycine | 2.0 | - | 3.0 |
| Indoleacetic acid | 0.3-3.0 | - | - |
| Kinetin | 1.0-2.0 | 0.1 | - |
| 2,4-Dichlorophenoxyacetic acid | | 0.1-1.0 | - |

***pH=5.7. **pH=5.5.**

## Media Constituents

The major constituents of a culture medium include (i) inorganic nutrients, (ii) organic nutrients, (iii) growth hormones and (iv) agar. Among the inorganic nutrients, besides C (carbon), H (hydrogen), O (oxygen), additional 12 elements (N, P, S, Ca, K, Mg, Fe, Mn, Cu, Zn, B, Mb), which are essential for plant growth, are included in culture media. Six of these (N, P, S, Ca, K, Mg) are major or macroelements (conc. above 0.5 mmol) and the remaining six are minor or microelements. The various culture media differ mainly in quantity rather than in quality of these elements. The original White's medium (1943) though widely used earlier, was later found inadequate quantitatively for good callus growth, so that the currently used media are richer in mineral salts.

The organic nutrients can be broadly classified into (i) nitrogen and (ii) carbon sources, the former mainly including vitamins ( B1, B3, B5, B6) and amino acids and the latter mainly including sucrose, glucose and fructose, or other carbohydrates (including starch). Other complex nutrients of undefined composition include casein hydrolysate (CH), coconut milk (CM), corn milk, malt extract (ME), tomato juice (TJ), and yeast extract (YE). It is, however, recommended to avoid their use and replace each by a single amino acid, so that the quantity and quality of constituents may not vary at different times.

The growth hormones included in culture media involve (i) auxins, (ii) cytokinins and (iii) gibberellins. The auxins ae mainly used to facilitate cell division and root differentiation. Commonly used auxins are IBA (indole-3-butyric acid), NAA (napthaleneacetic acid), NOA (naphthoxyacetic acid), p-CPA (para-chlorophenoxyacetic acid), 2, 4-D (dichlorophenoxy-acetic acid) and 2,4,5-T (trichlorophenoxyacetic acid). IBA and NAA are widely used for rooting and (in combination with cytokinins) for shoot proliferation. 2,4-D and 2,4,5-T are very effective for the induction and growth of callus. The cytokinins facilitate the cell division and differentiation. Commonly used cytokinins include BAP (benrylamino purine), 2,ip (isopentenyl-adenine) and kinetin (furfurylamino purine). Among 84 known gibberellins, mainly $GA_3$ is used to induce plantlet formation from adventive embryos formed in culture.

Another important component of any culture medium is agar (a polysaccharide obtained from seaweeds to provide solid surface for growth (agar is not a nutrient), because in the liquid medium, the tissue will be submerged and die due to lack of availabililty of oxygen (agar is used at a concentration of 0.8-1.0%). Cells can also be grown in suspension cultures devoid of agar, but such cultures need to be aerated regularly either by bubbling sterile air or by gentle agitation.

## Media selection

In order to find a suitable culture medium for a plant material, one should start with a well known medium (MS or BS ; sec Tabel), and by a series of experiments, a new medium may be evolved. The most variable factors in the medium are growth regulators (auxins and cytokinins), which may be adjusted by using five concentrations of each (0, 0.5, 2.5, 5, l0µmol 1-1). The combination giving best results may be selected. Different concentrations of salts (1/2 or 1/4 that of MS salts) and sucrose (2-6%) may be used with the best combination of growth regulators to find their optimum concentration. Other approaches like "broad spectrum experiment" may be used, if the above approach does not help in finding a suitable medium (for details, see Bhojwani mid Razdan, 1983).

## Media preparation

Dry powdered media having required concentration of different constituents are now available commercially, so that this powder is dissloved in distilled water (10% less than the final volume). Agar, sugar and other desired supplements are added and the final volume is made up using distilled water. The pH is then adjusted (5.0 - 6.0) and medium is autoclaved. When standard medium is not satisfactory, the. required quantities of different components are weighed and dissolved.

The medium prepared as above is poured into the desired culture vessels ( 15 ml for 25 x 150 mm culture tubes; 50 ml for 150 ml flasks, which are plugged with non absorbent cotton wool wrapped in cheek cloth). The vessels are transferred to baskets and covered with aluminium foil and sterilized by autoclaving at 120°C for 15 min. If plastic culture vessels (unautoclavable) are used, the medium is autoclaved in 250 or 500 ml flasks. In this case, medium is allowed to cool to 60°C before pouring. In either case, culture vessels may be stored at 40°C and used whenever needed.

## Cell Culture

As described earlier in this chapter, although single cells could he cultured by Haberlandt as early as 1902, they could be induced to divide in culture only after several decades. During the last flew decades, the progress in this area has been so spectacular that cultured cells are being used in all areas of biology for a variety of purposes. The cell culture techniques will be briefly described in this section.

## Isolation of single cells

Single cells can be isolated either from cultured tissues or from intact plant organs, the former being more convenient than the latter. When isolated from cultured tissue, the latter is obtained by culturing an organised tissue (explant) into callus (callus is undifferentiated growth in solid form). The callus may be separated from explant and transferred to fresh medium to get more tissue. Pieces of undifferentiated calli are transferred to liquid medium, which is continuously agitated to obtain a 'suspension culture'. Agitation of pieces breaks them into smaller clumps and single cells, and also maintains uniform distribution of cells and cell clumps in the medium. It also allows gaseous exchange.

Suspension cultures with single cells can also be obtained from intact plant organs either mechanically (grinding the tissue followed by cleaning, filtration and centrifugation) or enzymatically (treating excised and peeled leaves with macro-enzyme). The datails of technique are available in an advanced book (Bhojwani and Razdan, 1983).

## Suspension cultures

Suspension cultures can be maintained in either of the following two forms: (i) Batch cultures are initiated as single cells in 100-250 ml flasks and are propagated by transferring regularly small aliquots of suspension to a fresh medium. (ii) Continuous cultures are maintained in a steady state for long period by draining out the used medium and adding fresh medium; in this process either the cells separated from drained medium are added back to suspension culture (these are called 'close continuous cultures') or the addition of medium is accompanied by the harvest of an equal volume of suspension culture (these are called 'open continuous cultures'). In the second case, a steady state of suspension culture is maintained for a long period. Despite advantages associated with continuous cultures , they are rarely used, since a lot of attention is needed and the required equipment may not be easily available.

The medium used for suspension cultures is usually the same as for callus growth except that agar is omitted. A manipulation of auxin/ cytokinin ratio may be desirable to achieve cell dispersion. The suspension culture needs to be regularly agitated to avoid clumping of cells and to provide aeration. A shaking speed of 31 - 150 rpm is satisfactory. Cell suspensions are often asynchronous, different cells being in different stages of cell cycle (G1, S, G2, M). Therefore, efforts have also been made to obtain synchrony in suspension cultures. Synchrony can be achieved by one of the following two methods: (i) Starvation: cells are starved of a nutrient ( e.g. a growth hormone) necessary for cell division, so that cells are arrested at G1 or G2. These arrested cells are supplied with the nutrient (after sometime) to obtain synchronous cell division. (ii) Inhibition inhibitors of DNA synthesis (e.g. 5- aminouracil, Fudr = flurodeoxyuracil, hydroxyurea and thymidine) have also been used for achieving synchrony of cell division. Due to inhibition of DNA synthesis, cells are arrested at G1, so that the removal of inhibitors is followed by synchronous division of cells.

**Bergmann's cell plating technique**

In this technique, free cells are suspended in a liquid medium (if cell aggregates are there, it is filtered). Culture medium with agar (0.6 - 1%) is cooled and maintained at 35°C in a water bath. Equal volumes of liquid and agar media are mixed and rapidly spread in a Petri dish, so that cells are evenly distributed in a thin layer, after solidification. The Petri dishes are sealed with parafilm and examined with inverted microscope to mark single cells (marking is done on outer surface of the dish).The Plates are incubated in dark at 25°C and cell colonies developing from marked single cells are used to obtain single cell cultures *(Fig.)*. Various other methods (e.g. filter paper raft-nurse technique; microchamber technique) have also been developed to grow individual cells.

## Applications

### Mutant selection

An important use of cell cultures is in mutant selection in relation to crop improvement. The frequency of mutations can be increased several fold through mutagenic treatments, and million of cells can be screened. Since selection of mutations will be exercised at cellular level, no chimeras will be obtained, which is sometimes a drawback of mutation breeding methods, where mutants are selected at the level of whole plants. Further, if haploid cells (like microspores) are used for cell culture, the problem of dominance not permitting the expression of mutation in M1 generation is also automatically overcome. A large number of reports are now available where mutants have been selected at the cellular level. The cells are often selected directly by adding the toxic substance against which resistance is sought in the mutant cells. Using this strategy, cell lines resistant to amino acid analogues, antibiotics, herbicides, fungal toxins (causing diseases) etc. have actually been isolated. Sometimes indirect selection is also exercised. For instance, nitrate reductase (NR) deficient cell lines are selected as chlorate resistant cells. Tobacco protoplasts selected for resistance to methionine sulfoximine (MSO) showed enhanced resistance to Pseudomonas tabaci, even though MSO is not the phytotoxin of P. tabaci. Cell cultures of Nicotiana were also selected for resistance against several herbicides like amitrole, sodium chlorate, chlorosulfuron and sulfomentron methyl. In potato, cell lines were selected, which were resistant to 5 methyltryptophan (5 MT), thus permitting accumulation of free tryptophan, phenylalanine and tyrosine. Salt tolerant and cold tolerant cell lines have also been isolated in the genus Nicotiana. Auxotrophic mutants are selected by microbial methods using the minimal medium, where only prototrophs will grow. The cell lines will be grown first on the minimal medium. Lack of growth on minimal medium will suggest the presence of an auxotroph, which can be grown on supplemented media to identify nutritional deficiency. The selected mutant cells need to be grown into whole plants, which will be multiplied and tested for the mutant phenotype. This mutant phenotype may or may not express at the level of whole plants. If it expresses in whole plant, its genetics can be studied using conventional methods of genetics.

**Production of secondary metabolites**

It is well known that plants are an important source for a variety of chemicals used for a variety of purposes including pharmacy, medicine and industry. In recent years, plant cell suspension cultures and immobilized cells (see Chapter) are being utilized for the production of these chemicals on a commercial scale due to the following advantages over extraction from whole plants: (i) The yield and quality of the product is more consistent in cell cultures because it is not influenced by the environment. (ii) The production schedule can be predicted and controlled in the laboratory or industry.

The most important chemicals produced using cell cultures are secondary metabolites, which are defined as 'those cell constituents which are not essential for survival'. These secondary metabolites include alkaloids, glycosides (steroids and phenolics), terpenoids and a variety of flavours, perfumes, agrochemicals etc. The yield of these chemicals in cell culture is though generally lower than in whole plants, it can be substantially increased by manipulating physiological and biochemical conditions. In some cases, cell cultures accumulate these secondary metabolites at levels higher (2-10 times) than those found in whole mother plants, from which cell culture has been prepared. Automation in cell cultures discussed in the next chapter, can be used for industrial production of secondary metabolites. However, sometimes immobilized plant cells are used instead of suspension cultures to increase the efficiency of production system.

**Table**

Plant species and secondary metabolites obtained from them using tissue culture leachniques.

| *Plant species* | *Product* | *Activity* |
|---|---|---|
| 1. Catharanthtrs roseus | (a) Vincristine | Antileulkaemic |
| | (b) Ajmalthine (serpentine) | Antiarrhythmic<br>Tranquilizer |
| | (c) Ajmalicine | |
| | (d) Serpentine | |
| 2. Chrysanthemum cinerariaefolium | Pyrethrim storage) | Insecticide (for grain) |
| 3. Cinchona officinalis | Quinine | Antimalarial |
| 4. Digitalis lanata | (a) Digoxin | Cardiac tonic |
| | (b) Reserpine | Hypotensive |
| 5. Dioseorea deltoidea | Diosgenin | Antifertile |
| 6. Jasminum sp. | Jasmine | Perfume |
| 7. Papaver somniferum | Morphine | Analgesic, sedative |
| 8. P. bracteatum | Codeine | Analgesic |
| 9. Thaumatococcus danielli | Thaumatin | Sweetener |
| 10. Datura stramonium | Scopolamine | Anti hypertension |
| 11. Atropa belladonna | Atropine | Blocking of cholinergic |
| 12. Coptis japonica | Berberine | Antibacterial |
| 13. Campatotheca accuminata | Camptothecin | Anticancer |
| 14. Nicotiana tabacum | (a) Nicotine | Ganglion blocker |
| | (b) Glutathione? | Cardiovascular agent |
| | (c) Ubiquinone-10 | |
| 15. Coleus Mond | Rosamarinic acid | Spice, antioxidant |
| 16. Morinda citrifolia (also Cassia tora) | Anthraquinones | Laxatives, dyes |
| 17. Glycyrrhiza glabra | Glycyrrhizin | |
| 18. Cephalotaxus harringtonia | Cephalotaxine | Antitumour |
| 19. Lithospermum erythrorhizon | Shikonines | Dye, pharmaceutical |
| 20. Stevia rebaudiana | Stevioside | Sweetener |

Some of the important secondary metabolites obtained from plants are listed in *Table* and plant cell cultures techinque can be profitably utilized to produce them at a commercial level.

## Biotransformations

Plant cell cultures are also being utilized for producing valuable products including secondary metabolites through biotransformation, a technique also utilized with the help of microbes (microbial transformation = see Chapter). In this technique, low cost precursors are used as a substrate and are transformed into value added high cost products. At least in some cases, this transformation is achieved more easily by cultured plant cells, than either by micro-organisms or by chemical synthesis. Following are some of the examples of the use of cell cultures in biotransformation:

(a) Suspension culture of Digitalis lanata can convert 'digitoxin' or 'methyldigitoxin' into medically important 'digoxin' or 'methyldigoxin', which is used for treatment of heart diseases. The conversion rate has been estimated to be as high as 15% in 24 hours and 70% in 7 days. (b) Datura cell cultures possess ability to convert 'hydroquinone' into 'arbutin' (used as diuretic and urinary antiseptic) through glycosylation. (c) Cell cultures of Stevia rebaudiana and Digitalis purpurea can convert steviol into 'steviobiocide' and 'stevioside' which are 100 times sweeter than cane sugar. Single cell cultures are also being used for production of single cell proteins (SCP).

## Tissue Culture

In plant tissue culture experiments, unless single cell cultures are absolutely necessary, more often we use explants (pieces of differentiated tissues) to initiate their growth in culture. As discussed in the previous chapter, this is also necessary to obtain callus from which single cell cultures are derived. During the last few decades, even for development of new crops or for improvement in the characteristics of existing crops (e.g. to obtain disease free crop, etc.), the techniques of tissue culture in combination with genetic manipulation have become important tools. For instance, as an alternative to vegetative propagation, shoot tip propagation was used commercially for rapid and consistent reproduction of elite or difficult to propagate genotypes.

When an explant from differentiated tissue is used for culture on a nutrient medium, the non-dividing, quiescent cells first undergo certain changes to achieve a meristematic state. The phenomenon of the reversion of mature cells to the meristematic state leading to the formation of callus is called 'dedifferentiation'. The component cells of callus have the ability to form a whole plant, a phenomenon described as 'redifferentiation'. These two phenomenons of dedifferentiation and redifferentiation are inherent in the capacity described as 'cellular totipotency', a property found only in plant cells and not in animal cells. In other words, while a differentiated plant cell retains its capacity to give rise to a whole plant, an animal cell loses this capacity of regeneration after differentiation. Although, generally, a callus phase is involved before the cells can undergo redifferentiation leading to regeneration of a whole plant, but rarely the dedifferentiated cells can give rise to whole plants directly without an intermediate callus phase. These aspects of tissue culture and its application in 'clonal propagation' and in the generation of hereditary variation called 'somaclonal variation' will be described in this chapter

## Cell and Organ Differentiation

### Cytodifferentiation

Since the undifferentiated callus cells are all parenchymatous in nature, the differentiation of these cells into a variety of cells is required during redifferentiation of these cells into whole plants. In this connection, the study of differentiation into vascular tissue has received maximum attention. It has been shown that 'auxin' and 'sucrose' play a major role in vascular differentiation and also that cytokinins and gibberellins promote differentiation into xylem tissue. At low concentration (0.5mg. $l^{-1}$), auxin stimulates xylem formation and there is inverse relationship between auxin concentration and the degree of xylem differentiation. This effect of auxin is dependent on the presence of sugar, since the relative amounts of xylem and phloem formed can be altered by varying the sucrose concentration. In some studies (e.g. in Syringe), 1% sucrose (with 0.55mg $l^{-1}$ of auxin or 1AA) induced little xylem formation; 2% sucrose favoured better xylem formation (with little phloem), 2.5-3% sucrose induced equal xylem and phloem differentiation and 4% sucrose induced only xylem formation (with little or no phloem). Sugars, other than sucrose (e.g. maltose, glucose, fructose, etc.) have also been used to stimulate vascular differentiation. Besides auxin and sucrose, cytokinins and gibberellins have also been shown to play a role in xylem differentiation.

### Organogenic differentiation

For the regeneration of a whole plant from a cell or from a callus mass, cytodifferentiation is not enough and there should be differentiation leading to organogenesis (organ formation = formation of stem, roots, flowers, etc.). This may occur either through 'shoot bud differentiation' or through 'somatic embryogenesis'. In the former, shoot buds (monopolar structures) are formed while in the latter, somatic embryos (bipolar structures) are formed both leading to regeneration of whole plants.

**Shoot bud differentiation.** Shoot bud differentiation is stimulated in plants by a variety of chemical and physical factors, but these factors vary widely for different plant materials, so that no general recipes can be prescribed. In this connection, the relative ratio of cytokinin (kinetin) to auxin (IAA) was considered important in determining the nature of organogenesis in tobacco with tissue (Skoog and Miller, 1957). It was shown that high level of kinetin caused bud initiation, while high concentration of auxin favoured rooting. However, there are other studies which suggested that the type of organ (root or shoot) formed was exclusively determined by the concentration of auxin (NAA) and not by the ratio of cytokinin to auxin. In most cereals, callus tissue exhibits organogenesis when it is transferred from a medium containing 2, 4-D to a medium lacking it or where 2,4-D is replaced by IAA or NAA.

The requirement of cytokinin and auxin for bud differentiation depends on the endogenous levels of these two classes of substances in the tissues used for culture. Thus, not only different plant materials differ in their ability to undergo shoot bud differentiation, but some tissues are incapable of bud formation in cultures due to special requirements, rather than due to lack of inherent genetic ability to do so. Shoot bud formation is also controlled by the size and source of the explant. The larger the explant (containing parenchyma, cambium and vascular tissue), the better are the chances of shoot bud formation, irrespective of auxin - cytokinin concentration. Similarly, depending upon the plant species, leaves or leaf fragments, root sections and inflorescence sections or cotyledons may be used as explants leading to the formation of adventitious buds.

**Somatic embryogenesis.** It has been demonstrated that embryo development is not the monopoly of egg cell and that any cell of the embryo sac or any other part of the sporophyte may give rise to an embryo. However, in nature, embryos can not develop outside the ovules, although in several species (Citrus, Mangifera), adventive embryos (embryos developing from sporophytic tissue) may develop from nucellar cells or integuments of the ovule outside the embryo sac. These embryos, however mature, only when pushed into the embryo sac, suggesting that embryo sac should have a special environment, which facilitates the development of embryo and the same should be made available in culture if somatic embryogenesis needs to be induced.

Somatic embryogenesis in culture was demonstrated for the first time in carrots (*Daucus caroler*), where bipolar embryos developed from single cells (Fig.). Further, the tissue derived from these somatic embryos spontaneously developed into embryos again. However, somatic embryogenesis is influenced by plant extracts, growth regulators and by the physiological state of calli. For instance, 2,4-D in the medium is least conducive for embryogenesis in plants like cabbage or carrot. Embryogenesis is easy in suspension cultures, which are derived from explants grown on auxin containing medium. On the other hand, cytokinins and gibberellins caused a partial or complete inhibition of growth of the potential embryonic cells. It has also been shown that nitrogen supplied in the form of $NH_4+$ ions or as casein hydrolysate (rather than as nitrate), stimulates somatic embryogenesis.

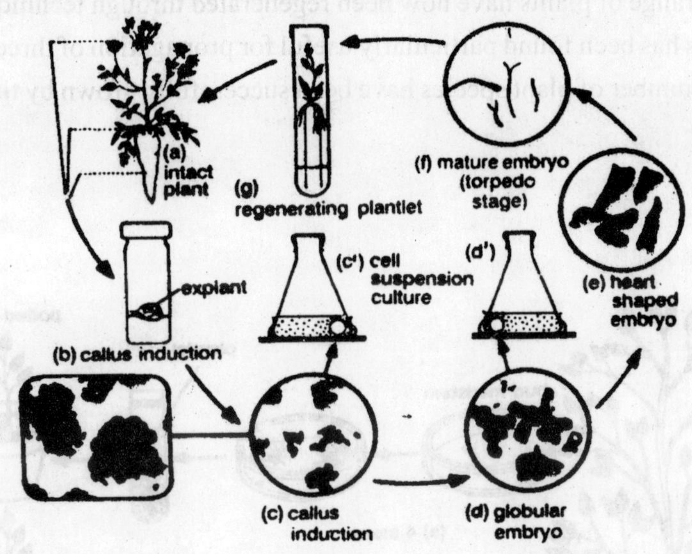


**Fig.** Schematic representation of somatic embryogenesis in carrot.

## Clonal Propagation or Micropropagation

It has been demonstrated that a variety of plant species can be conveniently propagated through the techniques of cell, tissue or organ culture. This is popularly described as clonal propagation or micropropagation. The major benefits of this method include the following; (i) rapid multiplication of superior clones and maintenance of uniformity; (ii) multiplication of disease free plants and (iii) multiplication of sexually derived sterile hybrids. In most cases, clonal propagation is achieved by placing sterilized shoot tips or axillary buds onto a culture medium that is sufficient to induce formation of multiple buds. This method has already been used to propagate a large number of marketable ornamentals. Following stages are involved in the method of clonal propagation : (i) Stage I involves establishment of tissue in vitro. (iii) Stage II involves multiplication of shoots (often media are not changed between stage I and stage II). (iii) Stage III concerns root formation and conditioning of propagules prior to transfer to the green house; this stage requires high intensity and alteration of media for promotion of root formation. (iv) Stage IV involves growth in pots followed by field trials. The number of seeps in micropropagation may sometimes be reduced to three or two as shown in Figure.

A wide range of plants have now been regenerated through technique of tissue culture. This has been found particularly useful for propagation of three species, so that a large number of plant species have been successfully grown by tissue culture (Table).

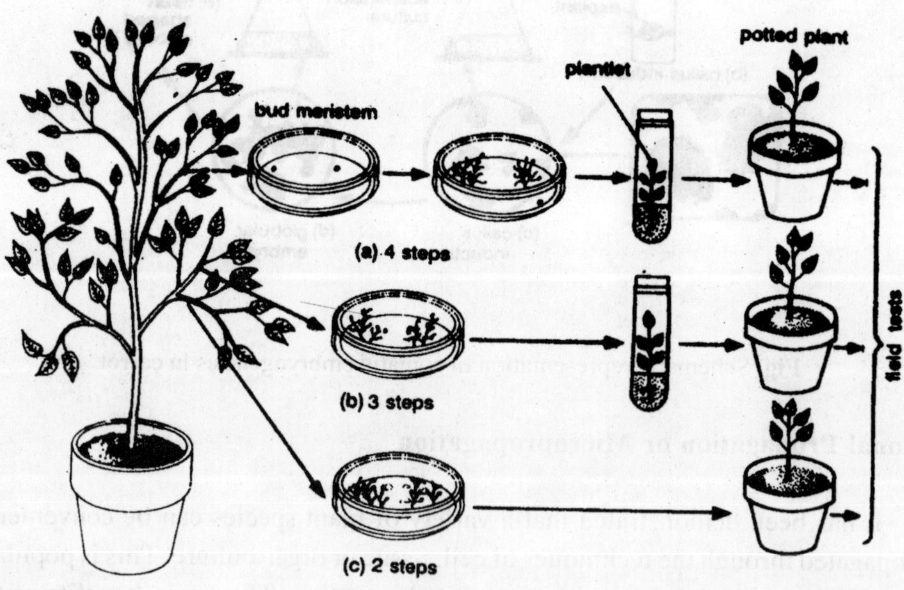


**Fig.** Three different methods (1-step. 3-step and 2-step) of micropropagation of plant material.

**Table**

A list of plant materials, which can be propagated by tissue culture (modified from Swaminathan, M.S., 1991)

| | | | | | |
|---|---|---|---|---|---|
| 1. | Alpittia | 25. | Cardamom | 49. | Gypsophila paniculate |
| 2. | Alstroemeria | 26. | Cattleya | 50. | Hydrangea |
| 3. | Anthurium | 27. | Crocusmia | 51. | Hamamelis |
| 4. | Asparagus | 28. | Cucumis sativus | 52. | Iris |
| 5. | Actinidia | 29. | Capsicum | 53. | Kalanchoe |
| 6. | Aglaonema | 30. | Coffee | 54. | Lily |
| 7. | Agapanthus | 31. | Dianthus | 5.5. | Maranta |
| 8. | Amaryllis | 32. | Dieffenbachia | 56. | Monstera |
| 9. | Amibia | 33. | Dendrobium | 57. | Males |
| 10. | Artemesia | 34. | Eryngium | 58. | Mimosa |
| 11. | Aster | 35. | Epidendrum | 59. | Narcissus |
| 12. | Arillata | 36. | Eucalyptus | 60. | Petunia |
| 13. | Allitan | 37. | Fiats benjamina | 61. | Philodendrons |
| 14. | Alocasia | 38. | Ficus elastic | 62. | Pineapple |
| 15. | Banana | 39. | Fiats h rata | 63. | Papaya |
| 16. | Begonia | 40. | Ficus robusta | 64. | Prumus |
| 17. | Betula | 41. | Ficus mini | 65. | Potato |
| 18. | Babaco | 42. | Ficus compacta | 66. | Polar |
| 19. | Bamboo | 43. | Fiats foliole | 67. | Rose |
| 20. | Calathea | 44. | Fragaria | 68. | Spathiphyllum |
| 21. | Cordyline australis | 45. | Gerbera | 69. | Syngonium |
| 22. | Chrysanthemum | 46. | Ginger | 70. | Turmeric |
| 23. | Cymbidium | 47. | Grape vine | 71. | Tomato |
| 24. | --------------- | 48. | -------------- | 72. | Vanilla |

## Bioreactors for micropropagation, and for bioactive compounds

In order to reduce the cost of plant tissue culture either for micropropagation or for production of bioactive compounds (see 'production of secondary metabolites' in the last chapter), automatic plants having bioreactors and bioprocessors are being developed. The production rate can be enhanced tens of thousands times, so that now elite plants are being either multiplied or used for the production of bioactive compounds by plant biotech companies at the commercial scale. This has become necessary, particularly in the developed countries, where the major part of the operational cost is due to labour. In India also at BARC, airlift bioreactors of 20 litre capacity have been fabricated and used for the production of several bioactive compounds.

In bioreactors, instead of a semi-solid agar medium, liquid culture media are used, which are prepared and replaced regularly through automatic devices in many cases. The growth conditions are also maintained automatically. For micropropagation, the organogenic or embryogenic tissue is multiplied in bioreactors and transferred to a bioprocessor where propagules of different sizes are separated from each other. These propagules then grow and develop into plantlets in separate culture vessels to be transplanted, stored or shipped.

## Synthetic seeds

In the conventional plant tissue culture for clonal propagation, storage and transportation of propagules for transplantation is a major problem. To overcome this problem, in recent years, the concept of synthetic or artificial seeds has become popular, where somatic embryos are encapsulated in a suitable matrix (e.g. sodium alginate), along with substances like mycorrhizae, insecticides, fungicides and herbicides. In India, this technique of synthetic seeds is being standardized and practiced for sandalwood and mulberry at BARC (Bombay) under the leadership of Dr. P.S. Rao.

Synthetic seeds have many advantages including the following: (i) they can be stored upto a year without loss of viability; (ii) they are easy to handle and useful as units of delivery; (iii) they can be directly sown in the soil like natural seeds and do not need hardening in greenhouse. The only limitation of synthetic seeds is the high cost of their production. However, this may go down in future, so that these synthetic seeds will become popular at the commercial scale in due course of time.

## Production of virus free plants

In certain crops, and particularly in ornamentals, it has become possible to produce virus free plants through tissue culture at the commercial level. This can be done by regenerating plants form cultured tissues derived from (i) virus free plants, (ii) meristems which arc generally free of infection, (iii) meristems treated with heat shocks (34-36°C) to inactivate the virus (Fig.), (iv) callus, which is usually virus free like meristems. Further, in callus or suspension culture, virus infection can be completely eliminated through repeated subculturing. During production of virus free plants, highly sensitive tests like ELISA (enzyme linked immunosorbent assay) can be employed to monitor virus free plants (for details, see next section).

Viruses have been eliminated from a number of economically important plant species (at least in 75 species) including the following: mottle virus has been eliminated from strawberry; potato virus X from potato; mosaic virus from cassava; and cauliflower mosaic virus from cauliflower. In all these cases, significant increase in yield has been obtained through production of virus free plants. These virus free plants, however, are not disease resistant, so that one will have to maintain stock plants to multiply virus free plants, whenever, required.

**ELISA for detection of viruses, fungi, bacteria, MLOs, mycotoxins and hormones.**

ELISA is based on the ability of low molecular weight antibodies to couple with enzymes to produce enzymatically active immunological conjugates. This allows the detection of immune reaction with histochemicalstaining techniques, because the antibody component is involved in immune reaction and the conjugated enzyme can be used for staining reaction utilizing appropriate substrate. Using this principle, ELISA was initially developed in 1971 independently by two groups (Engvall and Perlmann; Weeman and Schuurs). Later, Vollar et al. (1976) and Clark and Adams (1977) utilized ELISA for detection of virus infection. (In India, Usha M. Joshi of Mumbai also utilized it for detection of viruses). Several modification of original ELISA are mow known and are widely used for detection of viruses, fungi, bacteria, MLOs (mycoplasma like organisms), etc.

At least three different ELISA techniques are available for detection of plant viruses, which are diagrammatically represented in Figure. In all three techniques, a polystyrene or polyvinylchloride microtitre plate is used, which has wells to provide for a solid phase for the immune reaction. (i) DAS-ELISA. In the double antibody sandwich (DAS) ELISA, the wells are coated with antibody (or immunoglobulins = 1g) prepared from antisera. The test sample is then added to allow trapping of virus antigen by the coated antibody in the well. This is followed by the addition of enzyme labelled antivirus 1g, which will attach to the trapped antigen (Fig.). A substrate for the enzyme is now added for producing colour reaction. Peroxidase (OD 450) or alkaline phosphatase (OD 405) is used, which allow quantitative measurement through a spectrophotometric device (ELISA reader with automated, photometer is often used; Fig.). (ii) DAC-ELISA. In this direct antigen coating (DAC) technique, plant extracts prepared in a carbonate buffer are applied directly to the wells. In the second step, diluted unfractionated antiserum is added. 1g attatched to virus antigen is detected by the addition of enzyme-Ig conjugates (Fig.). This is the simplest procedure and can be completed within three hours. With 96 wells in a plate, as many as 16,000 seeds could be tested in a day. This procedure, however, is not suitable for quantitative estimations due to the use of crude extract. (iii) PAC-ELISA. This protein-A coating (PAC) technique involves four steps. The wells are coated with protein-A (lng ml-1), high dilution of unprocessed antisera were added, test samples were added and finally antisera (cross absorbed with healthy plant components to minimize non- specific reaction) were added (Fig.). This is followed by addition of the enzyme substrate for colour reaction.

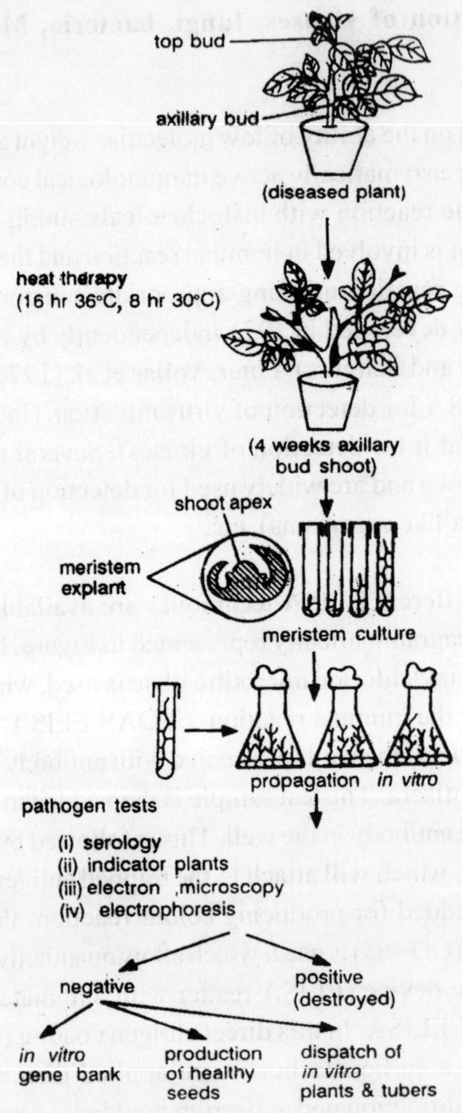


Fig. Different steps involved in elimination of pathogens in cultured plants: heat therapy is done after removing the terminal bud to allow the axillary bud grow during treatment the regenerated plants are tested for absence of the pathogen.

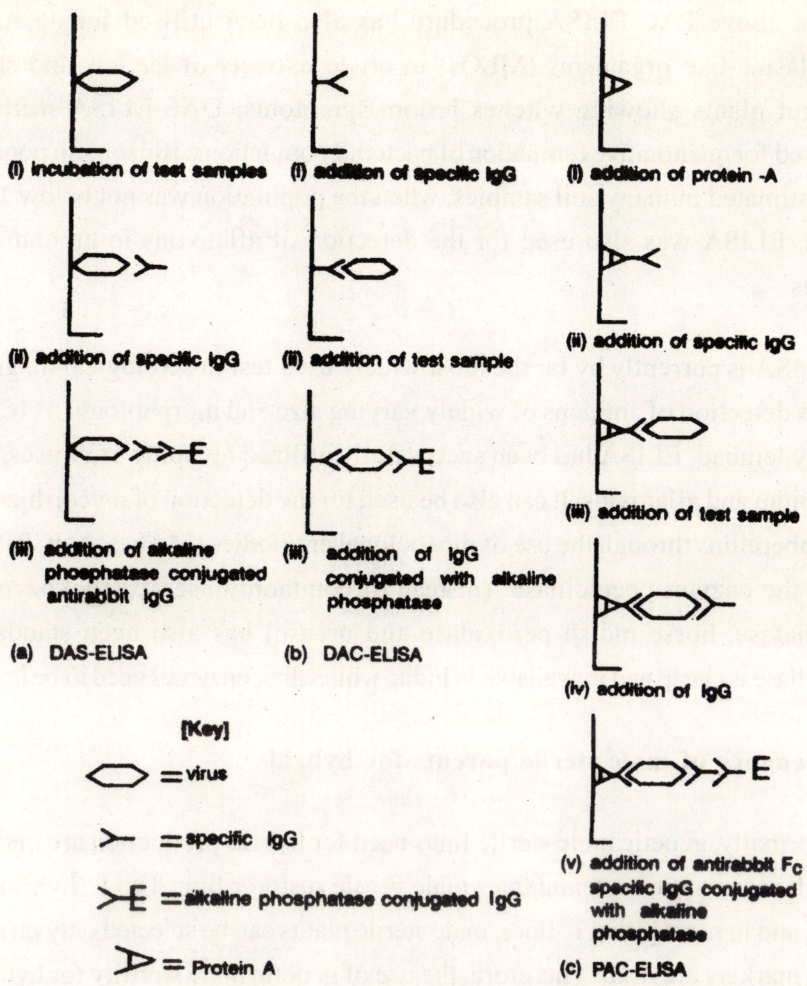


**Fig.** Different steps involved in ELISA test for detection of virus infection.

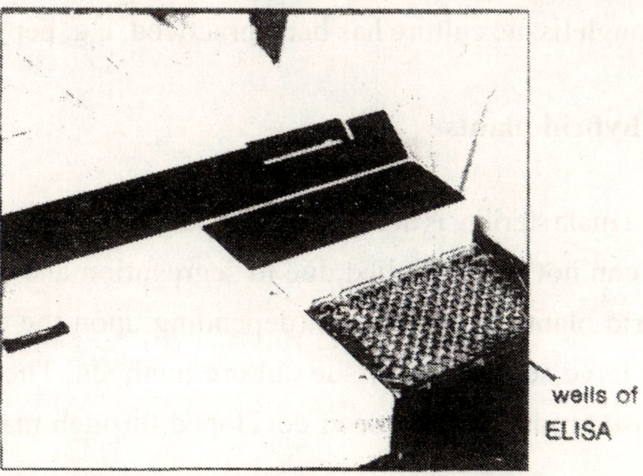


**Fig.** An ELISA reader for reading the results of an ELISA test.

The above PAC-ELISA procedure has also been utilised for detection of mycoplasma-like organisms (MLOs) in crude extracts of leaflets and stems of groudnut plants showing witches broom symptoms. DAS-ELISA method has been used for quantitative estimation of bacterial populations. Rhizobium populations were estimated in many soil samples, when the population was not below $10^3$ cells $g^{-1}$ soil. ELISA was also used for the detection of aflatoxins in groundnut seed samples.

ELISA is currently by far the most widely used test in serological diagnosis. It permits detection of antigens of widely varying size and morphology. At ICRISAT, near Hyderabad, ELISA has been successfully utilized for study of viruses, MLOs, Rhizobium and aflatoxins. It can also be used for the detection of mycorrhizal fungi, and gibberellins through the use of monoclonal antibodies (MAbs). At ICRISAT the use of the enzyme 'penicillase' (instead of commonly used costly enzymes like phosphatase, horse radish peroxidase and urease) has also been standardized. (penicillase is cheap and is available in India, while other enzymes need to be imported).

**Maintenance of male sterile parents for hybrids**

Normally, genetic male sterile lines used for hybrid production are maintained through crosses with a maintainer male fertile restorer line. The $F^1$ hybrid will be fertile, and in segregating $F^2$ lines, male sterile plants can be selected only on maturity unless markers are used. Therefore, the use of genetic male sterility for hybrid seed production is problematic. Therefore, in some cases, maintenance of genetic male sterile plants through tissue culture has been practiced, e.g. pepper.

**Propagation of hybrid plants**

In crops where male sterility is not available, production of hybrid seed is expensive and hybrid seed can not be multiplied due to segregation and loss of vigour in F, generation. Hybrid plants in such cases (depending upon the economics) can be propagated on a large scale using tissue culture methods. These hybrids may be cheaper than those produced fiom seed developed through manual emasculation and pollination.

**Somaclonal Variation**

For clonal propagation or micropropagation as discussed above, shoot tips or axillary buds are used for direct propagation on culture medium without intervening callus phase. It has been shown, in recent years that regeneration from callus, leaf explants or plant protoplasts leads to the generation of considerable variation, described as 'somaclonal variation'. This variation includes aneuploids, sterile plants and morphological variants, sometimes involving traits of economic importance in case of crop plants. This variation in cultures was ignored for long time, but during the last decade it received increased attention and excitement, in view of its potential in crop improvement programmes. The usefulness of this variability in crop improvement programmes was first demonstrated through the recovery of disease resistant plants in potato (resistance against late blight and early blight) and sugarcane (resistance against eye-spot disease, Fiji disease, and downy mildew). Not only phenotypic variability involving both nuclear and organel- DNA is observed in plants regenerated from callus, but variability even in chromosome number is observed in longtime cell cultures. This is exhibited in callus and in plants regenerated from the callus.

It may be necessary to remember that the above variation may be transient (epigenetic) or genetic ; only the latter is transmitted to the next generation and is thus important for crop improvement. Therefore, it may be necessary to study the transmission of variation to sexual progeny ($R_1$) to facilitate the estimation of its utility for improvement of a sexually propagated crop. In several crops $R_0$, $R_1$ and $R_2$ progenies were analysed for genetic analyses and 3:1 segregation leading to the isolation of true breeding variants was observed.

Although the details of the genetic basis of somaclonal variation in most crops are still unknown, variation in structure and number of chromosomes has been suggested to be one possible basis. Polyploidy, aneuploidy, translocations, inversions and deletions have been reported in several cases. Meiotic crossing over involving symmetric and asymmetric recombination could also be responsible for a part of the variation observed in the regenerated plants.

Somaclonal variation has actually been used in plant breeding programmes following the strategy outlined in Figure. A number of plant species, where useful somaclonal variation has been reported are listed in Table.

**Table**

A list of crop species, where desirable and heritable somaclonal variation has been reported.

| *Species* | *Characters, which were modified* |
|---|---|
| **A.Monocolyledorts** | |
| 1. Allittnt sativa | Bulb size and shape; clove no.; aerial bulbil |
| 2. Arena sativa | Plant ht.; heading date; awns |
| 3. Hordeum spp. | Plant ht.; tittering |
| 4. Lolium hpbrads | Leaf size; dower, vigour; survival |
| 5. Oryza sauna | Plant ht.; heading date; seed fertility: grain no. & wt. |
| 6. Sacchantnt officinanum | Diseases (eye spot, Fiji virus, downy mildew. leaf scald) |
| 7. Triticum aestivum | Plant & ear morphology; awns: gliadins; amylase; grain wt., yield |
| 8. Zea maps | T toxin resistance; male fertility; mtDNA |
| **B.Dicotyledons** | |
| 9. Lactuca sativa | Leaf wt., length, width, flatness and colour; |
| 10. Lycopersiccnr esculentum | Leaf morphology; branching habit; fruit colour; pedicle; male fertility; growth |
| 11. Medicago sativa | Muitifoliate leaves; elongated petioles; growth; branch no.; plant ht.; dry master yield |
| 12. Solanum tuberosum | Tuber shape; maturity date; plant morphology; resistance for early & late blight; photoperiod; leaf colour; vigour; height.; skin colour. |

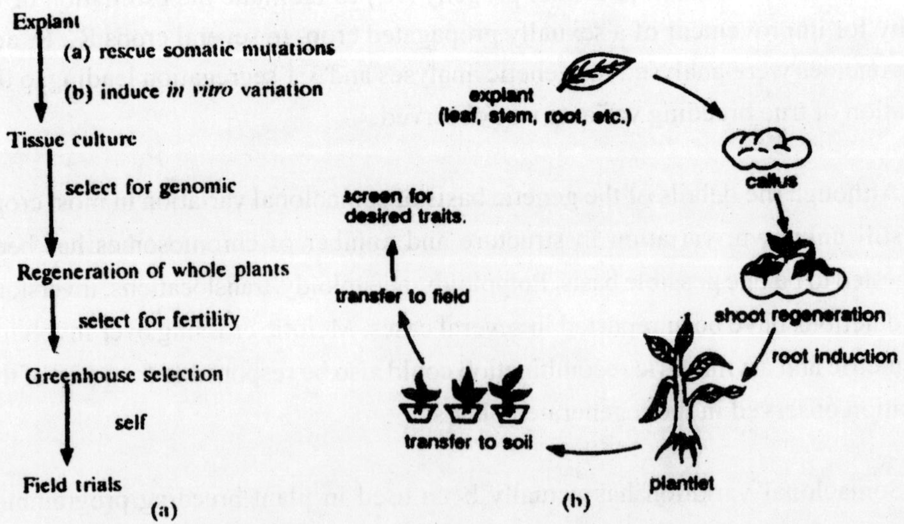


**Fig.** Different steps involved in generation and utilization of somatic variation (a) flow chart, (b) diagrammatic representation.

For the selection of useful generic somaclonal variation, two approaches have been followed : (i) Selection is exercised in cells cultured for different periods and screened for the derived traits (e.g. for resistance to specific herbicides, fungal toxins, pollutants, extremes of temperatures and salinity), and from these selected cells or cultures, plants are regenerated. (ii) In the second approach, selection is exercised at the phenotypic level in regenerated plants. The former approach of selection at the cellular level has the advantage of screening millions of cells with a relatively little effort and resources. But this method can not be used for traits like flower characters or grain yield. Further, the phenotype expressed by selected cells may not be expressed by regenerated plants. Therefore, the characters for which selection is exercised at single cell level need to be checked in the regenerated plants.

## Overcoming Crossing Barriers using Culture Techniques

In a crop improvement programme, sometimes it may be necessary to make crosses between two genera/species/varieties, which can not be conveniently hybridized using conventional methods due to a variety of crossing barriers. These crossing barriers may be pre-fertilization or post-fertilization, and the two will have to be dealt with differently.

### Pre-fertilization crossing barriers

The pre-fertilization barriers include situations or events before the fertilization and therefore, may be due to (i) differences in flowering times of the two parents, (ii) lack of stigma receptivity or pollen viability, (iii) failure of pollen tube to reach the ovule due to slow growth, (iv) cross- incompatibility due to any other unknown reason.

The above crossing barriers can be overcome by collecting pollen at the appropriate stage of development and using them for in vitro pollination. The pollination may be carried out on stigma, placenta or ovules of an excised ovary cultured on an artificial medium (under aseptic conditions). The choice between the organs used for pollination is made through a detailed study of crossing barrier. For instance, if lack of stigma receptivity or slow pollen tube growth are the crossing barriers, pollination can not be carried out on stigma, but instead placenta or ovules will have to be used for pollination. The bisexual flowers need to be emasculated and bagged before ovaries are excised for in vitro operations. After fertilization, the ovules or ovaries are transferred to a fresh medium for seed development.

In contrast to the above situation in plants (where ovule or ovary is used for in vitro fertilization), the in vitro fertilization in humans and other animals always involve fusion of an isolated egg with a sperm in test tube (consult Chapter). However, till recently, plant gametes could not be isolated and fused in a test tube. Only in July, 1993 Erhard Kranz and Horst Lorz from Hamburg (Germany ) reported in corn, the successful fusion of an egg cell and a sperm to form a zygote in a test tube. The fusion was facilitated by a short pulse of electricity and was described shotgun wedding. Alternative non-electrical methods for fusion of plant gametes are also being developed. A high quality culture medium containing early embryonic cells was also found essential for growth and differentiation leading to the production of fertile adult corn plants in a test tube. It is believed that this technique will provide the genetic engineers with a better tool for transfer of isolated, synthetic or manipulated genes into crop plants.

The technique of in vitro pollination and fertilization is also useful in overcoming self incompatibility, where selfing is desired but can not be achieved by conventional method of bagging bisexual flowers. As we know, self incompatibility does not allow fusion of male and female gamet from the same plant, even though the gametes are fertile. It is attributed to barriers at the level of pollen germination and/or pollen tube growth, so that fertilization can be affected, if in vitro pollination is carried out directly on the ovule or on the ovary after removing the stigma and sty.

**Post-fertilization crossing barriers : embryo rescue**

The term 'embryo rescue' is used for embryo culture, ovule culture at ovary culture, since in each case the objective is to rescue the embryo. During distant hybridization, often the embryo aborts at an early stage of development, so that no mature seed can be obtained. The problem can be overcome, if young hybrid embryos are excised and cultured on a synthetic medium. These cultured embryos can directly develop in young seedlings. When embryos can not be easily excised, whole ovules can be cultured and similarly when ovule is also too small to be removed whole ovaries may be cultured. These techniques of embryo culture, ovule culture and ovary culture are briefly described in this section.

**Embryo culture**—For embryo culture, embryos are excised from it mature seeds, usually under a 'hood' (laminar flow cabinet), which provides a clean aseptic and sterile area. Sometimes, the immature seeds are surface sterilized and soaked in water for a few hours, before the embryos are excised. The excised embryos are directly transferred to a culture dish culture tube containing synthetic nutrient medium.

The entire operation is carried out in the 'laminar flow cabinet' and the culture plates or culture tubes with excised embryos are transferred to a culture room maintained at a suitable temperature, photoperiod a humidity. The frequency of excised embryos that give rise to seedlings generally varies greatly and the medium may even have to be modified difficult cases. Most extensive use of this embryo culture technique has been made for making interspecific and intergeneric crosses within tribe Triticeae of the grass family (Poaceae). The hybrids raised through embryo culture have been utilized for (i) phylogenetic studies and genome analysis, (ii) transfer of useful agronomic traits from wild generà and species to the cultivated crops and (iii) to raise synthetic crops like triticale by producing amphiploids from the hybrids. It will be shown in Chapter, that the distant hybrids which can not be obtained even through embryo culture, can be obtained through protoplast fusion.

Embryo culture has also been used for haploid production through distant hybridization followed by elimination of the chromosomes of one parent in the hybrid embryos cultured as above. A popular example includes hybridization of barley and wheat with Hordeum bulbosum leading to the production of haploid barley and haploid wheat respectively. Haploid wheat planes have also been successfully obtained through culture of hybrid embryos from wheat x maize crosses.

**Ovate culture**—For a variety of difficult interspecific/intergeneric crosses involving members of the families Malvaceae, Fabaceae, Cruciferae, Solanaceae, etc., ovules after fertilization have been successfully cultured to obtain mature embryos/seeds. Depending upon when the embryo aborts, the ovules have to be excised any time from soon after fertilization to almost developed fruits, which may sometimes be lost due to premature abscission. However, ovule culture is mainly tried only in those cases, where embryo aborts very early, and embryo culture is not possible due to difficulty of its excision at a very early stage. In some cases, the medium may need to be supplemented with some fruit/vegetable juice to accelerate initial growth.

**Ovary culture**—In vitro rooting of the pedicle and a limited growth of ovaries, on a culture medium, were obtained rather early (La Rue, 1942), although technique for growing excised ovaries on culture medium was developed later in several plant materials, including tobacco, tomato, beans, etc. Subsequently, mature fruits containing seeds could be successfully obtained from ovaries excised from pollinated flowers (e.g. Cucumis, tomato). In our own country, at Delhi University (under the leadership of late Prof. P. Maheshwari), healthy fruits from excised ovaries could be successfully obtained in several plant materials including Iberis amara (an ornamental).

Ovary culture is often used either for in vitro pollination and fertilization (as discussed earlier) or for embryo rescue, when the embryo culture and ovule culture either fail, or are not feasible due to very small ovules. Interspecific hybrids, using ovary culture, have successfully been obtained in several genera including Brassica (B. campestris x B. olerncen). For interspecific or intergeneric crosses, ovaries are excised at the zygote stage or at the two celled proembryo stage and normal development is completed in vitro. The development of fruit may be promoted by application of chemicals like IAA, or by addition of coconut milk to the medium. Ovary culture, prior to fertilization, may also lead to the development of parthenogenetic haploids. This has been achieved successfully in materials, like wheat, barley and other crops.

Ovary culture has also been used for a study of the understanding of physiology of fruit development. For this purpose, specific chemical may be added and their effect observed by comparing the development with that in control. This is ordinarily not possible, when the treatment is given to the entire plant, because the chemical may undergo alterations, exercising a different effect on ovary growth and development.

## Other Uses of Tissue Culture

### Endosperm culture

Tissue culture methods are also used for culturing endosperm, which is unique firstly, in its function of supplying nutrition to the developing embryo and secondly, in being triploid in its chromosome constitution. Triploid plants are useful for production of seedless fruits (e.g. apple, banana, watermelon, etc.) and for production of trisomics for cytogenetic studies. Generally, these triploids are routinely obtained by crossing colchicine induced tetraploids with diploids followed by rescuing the triploid embryos. However, there may be strong crossability barriers, making it difficult to produce triploids from $4x \times 2x$ crosses, so that endosperm culture may be used as an alternative method for triploid production.

Extensive work on endosperm culture was initially done at Delhi University, where successful culture of endosperm was demonstrated initially in some parasitic

Angiosperms (families Santalaceae and Loranthaceae). The technique of endosperm culture involves the following steps : (i) the immature seeds are dissected under aseptic conditions and endosperms along with embryos are excised; sometimes mature seeds can also be used; (ii) the excised endosperms are cultured on a suitable medium and embryos are removed after initial growth; (iii) the initial callus phase is followed by 'embryogenesis' or 'shoot bud differentiation'; (iv) the shoots and roots may subsequently develop and complete triploid plants can be established for further use.

**Nucellus culture**

Nucellus from pollinated flowers can also be used for micropropagation, particularly in cases where adventive embryos develop from nucellar cells (e.g. Citrus). It can be grown on casein hydrolysate supplemented White's medium, giving a callus. The callus may then give rise to pseudobulbils, differentiating into embryoids, which can eventually develop into seedlings. Nucellus from unfertilized ovules may also develop emhryoids, when grown under suitable culture conditions. These embryoids can be excised and grown on medium supplemented with GA3, and plantlets can be obtained.

The seedlings obtained from the nucellas tissue are of parental type and thus may be important for maintaining purity of horticultural stocks. In Citrus, these may be better than those obtained from cuttings due to following reasons : (i) nucellar seedlings have tap root, while those from cuttings have lateral roots ; (ii) the vigour lost due to repeated propagation by cuttings is restored in nucellar seedlings , (iii) nucellar seedlings are disease free.

**Germplasm storage**

Conservation of plant genetic resources or germplasm by several national (National Bureau of Plant Genetic Resources = NBPGR) and international organizations (International Board of Plant Genetic Resources = IBPGR, now renamed as 'International Plant Genetic Resources Institute = IPGRI) has become a thrust area of biotechnology in recent years. For this purpose, various strategies for storage of germplasm without losing its viability have been developed and are being utilized. Germplasm can be stored in a variety of forms including seeds, buds, protoplasts, cells, tissues etc. Welt organized plant parts like shoot tips and plantlets in culture that can also be used for long term storage.

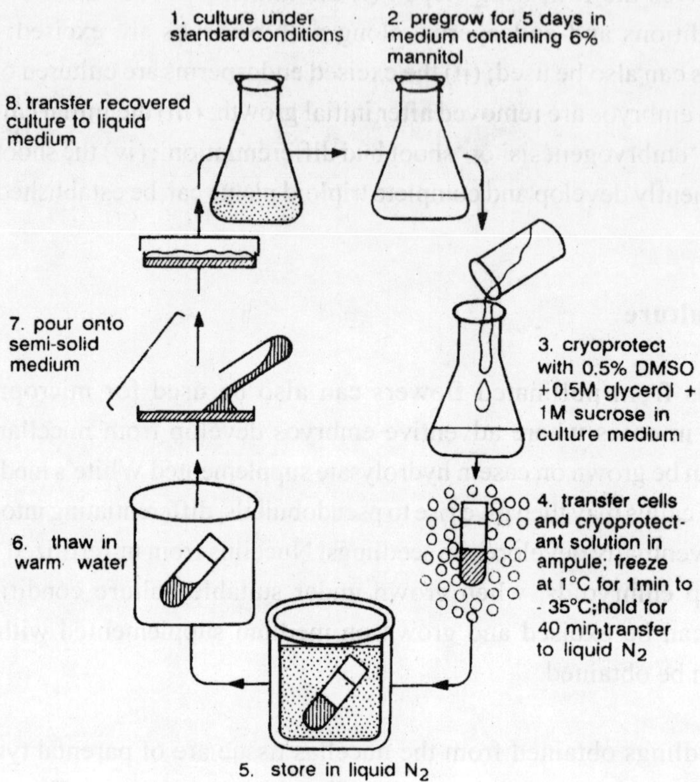


**Fig.** A typical cryopreservation procedure for cell culture; alternative treatments are possible at steps 2-4 and 6-8.

The techniques involving storage of tissues in culture may use the germplasm in the growing stage and their growth may be suspended by any one of the following methods: (i) lowering the temperature, (ii) adding chemical retardants or hormones and (iii) reduction in oxygen concentration. Such methods, where limited growth of cultures is allowed remains effective for about a year, so that periodic renewal of the cultures is required. However, the most popular and effective method for long term storage of cell cultures, involves storage at a very low temperature using liquid nitrogen. This is described as cryopreservation, which allows storage for virtually indefinite periods. Cryopreservation of isolated plant protoplasts has been recommended in many cases, since survival level of 75% has often been realized in some cases (e.g. Alums carom). DMSO (0.5 M or 10%) or some other medium

(DMSO + glycerol + sucrose) is used for step-wise freezing leading to storage in liquid nitrogen ($N_2$). The germplasm can be used whenever required by

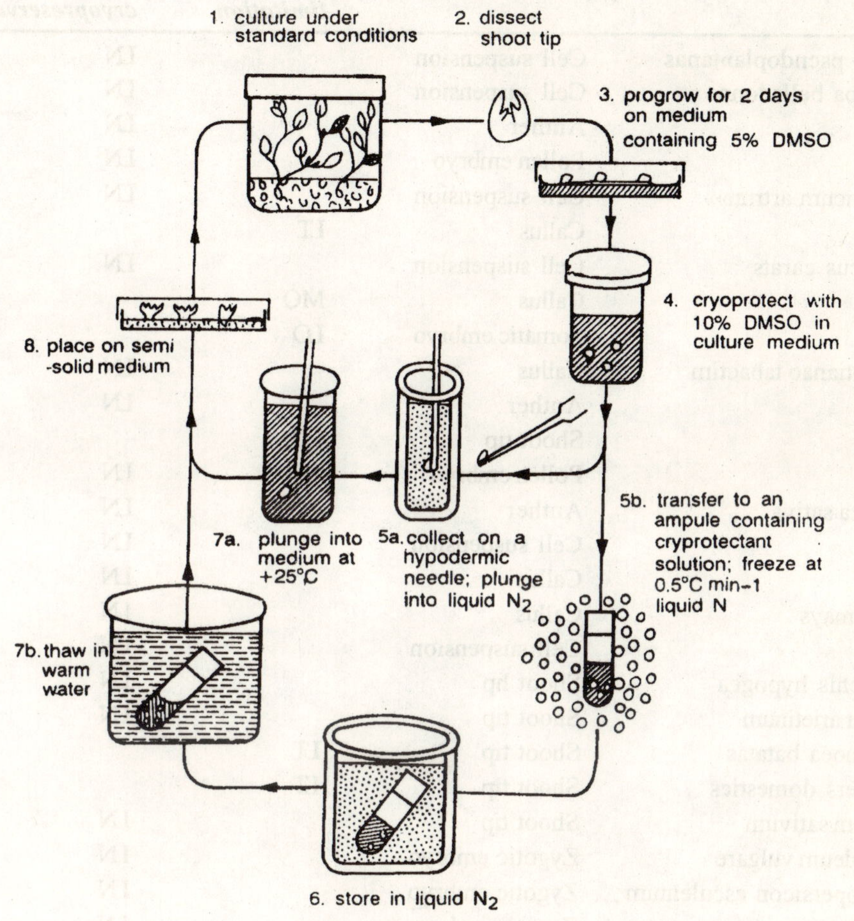


**Fig.** Standard procedure for cryopreservation of shoot tip; variation can be possible at several steps.

rapid thawing (Figs.). Pollen required for hybridization can also be cryopreserved. For cryopreservation, cells in early exponential phase have been found to be the most highly freeze tolerant, while thus in lag phase or stationary phase are susceptible to some freezing injury. This susceptibility to freezing injury actually depends on size of individual cells and also on the size of aggregate, or clumps used for cryopreservation. Callus cultures, cell suspensions, zygotic embryos (from wide crosses) and somatic embryos, all can be used for cryopreservation (Table).

**Table**

Examples of storage of plant tissue culture (condensed from Withers. 1983).

| *Species* | *Callure system* | *Conditions for storage* | |
|---|---|---|---|
| | | *By growth limitation* | *By cryopreservation* |
| Acer pscndopla»tanas | Cell suspension | | LN |
| Atropa belladonna | Cell suspension | | LN |
| | Anther | | LN |
| | Pollen embryo | | LN |
| Capsicum artrnn»» | Cell suspension | | LN |
| | Callus | LT | |
| Dancus carats | Cell suspension | | LN |
| | Callus | MO | |
| | Somatic embryo | LO | |
| Nicotianao tabactim | Callus | | LN |
| | Anther | | LN |
| | Shoot tip | LO | |
| | Pollen embryo | | LN |
| Oryza sativa | Anther | | LN |
| | Cell suspension | | LN |
| | Callus | | LN |
| Zea mays | Callus | | LN |
| | Cell suspension | | LN |
| Arachis hypogea | Shoot hp | | LN |
| Cicer arietinum | Shoot tip | | LN |
| lpomoea batatas | Shoot tip | LT | |
| Maters domestics | Shoot tip | LT | |
| Pisum sativum | Shoot tip | | LN |
| Hordeum vulgare | Zygotic embryo | | LN |
| Lycopersicon esculentum | Zygotic embryo | | LN |
| Triticum aestivum | Zygotic embryo | | LN |

LN, liquid nitroge; LT, law temperature; MO, mineral oil; LO, low oxygen.

## Status of Tissue Culture Technology in India

India is reported to have one of the largest groups of tissue culture scientists in the world. Mostly, the research is directed towards the development of improved plants for agriculture, horticulture and forestry using tissue culture methods. Another use of tissue culture methods, in some laboratories, involves studies in the areas of developmental biology, biochemistry, physiology, genetics wind molecular biology to answer questions of fundamental nature.

**The Department of Biotechnology (DBT),** New Delhi, is playing a special role in promoting research in the area of plant tissue culture, although the support is mainly directed towards biomass production (fuel, fodder, wood) and cardamom production. Several laboratories are being assisted by DBT and other funding agencies to carry out tissue culture research, the details of which are summarised in Table.

**Table**

Status of tissue-culture technology in India—semi-commercial or laboratory scale; other tissue-culture applications. (Front Curr. Sci. 1991)

| ***Laboratory/company*** | ***Plant species*** | ***Remarks*** |
|---|---|---|
| 1. Botany Department, Delhi University, Delhi | Flowering plants, endangered plants Environment (DOE) Bamboo | Sponsored by Department of Biotecnology (DOE) for evaluation teals |
| 2. Biotechnology Centre. IARI, New Delhi | Mustard<br>Rice | Somaclonal variation for yield<br>From haploids undergoing trims also research on genetic engineering |
| 3. Bhabha Atomic Research Centre, Bombay | Oil palm sandalwood<br>Medicinal plants | Oil palm micropropagation sponsored by DBT<br>For producing secondary products |
| 4. Cardamom Reserarch Istitute, Kerala | Cardamom* | For domestic market |
| 5. Central Institute of Medicinal and Aromatic Plants, Lucknow | Medicinal plants; endangered species | Somaclone of lemon grass isolated, sponsored by DOE, |
| 6. Deejay Hatcheries, Bangalore | Banana, cardamom | For domestic market |
| 7. Excel India Ltd. Bombay | Orchids | For domestic market |
| 8. EID Parry & Co. Madras | Sugarcane | — |
| 9. Grasim Forest Research Institute, Harihar | Eucalyptus spp., bamboo | For conducting trials |
| 10. National Botanical Research Institute, Lucknow | Dioscorea, Populus, Citrus sp. | — |

*(Table Contd.)*

| Laboratory/company | Plant species | Remarks |
|---|---|---|
| 11. National Chemical Laboratory, Pune | Cardamom, turmeric, pomegranate, ginger, sugarcane | Micropropagation for production of somaclones |
| | Turmeric | High curcumin |
| | Cotton, papaya | For secual hybrids by embryo |
| | Custard, apple | rescue Haploids |
| | Teak, eucalyptus, bamboo | Pilot-plant facility sponsored by DBT |
| | Salvadora | |
| 12. National Bureau of Economically important Activity at National Facility for Plant Genetic Resources, New Delhi | crops like timber crops. bulbous crops, spices, fruit crops, medicinal and aromatic plants and endangered plant species | Plant Tissue Culture Repository(NEPTCR), Sponsored by DBT |
| 13. North-Eastern Hill University, Shillong | Pitcher plant | — |
| 14. P.P. Pocha and Sons, Pune | Ferns, gerberas | For domestic market |
| 15. Rashtriya Chemicals and Fertilizers, Bombay | Orchids, sugarcane | For domestic market |
| 16. Regional Research Laboratory, Jammu | Medicinal plants | For isolation of secondary metabolites |
| 17. Rubber Research Institute, Kerala | Rubber | For their plantations |
| 18. Southern Petrochemical Industries Corporation, Madras | Coffee | — |
| 19. Sugarcane Research Institute, Coimbatore | Sugarcane | Somaclones released |
| 20. Tata Tea Ltd., Kerala | Eucalyptus globulus, tea | For their plantations |
| 21. Tata Energy Research Institute, New Delhi | Brassica<br>Forest trees | Embryo resuce, transformation Pilot-plant facility DBT |

*Process released through NCL

Besides laboratories, several private companies are also engaged to commercial production of many plant species (Table).

**Table**

Status of tissue-culture technology in India—commercial scale production. (from Curr. Sci, 1991)

| | *Laboratory/company* | *Plant species* | *Remarks* |
|---|---|---|---|
| 1. | Bio Tissue Calture Lab., Hyderabad | Pomegranate*, banana etc. | — |
| 2. | Harrison's Malayalam Ltd., Bangalore (RPG Group) | Vegetable crops (application. including micropropagation) | Biotechnology laboratory for micropropagation and other aspects of tissue culture with foreign collaboration |
| 3. | Hindustan Lever, Bombay, | Cardamom, sugarcane | For domestic market and trials |
| 4. | Indo American Hybrid Seeds, Bangalore | Banana, foliate and flowering plants. cardantom, etc. | For domestic and export market |
| 5. | Nath Seeds. Aurangabad | Ornamentals. etc. | Foreign collaboration for domestic and export market |
| 6. | Padumjee Pulp and Paper Mills Ltd, Pune | — | A commercial laboratory is under consideration |
| 7. | Reliance Co. India. Pune | — | A commercial laboratory is under consideration |
| 8. | A.V. Thomas and Co., Cochin | Banana, foliage and flowering plants, cardamom*, orchids | For domestic and export market; a laboratory to different aspects of biotechnology being set up |
| 9. | Tree Tech Corporation, Haryana | Tree species | Foreign collaboration |
| 10. | Unicorn Biotech, Hyderabad | Cardamom*, ornamentals, banana | With foreign collaboration |

*Process released through NCL

## Micro-ornamentals with foreign collaboration

There is widespread interest in micropropagation in India, with several private companies involved in this venture at a commercial scale. In mid 1980's, with the help of National Chemical Laboratory, Pune, a private company **(A.V. Thomas and Co. = AVT, Cochin)** established its first small scale production laboratory at Manalaroo in Kerala for cloning cordamom plants from its own farm collections. AVT has also established later a modern and well equipped laboratory in Cochin, with an annual turnover capacity of around eight million ($8x10^6$) plants. These pioneering efforts by AVT led to the development of Ernakulam as the first 'biotechnology district' in India. Consequently, several other entrepreneurs entered the field (Table). Another private company based in Bangalore is Indo-American Hybrid Seeds (IAHS), which is involved in the production of hybrid flower and vegetable seeds, with a turnover of 10 million plants per annum. They are also maintaining a tissue-culture bank for preservation of breeding lines of vegetables like tomato, cabbage, onion, chilly, etc. They also have diagnostic kits for screening plants which are free of major viral, fungal and bacterial diseases.

The department of biotechnology also initiated a project for biomass production using tissue culture technology. Fourteen (14) nationally important tree species have been identified and seven R & D projects oil selected forest tree species are in progress. Pilot plant units have been established (by DBT) at NCL, Pune and Tata Energy Research Institute (TERI), New Delhi. These units will serve as an intermediate slop in the transfer of tissue culture technology from the laboratory to the field. DBT has also funded a project to compare the performance of high yielding tissue culture clonal cardamom plantlets in a 100ha area. Under this project, AVT multiplies buds from elite cardamom selections identified by Cardamom Research Institute. A National Facility for Plant Tissue Culture Repository (NFPTCR) has also been established at National Bureau of Plant Genetic Resources (NBPGR), New Delhi. This repository is using tissue culture to conserve accessions of economically important crops like tuber and bulbous crops, spices, fruit crops, medicinal and aromatic plants and endangered plant species.

## Somaclonal and gametoclonal work in India

Somaclonal and gametoclonal variation technology has been used for developing new cultivars (for gametoclonal variation, consult next chapter). New cultivars have been obtained in sugarcane (Sugarcane Research Institute = SRI, Coimbatore), lemon grass (Central Institute of Medicinal and Aromatic Plants = CIMAP, Lucknow), rice (Indian Agricultural research Institute = IARI, New Delhi), Brassica (IARI, TERI, New Delhi), turmeric and custard apple (NCI-, Pune), etc. However, in some of these cases, the genetic basis of the variation introduced, will have to be confirmed using inheritance studies.

## Secondary metabolites research in India

For enhancement of the production of secondary metabolites, systems are being developed at Bhabha Atomic Research Centre (BARC), Bombay, Regional Research Laboratory (RRL), Jammu and at Botany Department, University of Jaipur. These will increase their activity in future.

Besides these, tissue culture facilities are also being created for recombinant DNA technology, fusion, protoplast culture and embryo rescue, etc.

# TWELVE

# Protoplast Culture

The isolation, culture and fusion of protoplasts. A more recent achievement is the manipulation and regeneration of these cultured or fused protoplasts into whole plants. Since in plant cells, the plasma membrane is bound by a rigid cellulose wall (unlike animal cells), it has been relatively difficult to handle plant cells. Only in 1960, it was demonstrated by E.C. Cocking at the University of Nottingham (U.K.), that naked cells called protoplasts can be obtained through enzymatic degradation of cell walls. This led to significant developments in the field of somatic cell genetics in higher plants. Cultured protoplasts can be used not only for somatic cell fusions, but also for taking up foreign DNA, cell organelles, bacteria and virus particles. In view of this, the isolation and culture of protoplasts has become a very important area of research, within the realm of plant biotechnology. The essential ingredients of the technique include isolation of protoplasts, culture of protoplasts, introduction of foreign DNA into protoplasts, raising whole plants from cultured protoplasts and fusion of protoplasts leading to somatic hybridization. These aspects of plant tissue culture will be discussed in this chapter.

## Isolation of Protoplasts

### Mechanical method

In mechanical method, cells are kept in a suitable plasmolyticum (in plasmolysed cells, protoplasts shrink away from cell wall) and cut with a fine knife, so that protoplasts are released from cells cut through the cell wall (Fig.), when the tissue is again deplasmolysed. This method is suitable for isolation of protoplasts from vacuolated cells (e.g. onion bulb or scales, radish roots). However, this method gives poor yield of protoplast and is not suitable for isolating protoplasts from meristematic and less vacuolateil cells. The mechanical method, though, was used as early as 1892, is now only rarely used for isolation of protoplasts.

## Enzymatic Method

The enzymatic method is almost invariably used now for the isolation of protoplasts, since it gives large quantities of protoplasts, where cells are not broken and osmotic shrinkage is minimum. However, sometimes mechanical and enzymatic methods are combined, where cells are first separate mechanically and later used for isolation of protoplasts through enzymatic treatment.

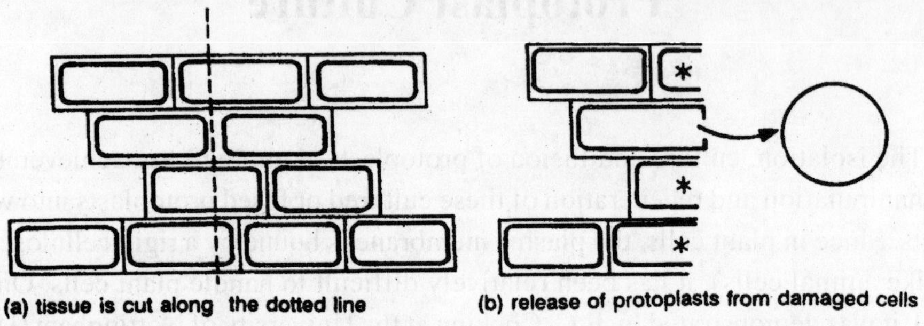


**Fig.** Principle of mechanical isolation; when tissue is cut along the dotted line, with say a razor blade, some protoplasts will be released.

The protoplasts can be isolated from a variety of tissues including leaves, roots, in vitro shoot cultures, callus, cell suspension and pollens. Young cell suspensions are particularly ideal for isolation of protoplasts in large quantities. However, the most commonly used organVF are leaves which can be employed for isolation of protoplasts using the following steps (Fig.) : (i) sterilization of leaves, (ii) peeling off the epidermis, (iii) enzymatic treatment and (iv) isolation and cleaning of protoplasts.

Fully expanded leaves are obtained from about 10 weeks old plant (in tobacco) and are sterilized by first dipping them into 70% ethyl alcohol for about a minute and then treating them with 2% solution of sodium; hypochlorite for 20-30 minutes. The leaves are then rinsed three time with sterile distilled water and subsequent operations are carried out under aseptic conditions under Laminar air flow chamber. The lower epidermi of the sterilized leaves is carefully peeled off and the stripped leaves are cut into small pieces. The peeling operation is easy, if the water supply is limited before excision or if leaves are allowed to become flaccid after sterilization. Mesophyll protoplasts can be obtained from these peeled leaf segments, while those for epidermis are obtained from the peeled epidermis. In cereals, where it is difficult to peel off the epidermis, leaves are cut in long strips and used with enzyme mixture. When the starting material is an in vitro shoot culture, the sterilization step can be omitted and leaves can be cut into pieces under aseptic condition and used directly for enzymatic treatment (Fig.).

From the peeled leaf segments, obtained as above, the protoplasts can be isolated using any one of the two methods : (i) direct (one step) method, in which treatment with macroenzyme (or pectinase) and cellulase is done simultaneously, or (ii) sequential (two step) method, in which cells are first isolated using macroenzyme and cells are then treated with cellulase to isolate protoplasts. In both these methods, peeled leaf segments are placed with their lower surface downwards in a Petri dish containing the enzyme mixture. The enzyme mixture in direct method consists of 0.5%P macroenzyme + 2% cellulase in 13% sorbitol or mannitol at pH 5.4, while

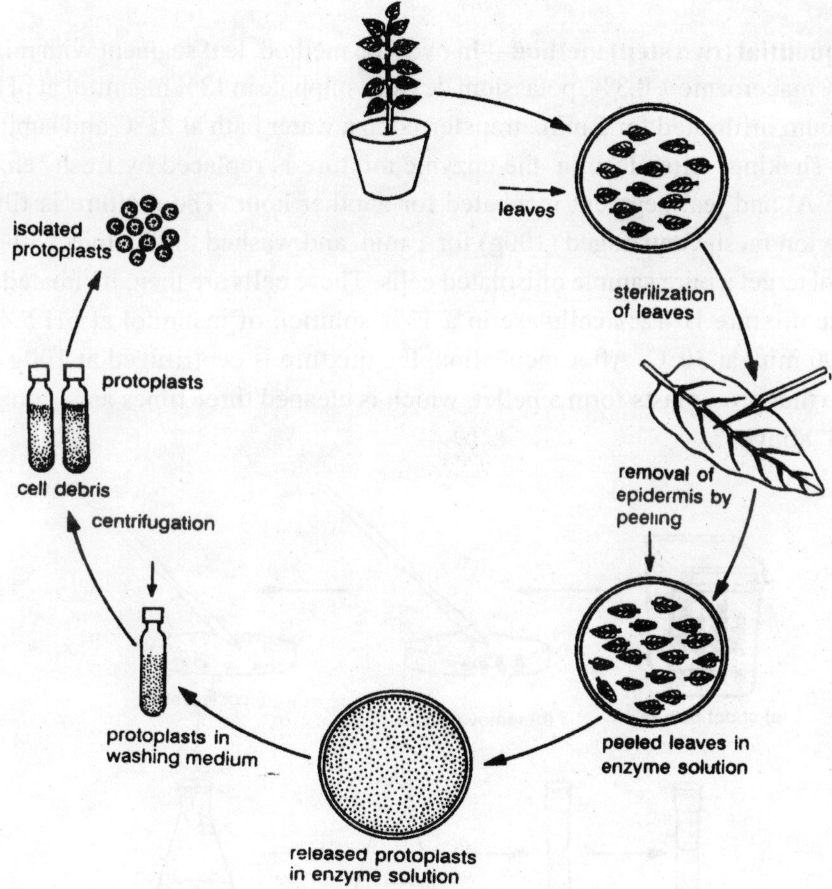


**Fig.** Isolation of mesophyll protoplasts using leaves with epidermis peeled off.

in sequential (or two step) method, two enzyme mixtures ('mixture A' any 'mixture B'; see later) are used one after the other. While the yield r protoplasts is higher in one step method (since both palisade and spongy mesophyll tissues are used in this method, while only palisade is used in two step method), the protoplasts isolated by two step method are used for culture studies.

**Direct (one step) method**—In one step method, the leaf segments are incubated overnight (15-18h) with enzyme mixture at 25°C and released gently to liberate the protoplasts. The mixture is filtered through a fitT wire gauze to remove leaf debris, transferred to 13 x 1000 mm screw capped tubes and centrifuged at 100g for 1 min. The protoplasts form a pellet and supernatant removed. The process is repeated three times and protoplasts washed with 13% sorbitol solution, which is later replace by 20% sucrose solution and centrifuged at a speed of 200g for 1 mir. The cleaned protoplasts, which will now float (debris settles down), can be pipetted out and bulked.

**Sequential (two step) method**—In two step method, leaf segment with mixture A (0.5% macerozme + 0.3%, potassium dextran sulphate in 13% mannitol at pH 5.8) are vacuum infiltrated for 5 min., transferred to a water bath at 25°C and subjected to slow shaking. After 15 min. the enzyme mixture is replaced by fresh 'enzyme mixture A' and leaf segment incubated for another hour. The mixture is filtered using nylon mesh centrifuged (100g) for 1 min. and washed three times with 13% mannitol to get a pure sample of isolated cells. These cells are then, incubated with 'enzyme mixture B' (2% cellulase in a 13% solution of mannitol at pH 5.4) foı above 90 min. at 30°C. After incubation, the mixture is centrifuged at 100g for 1 min., so that protoplasts form a pellet, which is cleaned three times as in 'one step method' above.

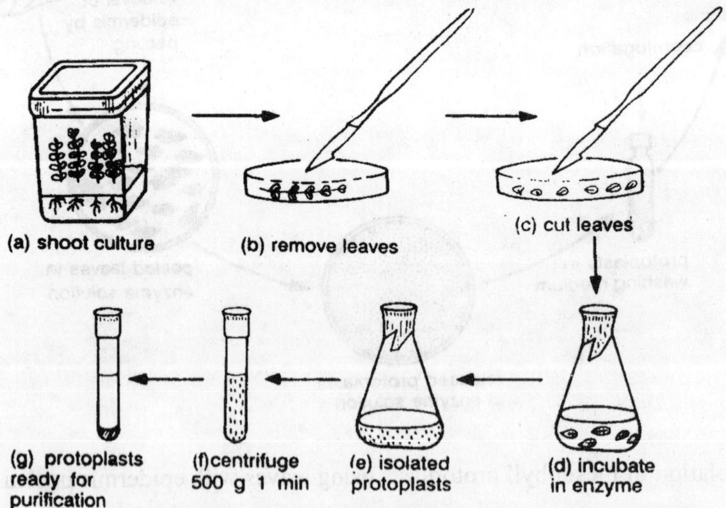


Fig. Diagram of the principle steps in enzymic isolation of protoplasts; leaves are removed and cut into fine steps incubated in enzyme mixture.The enzyme is then removed and the protoptasts are ready for purification.

## Purification of Isolated Protoplasts

The protoplasts isolated as above are present in the media together with a range of cell debris and broken cell organelles. A number of methods are available for purification of protoplasts from this mixture. Only two commonly used methods will be briefly described in this section.

### Sedimentation and washing

In this method, the crude protoplast suspension is centrifuged at low speed (50-100 g/5 min.). The intact protoplasts form a pellet and supernatant containing cell debris can be pipetted off. The pellet is gently resuspended in fresh culture media plus mannitol and rewashed. This process is repeated two or three times to get relatively clean protoplast preparation (Fig.).

### Flotation

Protoplasts being lighter (low density) than other cell debris, gradients may be used, which will allow the protoplasts to float and the cell debris to sediment. A concentrated solution of mannitol, sorbitol and sucrose (0.3-0.6M) can be used as a gradient and crude protoplasts suspension may be centrifuged in this gradient at an appropriate speed. Protoplasts can be pipetted off from the top of the tube after centrifugation. This method causes less loss or damage than the sedimentation and washing method. Babcock bottle shown in Figure is also used by some workers for flotation, since it facilitates removal of protoplasts.

## Viability and Plating Density of Protoplasts

### Testing protoplast viability

Protoplast viability could be tested in the beginning of protoplast research by a variety of methods including the following : (i) presence of cytoplasmic streaming, (ii) exclusion of Evans Blue dye; (iii) change in protoplast size due to change in the level of osmoticum (osmoticum is a solution causing change in osmotic pressure); (iv) presence of photosynthetic and respiratory activity. However, protoplast viability is now most frequently tested using several dyes including fluorescein diacetate (FDA), phenosafranin and calcofluor white (CFW). Following are some details: (i) As FDA accumulates within the plasma membrane, viable protoplasts fluoresce green/white. It should be examined within 5-15 minutes after the FDA treatment, after which FDA dissociates. (ii) Phenosafranin (0.1%) detects dead protoplasts, which turn red in its presence, the viable protoplasts remaining unstained even after 2 hours in the stain solution. (iii) CFW (0.1% v/v) detects onset of cell wall regeneration around plasma membrane of viable protoplasts in the form of a ring of fluorescence.

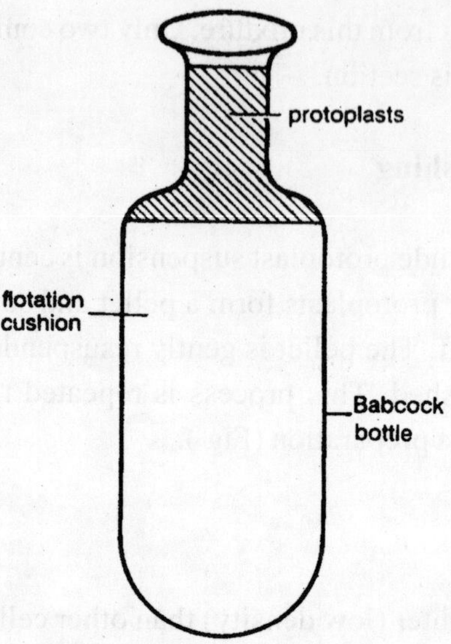


**Fig.** Flotation of the protoplasts to a Babcock bottle greatly eases the removal of purified protoplasts from the sUcrcOsc cushion,

## Minimum plating density (mpd) and culturing individual protoplasts

A minimum plating density of protoplasts is required for growth to begun because protoplasts enrich the culture medium due to cell leakage. The mpd for tobacco is $5 \times 10^3$ to $1 \times 10^5$ protoplasts/$cm^3$ (protoplasts/$cm^3$ can be estimated using a **haemocytometer**). However, due to requirements of genetic engineering experiments, there is a need to culture individual or fewer protoplasts. Following approaches are used for this purpose :

**(i) Small volume cultures** (less than 1$cm^3$ ) may be used to reduce the number of protoplasts, although mpd is maintained. Hanging droplets (2$\mu$1) and other similar techniques (e.g. multidrop array) have been used to reduce the volume. (ii) Conditioned media (in which plant cells were already grown), after filtering, can be used for growing isolated protoplasts at lower densities. (iii) Feeder layers of protoplasts can be prepared by plating solid media with protoplasts followed by irradiation (which should inactivate but not kill these protoplasts). Protoplasts at lower density (5-50 protoplasts/$cm^3$) can now be plated on this feeder layer.

## Protoplast Culture and Regeneration of Plants

The culture methods for isolated protoplasts are similar to those for single cells earlier discussed in Chapter (Fig.). The culture method selected also depends upon the objective of the experiment. For instance protoplasts can be cultured following 'Bergmann's cell plating technique'. Both semi-solid medium and liquid medium can be used, although the liquid medium is preferred. Protoplasts can be suspended in a liquid medium in Erlenmeyer flasks without shaking and can be cultured in small quantities in 'hanging drops' or in 'microchambers'. 'Multidrop array (MDA) techniques may also be used (consult Plant Protoplasts' by T. Bengochea and J. H. Dodds, 1986 for details). Within 2-4 days, protoplasts start developing cell walls, which can be detected by staining with 0.1% calcofluor white (CFW) fluorescent stain.

While the presence of a proper wall is essential for a regular division, it is not always a prerequisite for nuclear division; not all protoplasts with cell walls formed, embark upon division. The protoplasts, which are capable of dividing, undergo first division within 2-7 days and form multicellular colonies after 2- 3 weeks. After another two weeks, these colonies can be treated as standard tissue cultures. From these colonies or tissues in culture, plants can be regenerated, although success in this effort has been achieved only in a few species.

The first step in regeneration of plants involves the transfer of callus to a medium capable of initiating differentiation and it behaves just like the callus derived from cells. For instance, calli differentiate into shoots within 3-4 weeks after transfer to a medium containing IAA (4mg/1) and cytokinin (2mg/1). In this medium, roots may also be formed, failing which they can be induced by transfer of shoots to a basic White's medium. Subsequently, the plantlets may be transferred to pots. Different steps from isolated protoplasts to regeneration of plants are depicted in Figure. However, the variation in regenerated plants derived from protoplasts is much greater than that observed in plants regenerated from a meristem, a stem segment or a leaf disc (Fig.).

## Isolation of Subprotoplasts

Subprotoplasts do not contain the entire contents of plant cells and include the following : (i) cytoplasts lack a nucleus and each contains entire cytoplasm of a cell or a part thereof; these are often used for hybrid production through somatic hybridization (consult), and may sometimes also lead to the transfer of a part of the genome; (ii) karyoplasts or miniprotoplasts contain a nucleus surrounded by some cytoplasm and the original outer plasma membrane; (iii) microprotoplasts contain only a few of all chromosomes, and a fraction of the cytoplasm; (iv) microplasts contain only a fraction of cytoplasm and outer membrane.

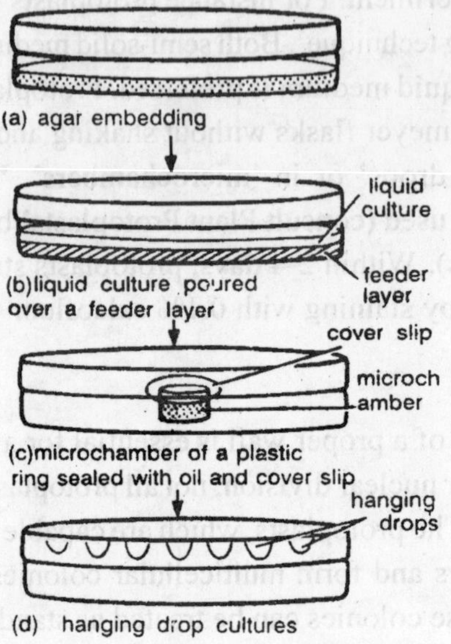


Fig. Different culture methods: (a) agar embedding, (b) liquid culture poured over a feeder layer of medium or dedium and inactivated protoplasts of another species, (c) microchamber constructed of a small plastic ring with oil and a cover slip, (d) hanging drop cultures.

**Cytoplasts** can be prepared by one of the following methods: (i) X-ray treatment or laser beam microsurgery of freshly isolated protoplasts; the nucleus is inactivated by such treatment. (ii) Cytoplasts occur spontaneously in the pericarps of ripening fruits as shown in tomato. These cytoplasts can be isolated by filtration of the juice through sieves followed by sedimentation and centrifugation. (iii) Enucleated protoplasts may also be obtained from cultured protoplasts due to phenomenon of 'budding' ( shaking of protoplasts may lead to separation of budded subprotoplasts). These may sometimes represent microprotoplasts. (iv) Cytoplasts and miniprotoplasts are also obtained through plasmolysis of elongated cells (Fig.).

## Protoplast Fusion and Somatic Hybridization

Protoplast fusion or somatic hybridization is one of the most important uses of protoplast culture. This is particularly significant for hybridization between species or genera, which can not be made to cross by conventional method of sexual

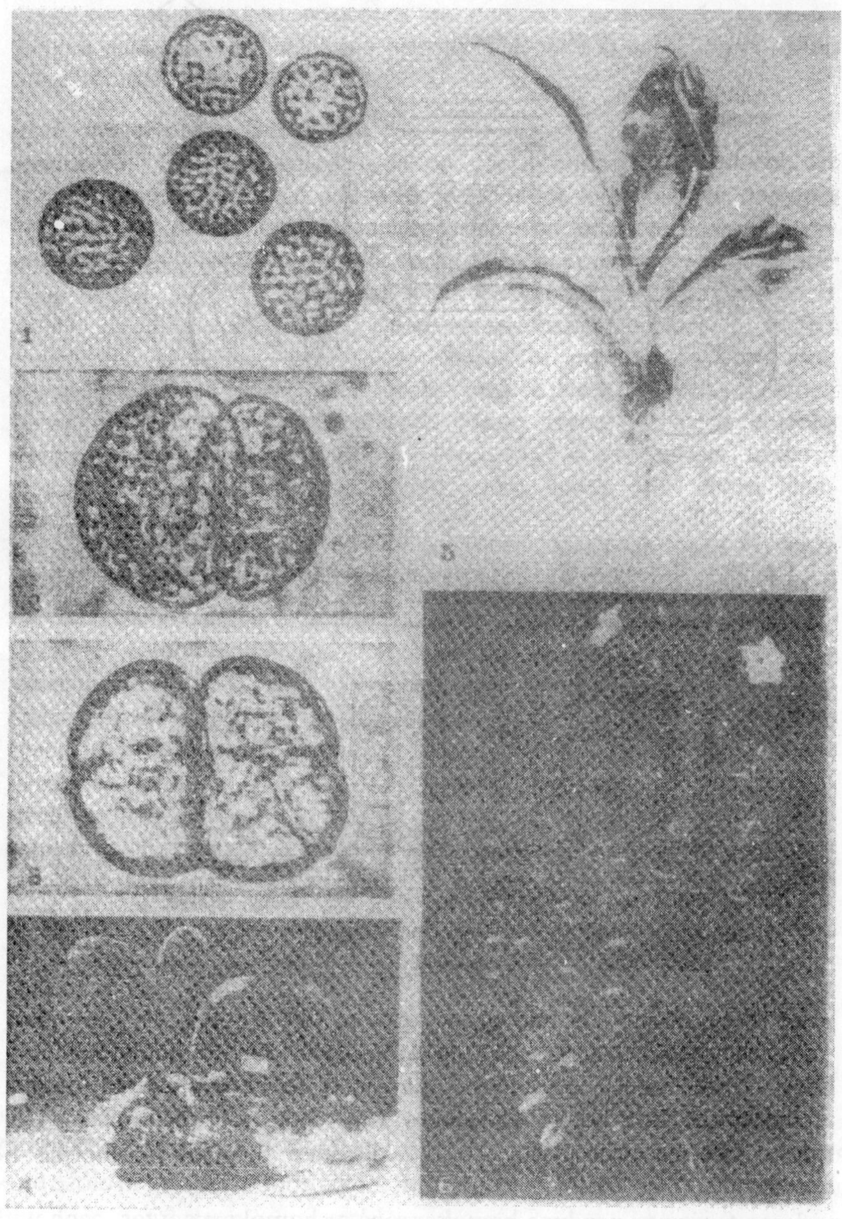

**Fig.** Regeneration of Petunia parodii plants from isolated mesophyll protoplasts.

hybridization. It was successfully achieved first in animals and only later in plants, its significance has been realized fully in plants because the hybrid cells can be induced to regenerate into whole plants.

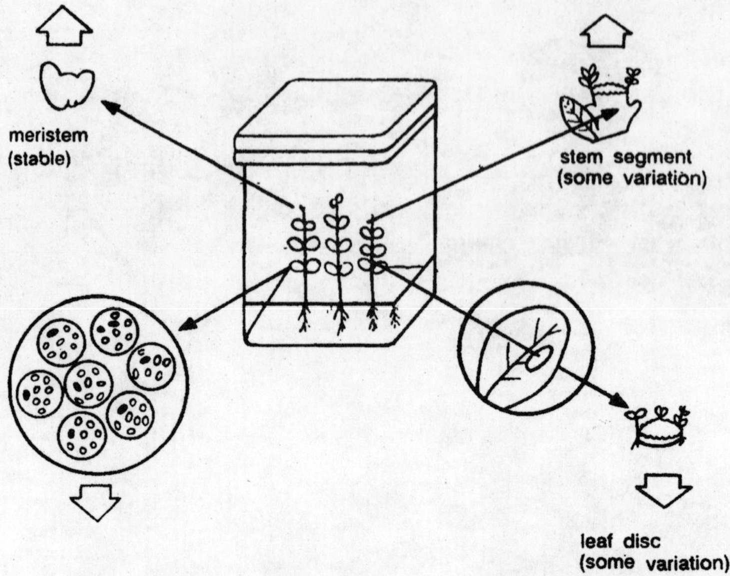


**Fig.** Diagrammatic representation of different regeneration methods with potato; meristem culture, stem explants, leaf discs and protoplasts can be regenrated with different degrees of genetic fidelity.

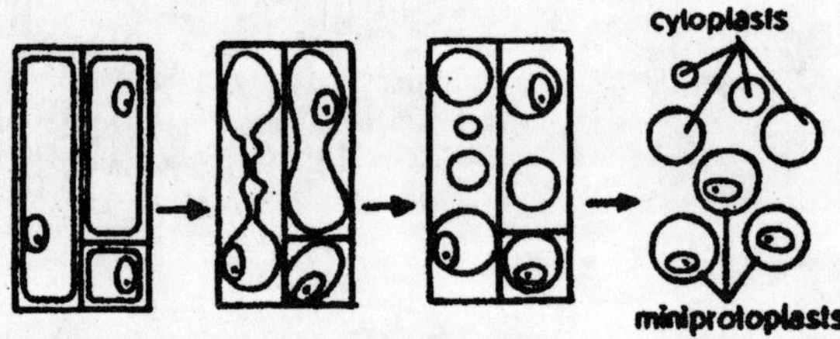


**Fig.** When plasmolysed cells are enzyme treated they will often geld a mixture of protoplasts, miniprotuplasts and cytoplasts (cy).

## Techniques of protoplast fusion

Spontaneous fusion—During isolation of protoplasts for culture, when enzymatic degradation of cell walls is affected, some of the protoplasts, lying in close proximity, may undergo fusion to produce homokaryons or homokaryocytes, each with 2-40 nuclei. The occurrence of multinucleate fusion bodies is more frequent, when protoplasts are prepared from actively dividing cells. This spontaneous fusion, however, is strictly intraspecific. However, spontaneous fusion of protoplasts can also be induced by bringing protoplasts into intimate contact through micromanipulators or micropipettes. There seems to be a correlation between the size of the leaf and the percentage of protoplasts undergoing spontaneous fusion; protoplasts from young leaves are more likely to undergo this fusion.

Induced fusion—Somatic hybridization is generally used for fusion of protoplasts either from two different species (interspecific fusion) or from two diverse sources belonging to the same species. To achieve this objective, spontaneous fusion may be of no value, and induced fusion requiring a suitable agent (fusogen) is necessary. In animals, inactivated Sendai virus is needed to induce fusion. In plants, however, the inducing agent first brings the protoplasts together and then causes them to adhere to one another for bringing about fusion. During the last two decades, a variety of treatments have been successfully utilized for fusion of plant protoplasts (Fig.). These treatments particularly include the following : $NaNO_3$, high pH with high $Ca^{++}$ ion concentration and polyethylene glycol (PEG).

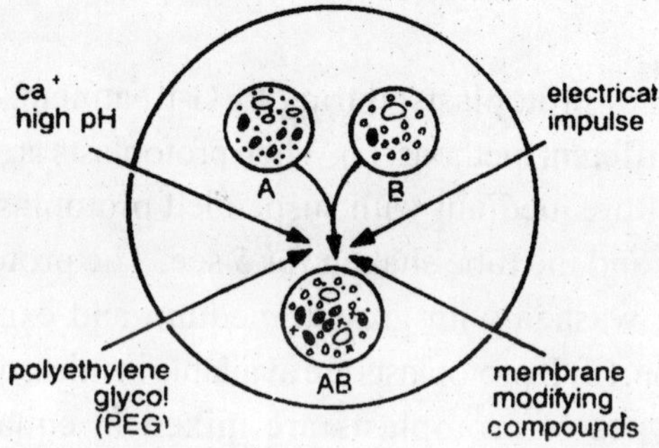


**Fig.** Different experimental methods used for induction of protoplast fusion

(a) $NaNO_3$ treatment—This method was successfully utilized for fusion of protoplasts from root tips of oat and maize seedlings but is not preferred due to low frequency of fusion, particularly when highly vacuolated mesophyll protoplasts are used. The method involves the following steps (Fig.) : (i) isolated protoplasts are suspended in an aggregation mixture (5.5% sodium nitrate in 10% sucrose solution). This mixture works as a fusion inducing mixture and causes fusion on incubation (water bath maintained at 35°C). In order to obtain a higher frequency of fused protoplasts, the mixture may be centrifuged and the pellet resuspended and incubated for one or more additional cycles. (ii) Finally the mixture is replaced by a liquid medium and the protoplasts in this mixture are incubated again; the cycle may be repeated once or twice before plating the protoplasts on a solid medium. The fusion of protoplasts may be monitored at different steps through examination under an inverted microscope.

(b) Treatment with calcium ions (Ca++) at high pH—This method involves spinning (centrifugation) the protoplasts in a fusion inducing solution (0 . 05M CaCI2 2H,0 in 0.4M mannitol at pH 10 . 5) for 30 min. at 50g, after which the tubes are placed in a water bath (37°C) for 40-50 min. This leads to fusion of 20-50% of the protoplasts. The details of the protocol are described by Bhojwani and Razdan (1983). The method has been found superior to other methods in some cases, but high pH may be toxic in other cases.

(c) Polyethylene glycol (PEG) treatment—Since 1974, protoplast fusion has been successfully achieved in several crops, using polyethylene glycol (PEG) as a fusogene. The technique gives high frequency of fusion with reproducible results and involves low cytotoxicity. The technique can be used for fusion of protoplasts from unrelated plant taxa (e.g. soybean tobacco, soybean - maize, and soybean - barley), from unrelated animal taxa and also between those from animal and plant cells.

The agglutination of protoplasts, during PEG-treatment, can be brought about by the following two different methods: (i) When protoplasts are available in sufficient quantities, 1m1 of culture medium with suspended protoplasts is added to 1ml of 56% solution of PEG and the tube shaken for 5 sec. The protoplasts are allowed to sediment for 10 min, washed with growth medium and examined for successful agglutination and fusion. (ii) If protoplasts are available in microquantities, drop cultures can be used. Two types of protoplasts are mixed in equal quantities and 4-fi microdrops (100 It/ each) are placed in a small Petri plate and allowed to settle for 5-10 min. at room temperature. Two to three microdrops (50 Eel each) of PEG are added from the periphery in each Petri plate, which is incubated for 30 min at room temperature (24°C). This leads to agglutination of protoplasts. Sometimes, a cover glass is placed in the middle of Petri plate before protoplast suspension is poured. This avoids sticking of protoplasts to the floor of Petri plate and also makes it convenient to handle the protoplasts including their fixation, staining and examination. After PEG treatment, protoplasts are gradually washed and during this process, most of the fusion is achieved. PEG is then replaced by culture medium to allow growth of fused protoplasts.

(d) Electrical fusion—If protoplasts are placed into a small culture cell containing electrodes and a potential difference is applied, then the protoplasts will line up between the electrodes. If now an extremely short, square wave electric shock is applied, protoplasts can be induced to fuse (Fig.).

**Selection of fused protoplasts**

After the fusion treatment as above, the protoplast population consists of a mixture of parental types, homokaryons and heterokaryons, of which heterokaryons (potential source of future hybrids) often make only 0-5%-10% (Fig.). Efforts have been made to select the heterokaryons or the calli derived from them. Following methods have been used for selection : (i) The hybrids Nicotiana glauca x N. longsdorfii could grow on a medium (Nagata and Takebe, 1971), on which none of the parents could grow (Fig.). (ii) Hybrids of Petunia perrodii- (green) with P. hybrida, P. inflorata or P. parviflora (albino), could be selected in the form of green callus, which represented only hybrid cells; similar approaches were also followed in hybrids of Datum innoxia (albino) with other species of Datura or with Atropa belladona (wild type = green) (iii) The protoplasts of two parents may be labelled by different fluorescent agents, which will then enable the selection of hybrids, as done in the genus Nicotiana (Fig.).

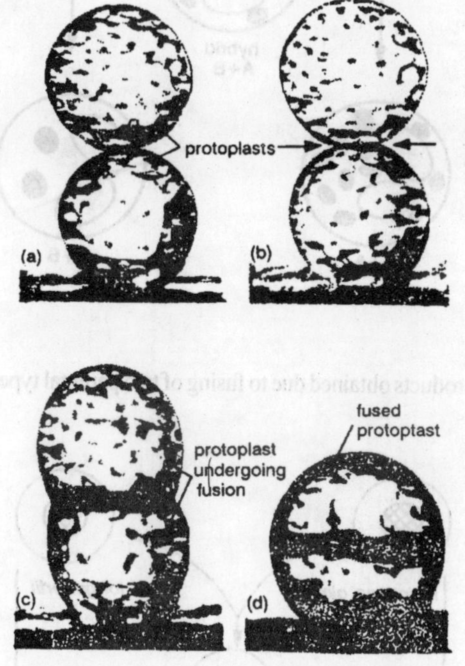


**Fig.** Protoplast fusion induced by the application of electric shock.

## Chromosome status of somatic hybrids

Various somatic hybrids successfully obtained through protoplast fusion are listed in Tables. The chromosome numbers of these hybrids indicate that only few have exact number expected in an amphiploid. In view of this, selection will also need to be applied at the cytological level, if true amphiploids need to be obtained.

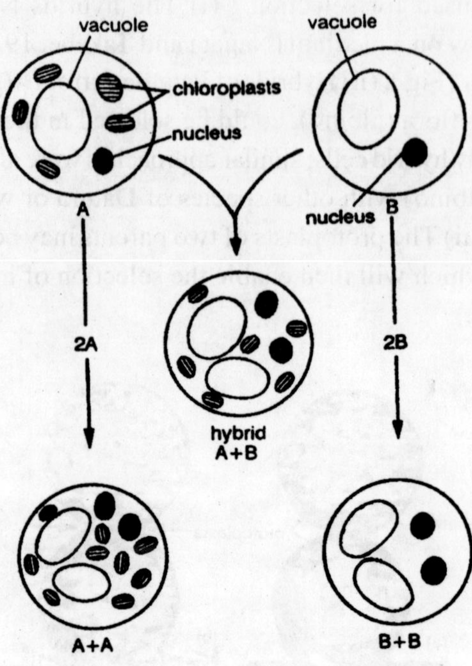


Fig. Different possible products obtained due to fusing of two parental types (A and B) of protoplasts

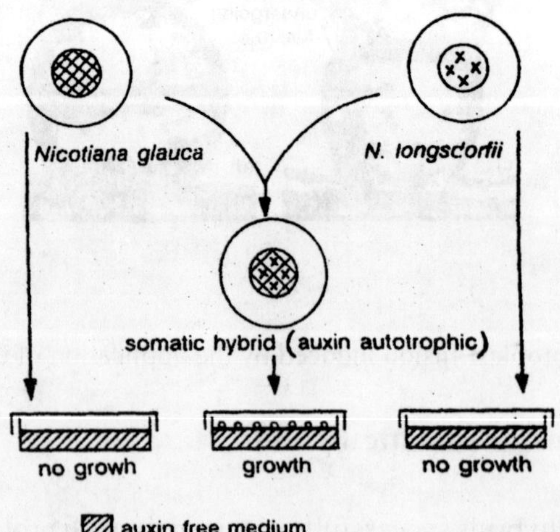


**Fig.** Auxin independent growth of hybrids of Nicotiana glatica and N longsdorffii; the two parental lines can not produce auxin and thus do not grow on auxin-free medium; the hybrid cells produce auxin and are thus able to grow and form callus

The variability in chromosome number in hybrids could be due to any one of the following reasons : (i) Multiple fusions give a higher chromosome number. In PEG induced and electro-induced fusions, about one-third of fusion products result

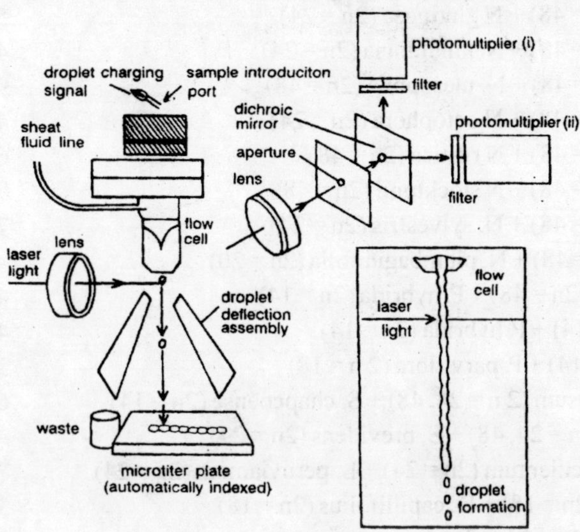


Fig. Simplified diagram of the principle of a fluorescently activated cell sorter; the droplet containing a single cell is electrically tested for its fluorescence property; by applying a charge to the plates, through which the droplet falls, it is possible to separate the parental protoplasts and the hybrids

from fusions between more than two protoplasts. (ii) Asymmetric hybrids result from fusion of protoplasts isolated from actively dividing tissue of one parent and quiescent tissue of the other parent. (iii) Unequal rates of DNA replication in two fusing partners may also give asymmetric hybrids. (iv) Somaclonal variation in cultured cells used for protoplast isolation may also lead to variation in chromosome number.

**Table**

Interspecific hybrids produced through protoplast fusion

| ***Parent species and their chromosome numbers number of hybrid*** | ***Chromosome*** |
|---|---|
| Brassica oleracea (2n = 18) + B. carnpeswis (2 n = 20) | wide variation |
| B. napus (2n = 38) + B. oleracea (2n = 18) | |
| B. napes (2n = 38) + B. nigra (2n = 16) | |
| B. napes (2n = 38) + B. carinata (2n = 34) | |
| B. napes (2n = 38) + B juncea (2n = 36) | |
| Nicotiana glauca (2n = 24) + N. longsdorffii (2 n = 18) | 56-64 |
| N. tabacum (2n = 48) + N. alata (2 n = 18) | 66-71 |
| N. tabacum (2n = 48) + N. glauca (2n = 24) | 72 |

*(Table Contd.)*

| *Parent species and their chromosome numbers number of hybrid* | *Chromosome* |
|---|---|
| N. tabacum (2n = 48) + N glutinosa (2n = 24) | 50-88 |
| N. tabacum (2n = 48) + N. knightiana (2n = 24) | 44 - 137 |
| N. tabacum (2n = 48) + N. mesophila (2n = 48) | 96 |
| N. tabacum (2n = 48) + N octophora (2n = 24) | 48 |
| N. tabacum (2n = 48) + N rustica (2n = 48) | 60-91 |
| N. tabacum (2n = 48) + N stocktonii (2n = 48) | 96 |
| N. tabacum (2n = 48) + N. sylvesfris (2n = 24) | 72 |
| N. tabacum (2n = 48) + N. plumbaginifolia (2n = 20) | – |
| Petunia parodii (2n = 48) + P. hybrida (2n = 14) | 44-48 |
| P. parodii (2n = 14) + P. hybrida (2n = 14) | 46 |
| P. parodii (2 n = 14) + P. parviflora (2 n = 18) | 31-40 |
| Solamum tuberosum (2 n = 24, 48) + S. chapcoense (2n = 14) | 60 |
| S. tuberosum (2 n = 24, 48) + S. brevidens (2n = 24) | – |
| Lycopersicon esculentum (2n = 24) + L. peruviamum (2n = 24) | 72 |
| Daucua carota (2n = 18) + D. capillifolius (2n = 18) | 36, 38 |
| Datura innoxia (2n = 24) + D. stramonium (2n = 24) | 46, 48, 72 |
| D. innoxia (2n = 24) + D. sanguinea (2n = 24) | 46, 72, 96 |
| D. innoxia (2n = 24) + D. candida (2n = 24) | – |

**Table**

Intergeneric hybrids produced through protoplast fusion.

| *Plant species and their chromosome numbers* | *New genus* |
|---|---|
| Raphanus satinets (2n = 18 ) + B. oberacea (2n = 18) | Raphanobrassica |
| Moricandia arvensis (2n = 24, 28) + B. oleracea (2n = 18) | Moricandiobrassica |
| Eruca saliva (2n = 22) + B. napus (2 n = 38) | Encobrassica |
| E. saliva (2n = 22) + B. juncea (2n = 36) | Erussica |
| Diplotaxis muralis (2n = 42) + B. napus (2n = 38) | Diplotaxbrassica |
| D. muralis (2n = 42) + B. juncea (2n = 36) | Diplotaxojuncea |
| Sinapis alba (2n = 24) + B. napus (2n = 38) | Sinapobrassica |
| S. alba (2n = 24) + B. oleracea (2n = 18) | Sinapo-oleracea |
| Nicotiana tabacum (2n = 24) + Lycopersicon esculenum (2 n = 24) | Nicotiopeisicon |
| N tabacum (2n = 24) + Petunia inflorata (2n = 14) | Nicotiopetunia |
| Solanum tuberosum (2n=24) + Lycopersicon esculentum (2n=24) | Solanopersicon |
| Daucus carota (2n = 18) + Petroselinum hortense (2n = 32) | Daucoselenium |
| Datura innoxia (2n = 48) + Atropa belladona (2n = 24) | Daturotropa |
| Oryza saliva (2n = 24) + Echinochloa oryzicola (2n = 24) | Oryzochloa |

**Table**

Intertribal somatic hybrids produced within the family Brassicaceae.

| | |
|---|---|
| Arabidopsis thaliana (2n = 10) + B. campestris (2n = 20)<br>(Tribe Sisymbrieae) (Tribe Brassiceae) | Arabidobrassica |
| Thlaspi perfoliatum (2n = 14) + B. napus (2n = 38)<br>(Tribe Lepideae) (Tribe Brassiceae) | Thlaspobrassica |
| Barbarea vulgaris (2n = 16) + B. tropes (2n = 38),<br>(Tribe Arabideae) (Tribe Brassiceae) | Barbareobrassica |

## Uses of protoplast fusion and somatic hybrids

Somatic hybrids for gene transfer—The family Solanaceae contains the most commonly used species for somatic hybridization. The genera from this family that have been often used for somatic hybridization include Nicotiana, Datura, Petunia, Solanum, Lycopersicon, etc. Other families include Umbellifere ae (e.g. interspecific hybrids in the genus Daucuscarro-and its relatives). In Rutaceae, Citrus has been combined with species from other genera. In Leguminoseae also, several genera including Medicago Trifolium and Lotus have been used for somatic hybridisation. Even among cereals, hybrids between rice (Oryza saliva) and Echinochloa oryzicola (barnyard grass) have been obtained. However, one of the most extensive programmes has been in progress in the family Brassicaceae at Uppsala Sweden. A collaborative programme in Brassicaceae is also underway in New Delhi, the collaborative institutes being IARI, TERI and Delhi University. Some of the interspecific somatic hybrids produced so far are listed in Table. Similarly intergeneric somatic hybrids are listed in Table and intertribal hybrids within Brassicaceac are listed in Table. The hybrids within Brassicaceae are also digrammatically represented in Figures.

Intergeneric fertile hybrids were obtained in several cases particularly within the tribe Brassiceae. These fertile hybrids were backcrossed to cultivated species followed by screening for economic traits like drought and insect resistance (transferred from Eruca), pathogen resistance (transferred from Sinapis), cytoplasmic male sterility or CMS (transferred from Diplotoxis), cold tolerance (from Barborea), and high concentration of nervonic acid, a lubricant (from Thalspi). Symmetric hybrids between tomato and potato, produced by protoplast fusion, have been shown to exhibit intermediate cold tolerance. In another example, substitution of Solarium acaule genome into S. tuberosum resulted in an appreciable increase of frost resistance.

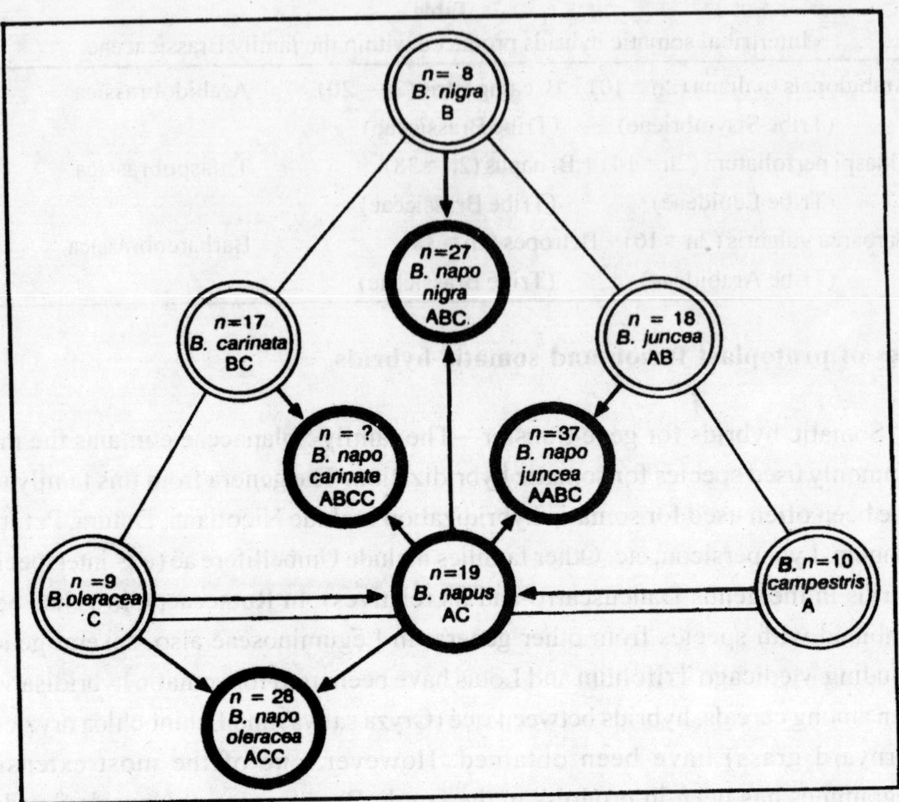


**Fig.** Interspecific somatic hybrids formed between different Brassica species; unbroken lines link amphidiploid Brassica species with their diploid progenitors; heavy arrows indicate which species are combined in somatic hybrids (represented by thick closed circles)

Asymmetry in somatic hybrids for gene transfer—As discussed above, in many somatic hybrids, the chromosome number has been found to be abnormal, often fewer than expected. For instance, in Erucobrassien about 60% plants had fewer than the expected chromosome number. Variability was also observed in Moricondiobrassica and other hybrids (see Table also). In many of these cases, chromosomes are eliminated or translocated thus facilitating limited alien gene transfer. An important example of use of such asymmetric hybrids in gene transfer, is the transfer of black-leg resistance to rapeseed (Fig). The disease blackleg, caused by the fungus Phoma lingum, is a world-wide problem for agricultural production of rapeseed and turnip rape. Fungal mycelia enter into stem, such that stem rots and plant dies by wilting. A toxin (sirodesmin PI) isolated from fungus can be used to select resistant material. To promote limited gene transfer, resistant donor lines were irradiated prior to fusion and hybridization, causing fragmentation and elimination of chromosomes, thus facilitating integration of resistance gene into the recipient line. The fusion products were allowed to develop into colonies before exposing them to selective toxin. Calli resistant to toxin were regenerated into plants and tested for resistance. Most of these plants exhibited resistance suggesting transfer during somatic hybridization.

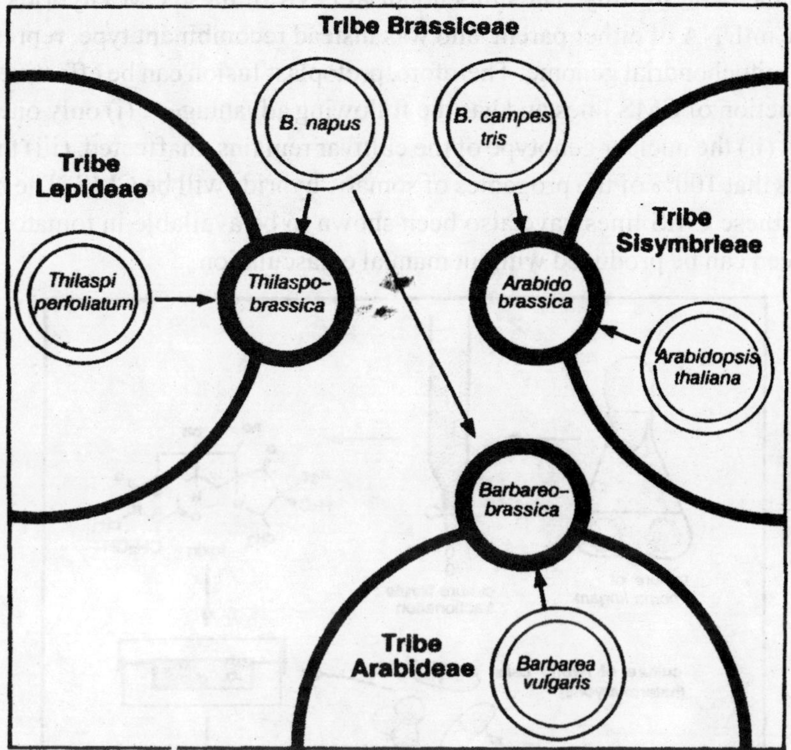


**Fig.** Intertribal somatic hybrids produced within the family Brassicaceae (hybrids shown by thick closed circles)

Somatic hybrids for cytoplasmic male sterility — Recently, methods have been developed to substitute the nucleus of one species into the cytoplasm of another species, whose mitochondria are inactivated. This type of substitution in some cases, led to generation of cytoplasmic male sterility (Fig.). For this purpose, in a recent report (published in August, 1992; PNAS, USA Vol. 89: 68-32) published by G. Melchers (from Germany) and his coworkers (from Japan), the two types of protoplasts, used for the production of somatic hybrids were treated differently as follows : (i) mesophyll protoplasts of tomato (Lycopersicon esccdennrm) were treated with iodoacetamide (IOA) to inactivate mitochondria and (ii) mesophyll protoplasts of S. tuerosum( = potato) were irradiated with y or x-rays to inactivate nuclei. The protoplasts were mixed in 1 : 1 ratio and induced to fuse using $Ca^{2+}$ and PEG, leading to the production of heterologous hybrids. Among the fusion products, some hybrid tomato plants were indistinguishable from the original cultivate with respect to morphology, physiology and chromosome number (2 n = 24), but exhibited various degrees of male sterility. The variation included (i) complete lack of pollen or malformation of anther; (ii) shrunken pollen, (iii) normal-looking stainable pollen that could not germinate. In five tomato cultivars, male sterility induced in this manner was inherited maternally over several generations. Therefore, it was obviously cytoplasmic male sterility. The mitochondrial DNA of these CMS hybrids did not resemble mtDNA of either parent, and was instead recombinant type, representing a hybrid mitochondrial genome. Therefore, protoplast fusion can be effectively used for production of CMS lines and has the following advantages : (i) only one step is required; (ii) the nuclear genotype of the cultivar remains unaffected, (iii) there are prospects that 100% of the progenies of somatic hybrids will be CMS. The restorer lines for these CMS lines have also been shown to be available in tomato, so that hybrid seed can be produced without manual emasculation.

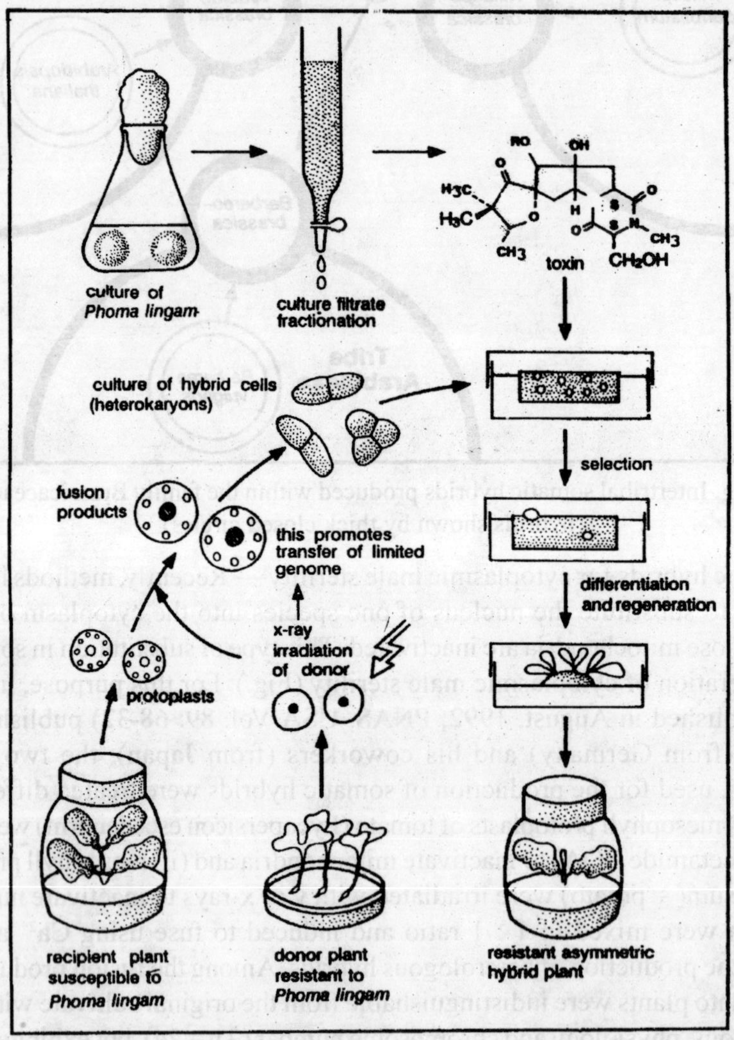


**Fig.** Transfer of resistance of Phoma lingam to susceptible plants, via production asymmetric hybrids between resistant and susceptible plants

## Cytoplasmic Hybrids or Cybrids

In sexual hybrids, the cytoplasm is derived from the maternal parent and in somatic hybrids, it is derived from both the parents. However, hybrids can be obtained,

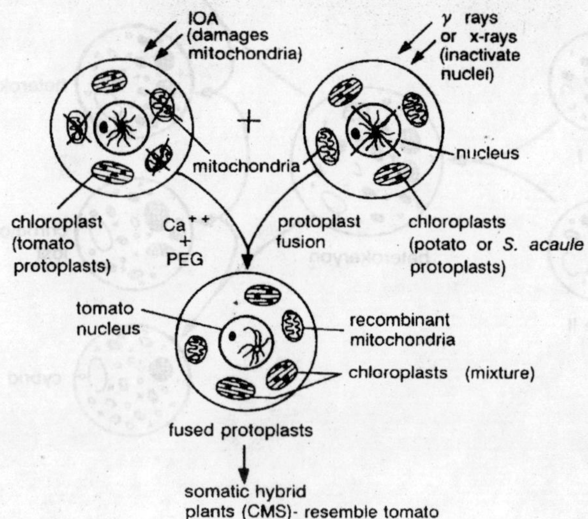


**Fig.** Generation of cytoplasmic male sterlity by fusion of mitochondrial-inactivated tomato protoplasts with nuclear-inactivated Solanum protoplasts

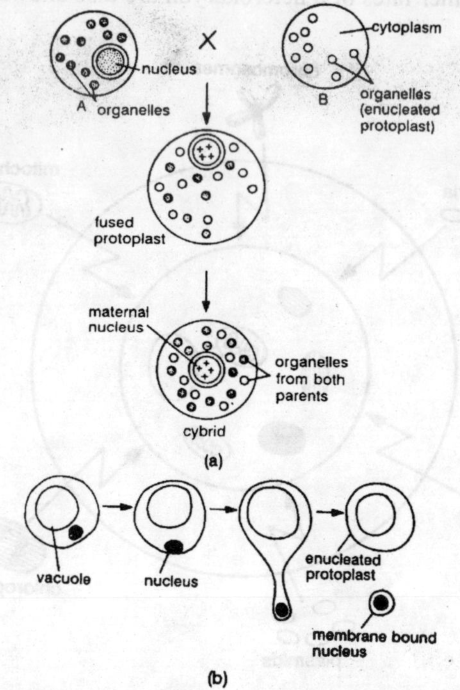


**Fig.** (a) Formation of a cybrid due to fusion of a normal protoplast (with a nucleus) with an enucleated protoplast; (b) protoplasts can be made enucleated by centrifugation

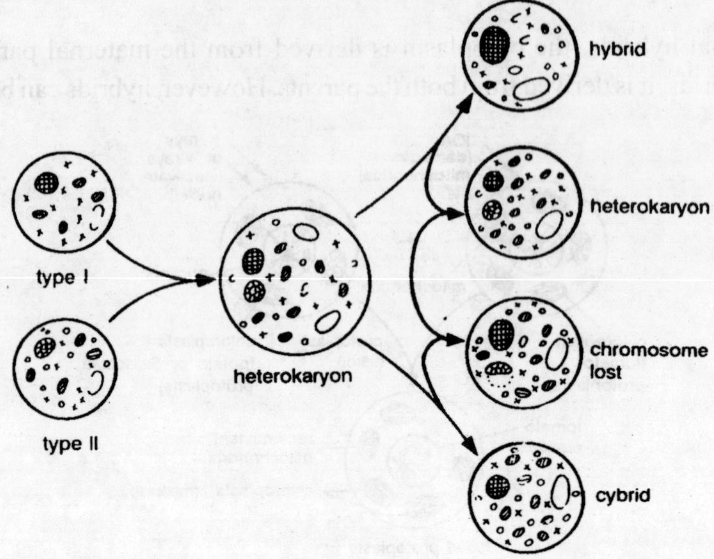


**Fig.** Formation of cybrid due to exclusion of a nucleus from the heterokaryon; other fates of a heterokaryon are also shown.

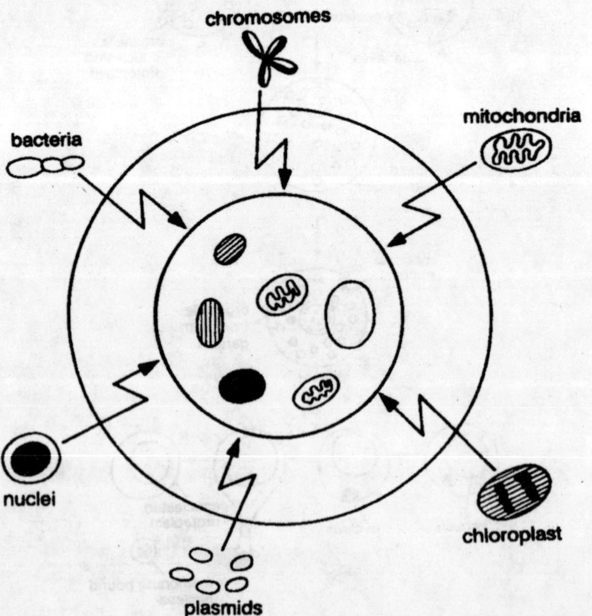


**Fig.** A diagrammatic representation of the wide range of biological components that can be taken up by isolated protoplasts

where nucleus is derived from one parent and cytoplasm is derived from both, thus producing cytoplasmic hybrids, which are also called cybrids_. These cybrids can be obtained using any one of the following methods : (i) fusion of normal protoplasts from one parent with enucleated protoplasts from the other parent (Fig.); enucleated protoplasts can be obtained by high speed centrifugation (20,000 - 40,000 for 45-90 min) of protoplasts or by irradiation treatment, (ii) fusion of normal protoplasts from one parent and protoplasts containing non-viable nuclei from the other; (iii) selective elimination of one of the nuclei from the heterokaryon (Fig.) or (iv) selective elimination of chromosomes of one parent at a later stage after fusion of the nuclei.

The technique of cybrid production has been utilized for transfer of cytoplasmic male sterility as has been successfully done in Nicotiana,Brnssica and Petunia. Other characters like streptomycin resistance have also been transferred using this technique. CMS lines with 'Ogura cytoplasm', a herbicide (atrazine) resistant line and lines with increased nectar production have also been obtained using cybrids in crop brassicas.

## Genetic Modification ( Transformation ) of Protoplasts

Since protoplasts lack a cell wall around its plasma membrane, they are suitable for uptake of chloroplasts, nuclei, microorganisms and naked DNA (Fig.). Although, some success has been achieved for the introduction of foreign chloroplasts, nuclei and microbes for genetic modification, the maximum work has been done and success achieved in the uptake of foreign DNA. The genetic transformation through uptake of foreign DNA involves uptake, integration, replication, expression and transmission.

## Haploids : Production and Uses

Haploids are the sporophytes with gametophytic chromosome number and have been produced in a number of plant species using different methods. Although, the significance of haploids in genetics and plant breeding has been recognized for long time, with the advent of new biotechnology it has received renewed emphasis, so that the production of haploids has become an important component of biotechnology programmes in different countries. Although, haploids could be produced following delayed pollination, irradiation of pollen, temperature shocks, clinical treatment and distant hybridization, the most important methods currently being utilized under biotechnology programmes include (i) anther or pollen culture and ovule culture and (ii) chromosome elimination following centerspecific hybridization (bulbosum technique). These techniques for production of haploids and the uses of haploids in genetic and plant breeding will be discussed in this chapter.

**Production**

**Anther and microspore culture**

One of the very popular methods for production of haploids is through culturing anthers or microspores on artificial culture medium. This leads to the growth of microspores into sporophytes. After the initial reports of successful production of haploids from anther culture in Datura (Guha and Maheshwari, 1966, 1967), haploids have been obtained in more than 150 species belonging to 23 families of angiosperms (Maheshwari *et al.*,1980). These include a wide variety of economically important species.

More often, anthers rather than microspures, are cultured, since the extraction and culture methods for microspores differ and have been successful only in a few species *(Datura inoxia, Nicotinnn sylvestris, N. tabacum, Oryza saliva, etc.)*. The following parameters have been recognized as particularly important for successful anther and microscope culture; (i) conditions of growth of donor plant; (ii) genotype of donor plant; (iii) the pretreatment; (iv) the developmental stage of anther/microspore; (v) the culture medium and the conditions during culture growth.

The different steps involved in anther and microspore cultures are shown in Figures and respectively. In either case, flower buds are brought to a laminar flow chamber and sterilized using appropriate chemical treatment before their dissection. While removing anthers from the Hover buds, care is exercised to avoid injury, because injury leads to development of callus, giving a mixture of diploids, haploids and aneuploids. As shown in the anthers are generally cultured on solid agar medium where they may directly give rise to embryoids or may lead to callus formation, before differentiation. The embryoids develop into haploid plantlets, which are colchicine treated to get diploid homozygous plants to be field tested for selection.

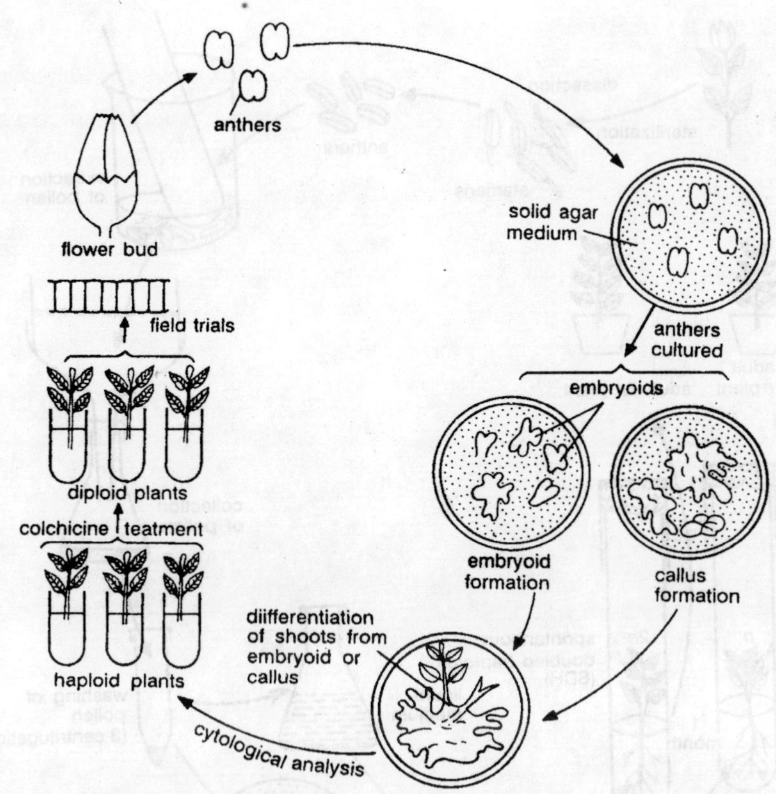


**Fig.** Different steps Involved in the production of haploid Plant, using anther culture.

Microspore culture may be preferred over anther culture, even though the degree of success is low in this case. Microspores are collected first using the following steps: (i) anthers, disssected as above, are taken in a sterile beaker containing liquid medium, and are pressed with a glass rod or syringe piston to allow the microspores squeeze out; (ii) the suspension with anthers and microspores is filtered through a nylon sieve, which allows only the microspores to pass through; (iii) the filtrate is centrifuged thrice, for 5 minutes each, at 500-800 rpm (the pellet resuspended in a fresh medium each time); (iv) the microspores are inoculated on a solid or in a liquid medium maintained at 25°C and 16/8 hr photoperiod. The microspores may develop directly into embryoids within 15 days or follow one of the several indirect paths to produce haploid plantlets (Fig). In anther culture as well as in microspore culture, spontaneous doubled haploids (SDH) are also obtained so that no colchicine treatment will be needed in such cases. Efforts are being made to increase the frequency of SDH, so that these can be multiplied and directly field tested.

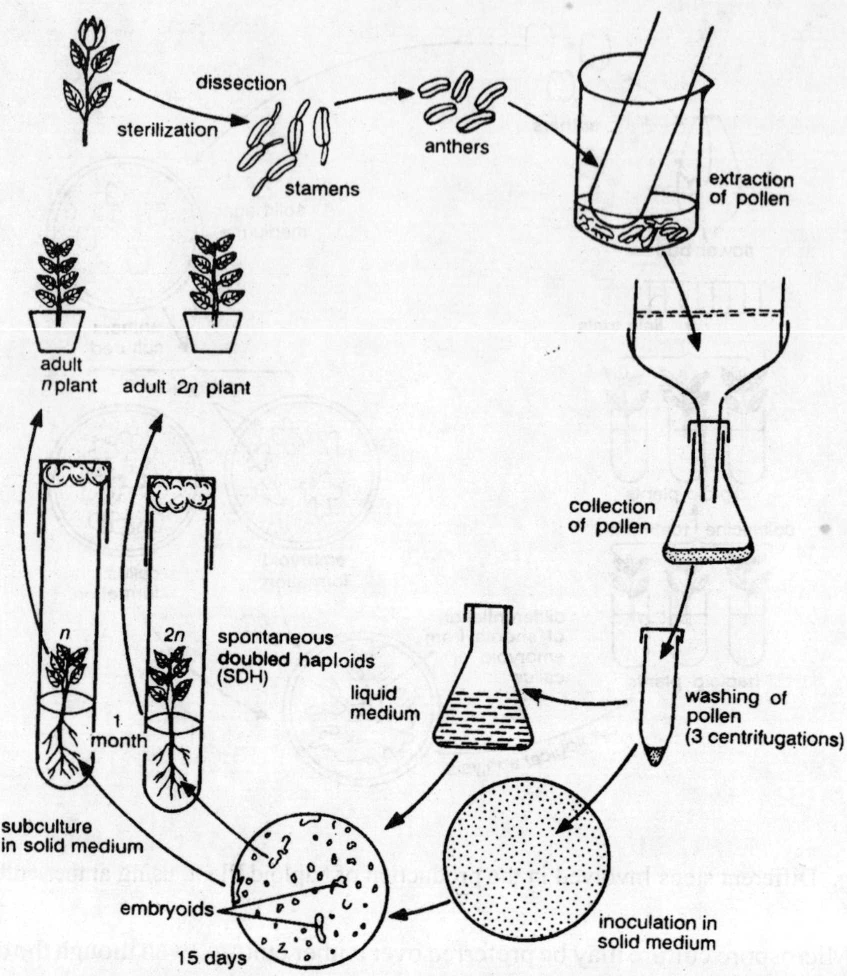


**Fig.** Different steps involved in the production of haploid plants using isolated pollen.

## Ovule culture

Haploid have also been successfully produced from cultured female gametophytes. Success in this connection was first achieved with some Gymnosperms like Zamia, Ephedra and sane Cycads. Although initial attempts for the production of haploids from cultured ovules in Angiosperms were unsuccessful, recently haploids could be obtained from cultured ovaries in a few crop species, like barley, wheat and tobacco. With more efforts in the future, haploids from ovule culture will be produced in a number of plant species. Further research is, therefore, necessary to improve our understanding about the following : (i) the events leading to induction of haploids in female gametophytes, (ii) the factors which control the in vitro development of proembryo into the fully organized plants and (iii) the differences in the growth patterns of in vitro development of unfertilized female gametophytes (in ovules) and the male gametophytes (in pollen).

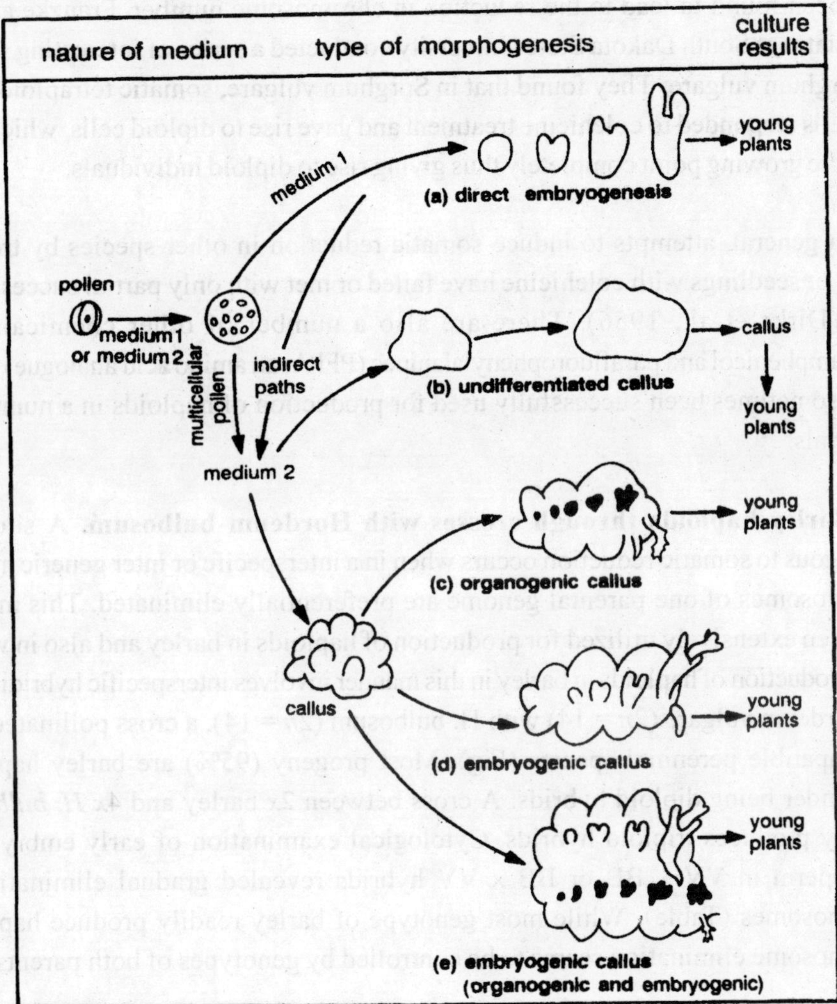


**Fig.** A direct and several indirect paths leading to production of haploid plants from isolated pollen grams.

## Chromosome elimination

**Somatic reduction and chromosome elimination**— Cases are known where either spontaneously or due to specific treatments, the chromosome number was reduced to half in the somatic tissues, a phenomenon described as somatic reduction or reductional mitosis. Swaminathan and Singh (1958) induced a haploid branch on a watermelon by irradiation of the seed used. This must have resulted by the reduction of chromosome number in the somatic tissue through an unknown mechanism (perhaps due to spindle organizer abnormalities). Similarly, colchicine treatment has also been found to lead to the reduction in chromosome number. Franzke and his associates at South Dakota State University conducted a series of interesting studies in Sorghum vulgare. They found that in Sorghum vulgare, somatic tetraploid (2n = 4x) cells responded to colchicine treatment and gave rise to diploid cells, which took over the growing point completely thus giving rise to diploid individuals.

In general, attempts to induce somatic reduction in other species by treating seeds or seedlings with colchicine have failed or met with only partial success as in flax (Dirks et al., 1956). There are also a number of other chemicals like chloramphenicol and parafluorophenylalaninec (PFPA, an amino acid analogue) which have sometimes been successfully used for production of haploids in a number of materials.

**Barley haploids through crosses with Hordeum bulbosum.** A situation analogous to somatic reduction occurs when in a interspecifc or inter generic hybrid, chromosomes of one parental genome are preferentially eliminated. This method has been extensively utilized for production of haploids in barley and also in wheat. The production of haploids in barley in this manner involves interspecific hybridization of Hordeum vulgare ($2n = 14$) with H. bulbosum ($2n = 14$), a cross pollinated, self incompatible perennial species (Fig). Most progeny (95%) are barley haploids, remainder being diploid hybrids. A cross between $2x$ barley and $4x$ *H. bulbosum* mainly produces triploid hybrids. Cytological examination of early embryo and endosperm in VV x BB or BB x VV hybrids revealed gradual elimination of chromosomes (Table). While most genotype of barley readily produce haploids, chromosome elimination seems to be controlled by genotypes of both parents.

Kasha and Kao (1970) summarised the findings for haploid production using the above technique as follows : (i) a very high frequency of haploid can be obtained and these haploids are not associated with any diploid hybrids; (ii) these haploids can be obtained with any cultivated barley and would resemble cultivated barley genome in reciprocal crosses. Contrary to these observations, Lange (1971) showed that haploids obtained due to chromosome elimination are associated with hybrids (Table). With embryo culture technique, he was able to obtain viable hybrids from several cross combinations, although frequency of hybrids was low except in case of VBB hybrids which were quite frequent.

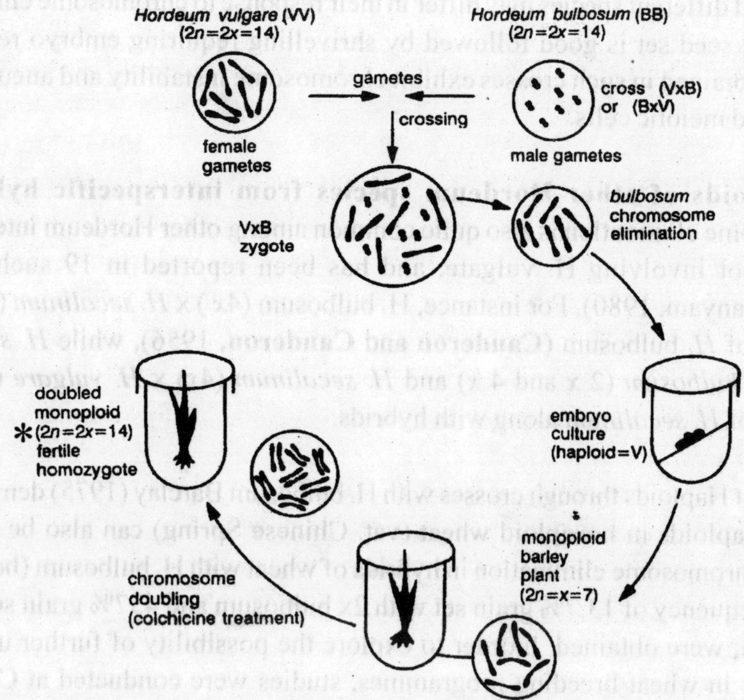


**Fig.** Scheme for production of monoploid and doubled monoploid barley plants using wide hybridization (bulbosum method involving elimination of *Hordeum bulbosum* chromosomes).

**Table**

Results of crosses between *Hordeum vulgare* (2*x* = VV 4*x* =VVVV) and *H. bulbosum* (2*x* =BB, 4*x* =BBBB)- (After Lange, 1971)

| *Cross* | *Progeny* | | |
|---|---|---|---|
| | *Hybrid* | : | *Haploid ratio* |
| VV x BB0 | 2 (VB) | : | 1(V) |
| BB x VV | 4 (BV) | : | 2 (V) |
| VV x BBBB | 145 (VBB) | : | 1 (V) |
| BBBB x VV | 12 (BBV) | : | 0 (V) |
| VVVV x BBBB | 11 (VVBB) | : | 15 (VV) |
| BBBB x VVVV | 5 (BBVV) | : | 3 (VV) |

Following is the subsequent summary of characteristics associated with chromosome elimination as later summarized by Kasha (1974), keeping in view the above results. (i) Elimination occurs in hybrid embryos following interspecific or intergeneric hybridization. (ii) Both haploids and species hybrids may be produced from a cross, and the haploids can have female or male origin, so that androgenetic haploids can also be produced. (iii) Genotypic variation may be found within parental species and different species may differ in their response to chromosome elimination. (iv) Initial seed set is good followed by shrivelling requiring embryo rescue. (v) Hybrids obtained in such crosses exhibit chromosome instability and aneuploidy in somaic and meiotic cells.

**Haploids of other Hordeum species from interspecific hybrids** — Chromosome elimination is also quite common among other Hordeum interspecific hybrids not involving H. vulgate, and has been reported in 19 such hybrids (Subrahmanyam, 1980). For instance, H. bulbosum (4*x* ) x *H. secalinum* (4*x* ) gave haploids of *H.* bulbosum (**Cauderon** and **Cauderon,** 1956), while *H. secalinum* (4*x*) x *H. bulbosum* (2 *x* and 4 *x*) and *H. secalinum* (4*x*) x *H. vulgare* (2*x*) gave haploids of *H. secalinum* along with hybrids.

Wheat Haploids through crosses with H. bulbosum Barclay (1975) demonstrated that the haploids in hexaploid wheat (vat. Chinese Spring) can also be produced through chromosome elimination in hybrids of wheat with H. bulbosum (both 2*x* and 4*x* ). A frequency of 13.7% grain set with 2x bulbosum and 437% grain set with 4x bulbosum, were obtained. Inorder to explore the possibility of further use of this technique in wheat breeding programmes, studies were conducted at Cambridge (U.K.) on the crossability of wheat with H. bulbosum (Snape et al., 1979). It was found that most of the European varieties did not cross, but four of the six Australian varieties were successfully crossed. A similar crossability pattern of wheat varieties was observed with rye also so that it was concluded that crossability with rye and H. bulbosum was controlled by same genetic factors. Through the use of chromosome substitution lines, it was also shown that 1B, 1D, 2D, 3A, 313, 4A, 4B and 6D in the variety Hope have genes promoting crossability, while 5A and 5B have gene reducing crossability. In view of the limitation of crossability with only few wheat varieties, *H. bulbosum* can not be extensively used for production of haploids in common wheat.

**Wheat haploids from wheat x maize crosses**—M.D. Bennett and his coworkers in Cambridge, U.K. (now at Kew, U.K.) demonstrated that wheat x maize crosses can be made successfully, and hybrid embryos can be rescued. Due to elimination of maize chromosomes, wheat haploid can be obtained.

**Mechanism of chromosome elimination**— The Mechanism of chromosome elimination in Hordeum hybrids was studied by **N.C. Subrahmanyam** and **K. J. Kasha** (1973) and by **M.D. Bennett** one of his associates (1976). The following conclusions were made : (i) normal double fertilization occurs in interspecific crosses as confirmed through cytological study; (ii) after fertilization, there is a gradual and selective elimination of H. bulbosum chromosomes from nuclei of endosperm as well as embryo cells, so that eventually haploid embryos are producecd. No evidence of the elimination of *H. vulgare* chromosomes was available.

Differences were, however, observed in the rates of chromosome elimination and these were attributed to genotypic differences in the stock, used in hybridization programmes. In 1970's, **M.D Bennett** and his coworkers, then at Combridge (U.K), had examined the possible cause of chromosome elimination and concluded that sudden shortage of protein in the developing embryo and endosperm, and the better ability of vulgara chromosomes to form spindle attachments relative to bulbosum chromosomes may be responsible for elimination of the bulbosum chromosomes. Other possible causes like differences in mitotic cycle congression during mitosis, etc. were ruled out by them.

It has also been demonstrated that the elimination of bulbosum chromosomes is under genetic control. Ho and Kasha (1975) used primary trisomics and monotelotirsomics in crosses with tetraploid f bulbosum and concluded that both arms of chromosome 2 and the short arm of chromosome 3 of H. vulgare are responsible for chromosome elimination although their effect may be neutralized or offset if sufficent dose of bulbosum chromosomes is available. The effect of these vulgar chromosomes, however, will be indirect.

## Detection or Identification of Haploids

### Morphology

A number of methods have been developed for the detection of haploid in a population of diploids. Although the most effective method is the use of genetic markers to be discussed in the next section, phenotyp effects can also be used for identification of haploids. The vegetative and floral parts and the cell size in haploids are reduced relative to diploid. In maize, Greenblatt et al. (1967) selected 45 seedlings out of 10,000 seedlings, on the basis of size (i.e. thin coleoptiles, etc.) and five of them were found to be monopoids showing some success in screening for haploidy. Stomata size, pollen size and pollen abortion can be other criteria to help initial screening, although in each case cytological confirmation will be essential.

### Genetic markers

Many genetic marker systems are used for recognizing the haploids. The basic principle involved is that the hybrids and nonhybrids should differ, so that any failure of fertilization and parthenogenetic development of embryo can be recognized in the progeny at the seed stage, seedling stage or adult plant stage. For instance, if a female parent in a cross carries a recessive marker and is pollinated by dominant stock, then any recessive progeny would be generally a maternal haploid, although possibility also exists of its being a diploid of one of the following types(i)diploid hybrids, with mutation or deficiency for the marker gene in male parent, (ii) diploid hybrids with suppression of dominant marker due to a suppressor gene or due to a disease, (iii) diploids which are maternal due to apomictic development.

If the recognition needs to be done at the seed stage, then the marker should express itself in the late stage of embryo development, such as aleurone colour marker. If screening is to be done at seedling stage, then the marker should clearly express in the seedling (e.g. the purple seedling colour). This technique of genetic markers has been extensively used in developing maize monoploids. Most extensively used markers include '$a_1$' for brown coloured aleurone, plant and root and *lg* for liguleless, their corresponding dominant markers being *'A'* for purple colour and *'Lg'* for liguled character. If the female parent carries *'a'* and/or *lg,* then the monoploids will be recognized by lack of purple colour and/or ligule, in crosses with male parent carrying *'A'* and/or *'lg'*. Another critical marker gene is *'R'* allele ($R^{nj\text{-}cudu}$), which produces deep pigmentation of aleurone, and thus allows screening at the dormant seed stage. Other markers included purple plumule (*A,* $Pu_1$, $Pu_2$), scutetllum colour (A, C, R genes, and $S_1$, $S_2$ and $S_3$ genes), and purple coleorrhiza ($Pc_1$, $PC_2$, $Pc_3$ and $Pc_4$). These markers did not prove to be equally useful, because the frequency of these markers in maize populations is otherwise also quite high. Endosperm markers, however, are not as useful because very often monoploid embryo in the seed is associated with triploid endosperm, so that the latter will exhibit the male dominant endosperm trait. It should also be noted that for the recognition of paternal or androgenetic haploids, the dominant marker should be present in the female parent so that the androgenetic haploids may be phenotypically different from the hybrid.

In most of the above cases when genetic markers are used, 90-98% of the progeny can be discarded rather early and a very high frequency of haploids are recovered in the remaining population. The above technique of using genetic markers is particularly useful for the development of inbred lines in maize, but in crops like barley and wheat which are self pollinated and where doubled haploids need to be used for cultivation, the availability and utility of genetic markers may be limited.

## Use in Plant Breeding Programmes

Among the different uses of haploids, their use in crop improvement is considered to be the most significant and stands out as the most important reason for emphasis on haploid research. Several extensive reviews have appeared during the last two decades, outlining the theory and the application of haploid breeding in general and also in particular crops (Kasha, 1974 ; Nitzsche and Wenzel, 1977; Clapham, 1977; Davies and Hopwood, 1980 ; Kasha and Seguin-Swartz, 1983; Choo et al., 1985; Peloquin et al, 1991). Some of the important crops, where haploid plants have been produced are listed in Table.

**Table**

Some commercially important crops where haploids have been produced

| | | | | | |
|---|---|---|---|---|---|
| 1. | Alalfa | 9. | Cucumber | 17. | Rye |
| 2. | Asparagus | 10. | Grape | 18. | Strawberry |
| 3. | Barley | 11. | Kale | 19. | Sugarcane |
| 4. | Broccoli | 12. | Peanut | 20. | Sunflower |
| 5. | Citrus | 13. | Pepper | 21. | Sweet potato |
| 6. | Clover | 14. | Potato | 22. | Tomato |
| 7. | Coffee | 15. | Rapeseed | 23. | Triticale |
| 8. | Corn | 16. | Rice | 24. | Wheat |

The methods used for utilizing haploids in plant breeding depend on the inheritance pattern observed in a crop, particularly when we use haploids derived from a polyploid crop. For instance a polyploids, particularly those which are self pollinated (e.g. tobacco), exhibit disomic inheritance, while autopolyploids (e.g. potato), particularly those, which are cross pollinated, exhibit polysomic inheritance. Uses of haploids for two types of crops will be discussed separately.

### Use in species with disomic inheritance

**Production of homozygous lines**—The most important use of haploids advocated, is their use in the production of homozygous lines, which may be directly used as cultivars or may be used in breeding programmes. For instance, a doubled haploid, 'Marglobe tomato' has been used commercially. Similarly, doubled haploids were used as inbred lines in hybrid maize programme. However, the chief limitation in the use of haploids for production of homozygous lines had been the low frequency, with which they were available. In recent years, the availability of anther culture technique and the use of Hordeum bulbosum, in barley and wheat have led to the production of haploids in high frequency. This has greatly accelerated the pace of research on haploids leading to several symposia on haploids arranged in recent years.

Hybrid sorting in haploid breeding—One of the essential steps in haploid breeding involves selection of superior plants among haploids derived from Fl hybrids through anther culture. It is popularly described as hybrid sorting and virtually means selection of recombinant superior gametes. The haploid method of breeding involving hybrid sorting is considered superior over pedigree and bulk methods, firstly because the frequency of superior gametes is higher than the frequency of corresponding superior plants in $F_2$ generation, and secondly because haploid breeding reduces significantly the time required for development of a new variety (see later). For instance, if we assume that the frequency of superior Fl gametes is one in one hundred, then the frequency of homozygous F2 plants derived from the fusion of two such superior similar gametes would be one in ten thousand. Therefore, smaller populations of doubled haploids derived from haploids will need to be handled.

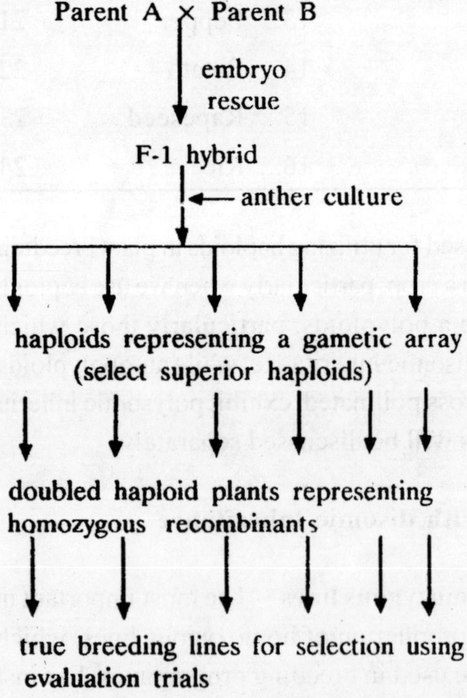


**Fig.** Different steps in hybrid sorting using anther culture technique.

The above technique of hybrid sorting has been successfully utilized, particularly in China and Japan, for development of new varieties in several crops (Table). In Japan, an excellent tobacco cultivar, **'F-211'** bred through anther culture, is resistant to bacterial wilt and has a mild smoking5 quality. In China, more than 100 rice varieties developed using the same, technique, gave an increased yield and, therefore, occupy significant areas of rice cultivation (Table).

Haploids have also been successfully uitlized for breeding of barley, maize, sugarcane, oilseed rape or Canola (B. napus), bell peppers and some other crops. In barley, at least two varieties (Mingo, Gwylan) have been produced in Canada using haploids derived through bulbosum technique of chromosome elimination. The varieties were superior for characters like disease resistance (against powdery mildew and barley yellow mosaic virus) and other yield contributing traits. Similarly, in maize superior inbred lines were developed through screening of haploids derived from anther culture in a hybrid. In sugarcane, selection among anther-culture derived

**Table**

Some high yielding crop varieties bred through hybrid sorting using anther culture.

| *Crop* | *Varieties developed* | *Attributes* |
|---|---|---|
| 1. Tobacco (Nicoriarra 2. tabocceon) | F 211, Tanyu 1, Tanyu 2. Tanyu 3 | Disease resistant, mild smoking |
| Rice (Oryza Sativa) | Tanfong I, Tanfeng 1, Tanyu 1, Huayu I, Huayu 2, Huayu 15, Hua- 03, HJa Be 702 Hua Ilan Zao, Zhonghua 8, Zhonghua 9, Zhonghua 10, Zhonghua 11, Tonghua I,'I'onghua 2, Nanhua 11, Chaohua Ai, Ta lie 78, Shin Shu, Xin Xiu, Xin Xion, Late Keng 76, Late Keng 959, zhe Keng 79-66, Huapei Shanyou 63 | Good quality, disease resistant, high yield |
| 3. Wheat (Triticum aestivum) | Huapei 1, Lunghua 1 Jinghua 1, Yunhua 1. Yunhua 2 | Rust resistant, cold resistant, high yield Large spikes, more tillers, resistant to rust & powdery mildew, lodging resistant |
| | Jungdan 2288 (winter wheat) | Large spikes, more tillers, resistant to rust & powdery mildew, lodging resistant |

**Table**
Performance of some rice varieties bred in China through hybrid sorting using anther culture.

| | *Variety* | *Cultivated acreage (ha)* | *Yield (t/ha)* | *Increase in yield over local variety (%)* |
|---|---|---|---|---|
| 1. | Xin Mon | $>3 \times 10^4$ | — | — |
| 2. | Huayu 1 | $1 \times 10^4$ | — | — |
| 3. | Huaya 15 | — | 8—11 | — |
| 4. | Late Keng 76 | $1 \times 10^4$ | 6.7 | 17 |
| 5. | Late Keng 959 | $>3 \times 10^4$ | 4 5—6.0. | 10 |
| 6. | Tonghua no. 1 and 2 | $2 \times 10^4$ | 6.8—7.7 | 10—15 |
| 7. | Hua Han Zao | $1.5 \times 10^5$ | — | 8.7—27.7 |
| 8. | Zhe Ken.g 79—66 | $1 \times 10^5$ | 6.05 | 8.1 |
| 9. | Zhonghua 8, 9, 10, 11 | $1.5 \times 10^5$ | 7—10 | 8.8—13.7 |
| 10. | Nanhua 11 | $1 \times 10^4$ | 6.12 | 7 4-10.1 |
| 11. | Chao Hua Ai | — | 6.5—8.3 | 18.8—51.7 |
| 12. | Huapei Shanyou 63 | $3 \times 10^5$ | — | 5.7 |
| 13. | Hua—03 | — | 6.0 | 10 (high protein) |

haploids led to the development of superior lines with tall stem and higher sugar content. In Brassica napus also, antherderived doubled haploids, produced in Canada, had low erucic acid and glucosinolate content. In another case, a haploid derived from twin seedlings in B. napus led to the development of a superior cultivar 'Maris Haplona'. In bell peppers, DH lines produced from anther-derived haploids, exhibited all shades of colour ranging from dark green found in one parent to light green found in the other parent of the hybrid employed.

Despite the above conspicuous achievements made through haploid breeding, we find that on a world wide level, the acreage occupied by varieties developed through haploid breeding is negligible. In India, no variety in any crop has so far been developed through haploid breeding. Even at an international level, the emphasis on haploid breeding has gone down in recent years, partly due to failure in achieving the expected targets and partly due to shift in emphasis towards plant breeding based on molecular mapping (see Chapter).

**A comparison of haploid breeding with conventional breeding**—A comparison of haploid method of breeding with conventional methods (pedigree method, bulk method, single seed discent or SSD) suggests that the time period required in haploid breeding is only 4-5 years, as against 10 years required in pedigree/ bulk method. (Table).

**Table**

A comparison of pedigree and bulk methods of conventional plant breeding with haploid breeding in a self pollinated crop like barley.

| *Years* | *Pedigree method* | *Bulk method* | *Haploid method* |
|---|---|---|---|
| 1 | Cross parents | | Cross parents |
| 2 | Grow Fl | Same as in pedigree method | Grow F1 ; derive haploids and DHs . (anther culture) |
| 3 | Grow F2 (2000-10000 seeds space planted); select for qualitative traits; harvest seed on single plant basis | | Dihaploids (DHs) are used for initial trials |
| 4-5 | Grow F3, F4 (plant to progeny rows from selected plants); select rows, plants | Harvest seed in bulk in F2-F5; no selection; only natural selection | Trials to verity performance ;. release of variety |
| 5-6 | Grow F5; initial trial (3 4 rows of each plant progeny); compare and select. | | |
| 7-9 | F6-F8; trials to verify performance | Space planting; select single plants; plant to progenies; yield trials | |
| 10 | Final trial and release of variety | Final trial and release of variety | |

**Production of spontaneously doubled haploids for breeding**—The utility of haploids in plant breeding eventually requires production of doubled haploids (DH) as discussed above. For this purpose, often efficient colchicine treatments are applied and colchicine doubled haploids (CDH) are obtained. In a system where haploids are detected through genetic markers, we can also identify the **spontaneously doubled haploids (SDH),** if doubling does occur spontaneously. This will avoid unnecessary colchicine treatment and the breeder can save time and plant material.

Although there are several reports of SDHs produced during haploid production, but at Svalov in Sweden, hap gene is being utilized, not only for the production of haploids, but also for the production of SDHs (**Hagberg and Hagberg,** 1987). It utilizes a marker gene system for mildew, resistance due to an allele (*ml*-o), which is recessive to susceptibility; allele (*Ml*—o). The breeding scheme used to produce spontaneously doubled haploids is shown in Figure. In such a programme, spontaneously doubled haploids (SDH) were three times as many as the normal haploids, which seems to be really promising. These SDH plants were found to be completely homozygous thus proving their maternal origin due to doubling of chromosome number in the original egg cell. The marker ml -o is suitable, because it allows screening at the seedling stage, but other gene pairs like K-k (hooded vs normal awns) or V-v (two row vs six row ears) can also be used in barley. It is expected that more work in this direction will be carried out in future to make SDH production system useful to the breeders and cytogeneticists.

**Use of haploids in species with polysomic inheritance**

Polyhaploids derived from polyploids exhibiting polysomic inheritance may be homozygous or heterozygous. For instance, dihaploids of potato may be *AA* or An or nn, depending upon the genotype of the tetraploid potato, which can be heterozygous for this gene *(AAArr, AAaa, Aaaa)*. Therefore, by doubling of haploids, homozygosity can not be always achieved in these cases, unless dihaploids are again used for production of monoploids. Therefore, the use of haploids in an autotetraploid crop like potato differs from that in crops like barley, rice, wheat, etc.

**Analytical breeding in potato**—In a crop like potato, the genetic improvement through crosses and selection may be achieved at the dihaploid level and improved dihaploids may then be used for chromosome doubling to get increased tuber size. This has been described as analytical breeding (Chase, 1963), which has the following advantages: (i) dismic inheritance in the dihaploids provides simpler inheritance ratios, ease of selection and smaller population size required to recover recombinants; (ii) dihaploids can be crossed with wild diploid species for introducing new germplasm; (iii) the dihaploids represent a level of homozygosity equivalent to three generations of self pollination in an autotetraploid.

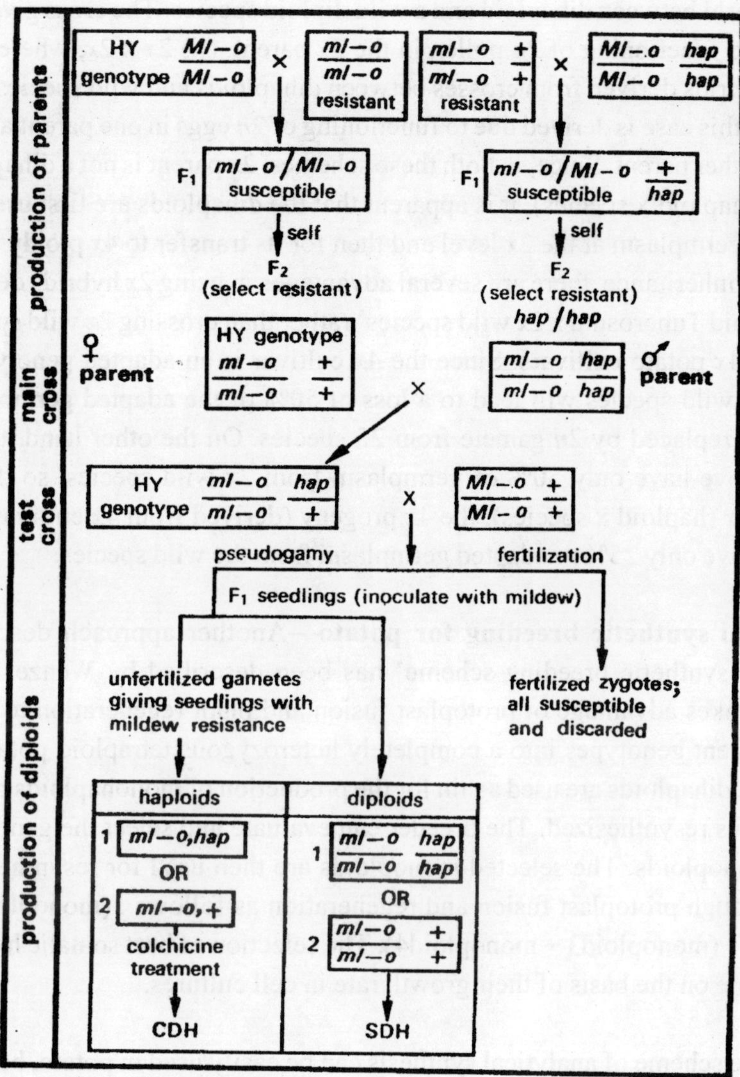


**Fig.** The use of hap gene in barley for the production of colchicine doubled haploids (CDH) and spontaneously doubled haploids (SDH) as used at Svalov. Sweden.

**Two breeding schemes using 2*n* gametes in potato**—In the above analytical method, where improved dihaploids are doubled by colchicine, there is reduced vigour of dihaploids due to inbreeding and expression of deleterious genes. This leads to failure in recovery of sufficient vigour in the tetraploids on chromosome doubling, because vigour depended on intra- and interlocus interaction in multiple allelic condition in the autotetraploids. To overcome this problem, following two breeding schemes were suggested by Mendiburu and Peloquin (1977): (i) 4*x* x 2*x*, where 4*x* is a cultivar and 2*x* is a hybrid between dihaploid and a wild diploid species. The cross gives 4 x progeny due to functioning of 2*n* pollen in the 2*x* parent. (ii) 2*x* x 2*x*, where both parents are hybrids derived from crosses between dihaploids and wild species. The 4*x* progeny in this case is derived due to functioning of 2*n* eggs in one parent and 2n pollen in the other parent. Since, in both these schemes, 2*x* parent is not a dihaploid, but a hybrid (haploid x species), it is apparent that the dihaploids are first used for enriching the germplasm at the 2*x* level and then for its transfer to 4*x* ploidy level. Due to dismic inheritance, there are several advantages in using 2*x* hybrids derived from '2*x* haploid Tuberosum x 2*x* wild species' rather than crossing 2*x* wild species directly with 4*x* potato cultivars. Since the 4*x* cultivar is an adapted genotype, a cross with 2x wild species will lead to a loss of 50% of the adapted germplasm, which will be replaced by 2*n* gamete from 25 species. On the other hand, the 2*x* hybrids as above have only 50% of germplasm from 2*x* wild species, so that in crosses 4*x* x 2*x* (haploid x species), the 4*x* progeny (derived from 4*x* cultivar x 2*x* hybrid) will have only 25% unadapted germplasm from the wild species.

**Analytical synthetic breeding for potato**—Another approach described as 'analytical synthetic breeding scheme' has been described by Wenzel et al. (1979). This takes advantage of protoplast fusion and plant regeneration to bring together different genotypes into a completely heterozygous tetraploid potato. In this approach, dihaploids are used again for the production of monohaploids before the tetraploid is resynthesized. The breeder can evaluate and select the genoomes present in monoploids. The selected monoploids are then used for resynthesis of tetraploid through protoplast fusion and regeneration as follows : (monoploid1 + monopolid2) + (monoploid3 + monoploid4). The selection of best somatic hybrids may be feasible on the basis of their growth rate in cell cultures.

The above scheme of analytical synthesis can be easily used in potato, because once the resynthesized tetraploids are regenerated, they can be multiplied via tubers for commercial cultivation. In other autotetraploid crops like alfalfa (Medicnko saliva) or orchardgrass (Dactylis glomerata), where reproduction is by seed, the above scheme will not be successful. However, the analytical breeding described earlier can be used. In alfalfa, where 2n gametes function, the two schemes suggested by Mendiburu and Peloquin (1977) for potato can also be used.

## Limitations to haploid breeding

As discussed above, haploid breeding has not given the desired and expected dividends, despite substantial investments made on haploid research during the last two decades. Following are some of the widely recognized limitations : (i) Haploids can not be obtained in high frequency required for selection; (ii) The cost- benefit ratio in haploid breeding is often not favourable, thus discouraging the use of haploid breeding despite its obvious advantages; (iii) Haploids will express recessive deleterious traits and deleterious mutations may arise during anther culture; (iv) Different ploidy levels may be available so that haploid status may need to be confirmed cytologically; alternatively, pollen culture may need to be used, which is more expensive and gives relatively low success rate.

## Other Uses

### Mutation research

In Chapter, we discussed the use of cell culture in induction and selection of mutations. The use of haploids provide with the following additional advantages in mutation research : (i) Due to the presence of only one set of chromosomes, even the recessive mutations are immediately expressed in haploids (unlike diploids, where the recessive mutations express only in segregating single plant progenies grown in M2 generation). (ii) Pollens can be plated on solid medium like single cells and therefore much bigger populations can be screened in the laboratory, without initially growing the plants in the field.

In several crops, desirable mutants have been isolated among haploids derived in culture. These desirable attributes included resistance to drugs (e.g. antibiotics), amino acid analogues, pathotoxins, salts, herbicides, chilling, diseases (caused by fungi, viruses and nematodes). Such mutants can be used for a variety of purposes leading to the development of suitable cultivars.

### Gametoclonal variation

In the previous chapter, we described somaclonal variation, which originates in tissue culture experiments and can be profitably utilized for improvement of economic traits in crop plants. Similar variation can also be observed among haploid plants (having gametic chromosome number), developing from cultures derived from anthers or microspores. Since this variation is available among plants with gametic chromosome number, it is popularly described as **gametoclonal variation.** The utilization of gametoclonal variation differs from that of somaclonal variation in the following respects: (i) recessive mutations will be immediately expressed, while somaclones need to be selfed in some cases to allow expression of variation; (ii) chromosome number needs to be doubled in selected gametoclones. The gametoelonal variation is actually utilized in hybrid sorting during the use of haploids for plant breeding purposes.

## Cytogenetics research

Haploids have been found to be useful in various areas of cytogenetics research, including the following : (i) production of aneuploids, (ii) determination of basic chromosome number and (iii) determination of the nature of polyploidy. For instance, in wheat, initially monosomics were produced by E.R. Sears by pollinating a haploid wheat plant by the pollen from a normal wheat plant. The monosomics in this case presumably resulted due to functioning of eggs with restitution nuclei deficient for any individual chromosome. In potato, haploids were used for the production of trisomics using one of the following procedures: (i) isolation of parthenogenetic aneuhaploid offspring following $4x \times 2x$ crosses; (ii) isolation of aneuploids from $3x \times 2x$ crosses, where $2x$ is a haploid. In both the schemes, a number of trisomics with $2n = 25$ could be obtained; (iii) in tobacco, several nullisomics were derived from haploids obtained from monosomics, which could not produce nullisomics in selfing due to sterility; (iv) haploids have also been used for the production of alien barley addition and substitution lines in a crop like wheat.

A study of meiosis in haploids also gives evidence for the original basic chromosome number in a species or a genus. For instance, in pearl millet (Penninetum typhoides), occurrence of some pairing (upto two bivalents in some cells) suggested that the basic chromosome number may be $x = 5$. Similar evidence is also available in maize, which is sometimes considered to be an archaic tetraploid. In potato also, chromosome behaviour of dihaploids ($2n = 2x = 24$) suggested that the base number is $x = 12$ and that potato is an autotetraploid. Chromosome behaviour of monoploids in potato ($2n = x = 12$) formed only univalents confirming this conclusion.

## Evolutionary studies

A comparison of dihaploids in potato with diploid wild species proved useful in deriving conclusions about the origin of cultivated potato. Since, production of 2n gametes is common both in dihaploids as well as in diploid species, it is suggested that Andigena, the progenitor of Thuerosum must have originated many times from different but related $2x$ populations in different locations through sexual polyploidization. Haploids in potato have also been used for crosses with diploid species to conduct genome analysis. The meiosis in hybrids and further generations suggested that wild species differ from cultivated potato in a small number of genes.

Potato dihaploids were also used for a study of relationship between potato and tomato. A study of pachytene chromosomes in potato dihaploids and tomato (diploids, $2n = 24$) suggested close evolutionary relationship between potato and tomato. Recently, potato dihaploids were used for preparing RFLP (restriction fragment length polymorphisms) maps in potato and for a comparison of RFLP maps in potato and tomato. The results indicated a high degree of conservation of DNA sequences (For details, see Chapter).

## Genetic studies

Dismic inheritance and reduced chromosome number in polyhaploids (e.g. dihaploids in potato) offer special advantage for their use in a variety of genetic studies. Using potato as an example, the uses of haploids in genetic studies can be listed as follows : (i) Using haploids for electrophoresis, a number of genes for several isozymes were determined for potato (Simon and Peloquin, 1980; Quiros and McHale, 1985). Progress has also been made in determining the number of genes for other traits like disease resistance and other agronomic traits (De Maine, 1984). (ii) Using haploids in potato, heritability was also determined for several traits like specific gravity, vine maturity, glucose concentration, tuber dormancy, total yield and tuber weight. (iii) The genotype of a tetraploid can also be determined by extracting and analysing the possible haploids, as has been done in potato for acid gel proteins, several isozymes and self incompatibility, because the recessive alleles are also unmasked in these haploids. (iv) Interlacus interactions could be studied using haploids in potato, because at a tetra-allelic locus ($A_1 A_2 A_3 A_4$) in potato, one expects six first order interactions ($A_1 A_2$ $A_1 A_3$, $A_1 A_4$ $A_2 A_3$ $A_1 A_4$, $A_3 A_4$), three second order interactions ($A_1 A_2 A_3$ $A_1 A_3 A_4$, $A_2 A_3 A_4$) and one third order interaction ($A_1 A_2 A_3 A_4$). As opposed to this, dihaploid will have only one interaction, which partly explains loss of vigour. (v) Recently dihaploids x wild species crosses were also used in various crops for the preparation of RFLP linkage maps.

# THIRTEEN

# Mapping of Genomes

One of the most important advances of molecular biology has been the detection and exploitation of naturally occurring DNA sequence polymorphisms represent The preparation of molecular maps in crop plants has been a priority area of research in plant biotechnology also. The utility of these maps for crop improvement programmes has been emphasized during 1980's and 1990's, so that molecular maps have already been prepared for all major crop species The techniques of the use of molecular markers have also proved useful for characterization of crop varieties and plant genetic resources THOUGH DNA fingerprinting. These aspects of plant biotechnology will be briefly discusses in this chapter.

## RFLP Genetic Maps in Plants

### Preparation of genetic maps using RFLP loci

The first step for the preparation of linkage maps based on RFLPs is to procure as many unique DNA sequence clones as possible and to determine which clones are effective in revealing polymorphisms. As discussed earlier, both cDNA clones (derived from mRNA) and genomic DNA (gDNA) clones have been used in this connection. In order to determine which clones are actually useful in determining RFLPs, the clones are subjected to an initial screening. The following critieria are used for selecting useful clones: (i) it should yield strong hybridization signal; (ii) it should reveal a small number of 'fragments and (iii) it should reveal polymorphism among homozygous lines (as in maize) or related species (as in tomato) which will be used for crosses to yield $F_2$ segregating population for linkage studies.

The second step is to make crosses between homozygous lines or related species showing polymorphism for the selected DNA clones. The $F_1$ hybrids are selfed to yield segregating $F_2$ population. Heterozygous parents like those in potato can also be used rarely for crosses to estimate recombination frequencies leading to construction of RFLP maps . The DNA samples extracted from parents, F1 hybrids and from individual plants of $F_2$ population can be used for collection of RFLP data using DNA clones (selected as above) as probes. A population of 50 $F_2$ individuals

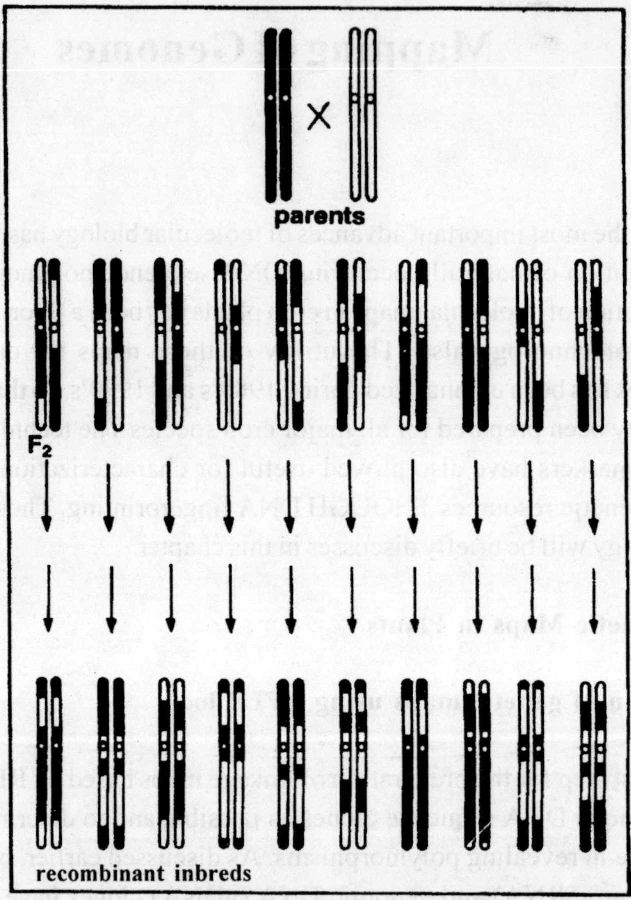


**Fig.** The production of recombinant inbred lines. This figure depicts a hypothetical diploid organism with a single pair of chromosomes. The original parents are themselves inbred lines and homozygous at all loci. Individuals of a large $F_2$ population produced from a mating of these inbreds are self-pollinated, in case of maize for at least six generations to produce new inbred lines.

has been shown to be adequate to give information on linkage among RFLPs, if any. Recombination frequencies can also be estimated from backcross populations and doubled haploids (DH) derived from pollen of $F_1$ plants. In maize, recombinant inbred

(RI) lines have also been produced from a number of crosses through continuous selfing or sib-mating in $F_2$ plants. These RI lines are comparable to RI lines used in mouse and are also comparable to doubled haploids (DH). They can be used in much the same way as the $F_2$ plants and offer following advantages: (i) They can be maintained indefinitely and supply unlimited DNA, unlike individual $F_2$ plants. (ii) They are produced by multiple rounds of meioses and, therefore, the linked genes have a better chance of recombination. The data on recombination frequencies are processed using a computer device with the help of a suitable software. One such software in common use for RFLP maps is MAPMAKER (other software, like LINKAGE was used in earlier studies). Once RFLP genetic and chromosome maps are available, if unmarked regions are available, chromosome walking or chromosome hopping or insertion targeting may be used, starting from a known RFLP locus. This will help saturation of the unmarked region or in reaching close to a gene of interest.

**Table**

List of available recombinant inbred Populations in maize.

| *Parents* | *No. of RI lines* | *Generations of inbreeding* |
|---|---|---|
| T232 x CM337[b] | 48 | 8 |
| CO159 x Tx303 [b] | -11 | 8 |
| CO159 x Tx303 | 160 | 5 |
| Mol⁻ x B73[b] | 44 | 7 |
| PA326 x ND300 | 74 | 5 |
| CK52 x A671 | 162 | 5 |
| CG16 x A671 | 172 | 5 |
| CH593—9x CH606 - 11 | 101 | 5 |
| CO220 x N28 | i73 | 5 |

[a]Generations of selfing beyond the $F_2$

[b]Extensive mapping has been done in these families.

The most extensive molecular maps using RFLPs, morphological markers and QTLs (quantitative trait loci;) have been prepared in maize and tomato. For maize the two inbred lines used initially were H427 and 761. For tomato, however, two different species (L. esculentum and L. hirsutum or L. esculentum and L . pennellii) had to be used due to limited variability available in L. esculentum. Segregation of codominant RFLP loci (1 : 2 : 1) could be detected in $F_2$ generation, and linkage, if any, could be detected by deviation from the expected segregation ratio. For the linked RFLP loci, recombination frequencies were also worked out using maximum likelihood analysis (MAPMAKER, Lander et al., 1987).

In preparing potato RFLP maps at Cologne in Germany, following two modifications were used to improve the technique of RFLP mapping: (i) only those restriction enzymes were used, whose restriction siles consisted of only four rather than six base pairs, so that the restriction sites will be more frequent and consequently the genomic DNA will be cleaved more frequently to release additional polymorphism; (ii) denaturing polyacrylamide sequencing gels were used instead of agarose gels to produce better resolution of DNA fragments, which differ in length by only a few base pairs.

Heterologous probes may also be used for RFLP mapping in plants. For instance, a number of DNA clones from maize have also been utilized for RFLP mapping in sorghum. Similarly, DNA clones belonging to tomato could also be successfully used for potato (also for pepper) and vice versa. The resulting RFLP maps of potato and tomato were found to be similar. Such studies also suggest that similar RFLP loci occur in related species and can also be used, therefore, for phylogenetic studies.

**Table**

A Vsl of some crops where RFLP maps have been prepared.

| *Crop* | *Gametic chromo-some no. (n)* | *Number of RFLP loci* | *Map length (cM)* | *Genome size (bp)* | *Resoultion bp/cM* |
|---|---|---|---|---|---|
| 1. Maize (Zea mars) | 10 | 338 | 500 | $700x10^5$ | $1.4 x 10^5$ |
| 2. Tomato (Lycopersicon escttlertaarr) | 12 | >1000 | 1300 | $715 x 10^6$ | $00.551x 10^6$ |
| 3. Potato (Solamun tuberosum) | 12 | 350 | 690 | $750\text{-}1000x10^6$ | $>1.0 x 10^6$ |
| 4. Rice (Oryza sativa) | 12 | >600 | 1389 | $600x 10^6$ | $0.432x10^6$ |
| 5. Wheat, D genome (Aegilops squat-) | 7 | 144 | 1554 | $500x10^6$ (approx.) | $.322x10^6$ |
| 6. Lentil (Lens culirtaris) | 7 | 20 | 333 | $500x10^6$ (approx.) | $1.5 x 10^6$ |
| 7. Barley (Hordeum vulgare) | 7 | >200 | 1500 | $5.2x10^9$ | $3.5 x 10^6$ |

## Mapping of major genes of economic value using RFLPs (gene tagging)

The RFLP markers can be used to map positions of major genes as well, as polygenic traits (also called quantitative trait loci or QTLs). Once these genes are mapped on RFLP maps, the molecular markers (RFLPs) linked to specific genes or gene clusters (gene tagging) can be used in selection schemes for plant breeding. Molecular markers mapped very close to important genes can also be used for isolation of these genes.

**Use of near isogenic lines (NILs) for gene tagging**— In recent years, near isogenic lines (NILs) have been utilized to identify linkages between molecular and conventional phenotypic markers. An NIL is produced when a conventional marker, like that for a disease resistance is transferred from a donor parent (DP) into a recurrent parent (RP) and the genotype of the recurrent parent is restored through

**Table**

Some examples of RFLP linked mapping of major genes of economic value to some crops.

| *Crop* | *Gene located on RFLP map* | *Lines use* | *Reference* |
|---|---|---|---|
| 1. Tomato | (a) Tm-*2a* locus for resistance to TMV | Near isogenic lines (NIIs) | Young et al., 1988 |
| | (b) Gene 12 and 11 for resistance against *Fusarium oxysporium* | Near isngenic lines (NIL-s) and BC populations | Sarfatti et al., 1989 1991 |
| | (c) Gene Mi for resistance against root knot nematode (Melodogyne spp. ) | Near isogenic lines (NILs) | Klein-Lankhorst et al., 1991 |
| | (d) Gene Sm for Stemphylium resistance | Near isogenie lines (NILs) | Behare et al., 1991 |
| 2. Potato | (a) Gene for resistance against cyst nernatode (Globodera rostochicusis ) | Two diploid clones | Barone et al., 1990 |
| | (b) Two genes (Rx1, R2) for resistance against potato virus × (PVX) | Two $F_1$ populations | Ritter et al., 1991 |
| 3. Maize | Gene for resistance against maize dwarf mossaic virus (MDMV) | Near isogene lines (NILs) | McMullen and Louie, 1989 |
| 4. Barley | *ml-o* locus for resistance against *Erysiphe graminis* f sp. hordei | Near Isogonic lines (NILs) and two subspecies of *Hordeum vulgare* | Hinze at al., 1991 |
| 5. Sugarbeet | Gene for resistance against a nematode (*Heterodera schachtii*) | Addition for *Beta procumbens* chromosome | Jung et al., 1990 |
| 6. Rice | Two genes for blest resistance, *Pi-2(t)* ant *Pi 4(t)* | Near isogenic lines (NILs) | Yu et al., 1991 |

5-7 backcrosses, retaining the conventional marker in each such backcross. Although an NIL, as the term near isogenic indicates, will contain alleles derived from DP at several loci on different chromosomes, due to selection of conventional marker, about 50% of such alleles are expected to be present on the chromosome segment that carries the conventional marker introgressed into RP. It has been calculated that if DP and RP differ for 100 molecular markers, NIL will retain four of them in a species with n = 20 (each chromosome with 5(l cM length), and of these four, two or three will be present on the marker chromosome. Therefore, allelic contrast between RP and its NIL, and allelic equality in the corresponding DP and NIL for a molecular marker suggests linkage of molecular marker with the conventional marker. This will also give an estimate of the linkage distance between the conventional and molecular markers. This technique has been successfully employed for tagging genes of economic value in several crops including, tomato, potato, maize, barley, rice and sugarbeet. Some examples of RFLP linkage mapping of genes of economic value are the given in Table.

## Identification and mapping of quantitative trait loci (QTLs) using RFLPs.

Many quantitative trails of economic value are under polygenic control and are selected directly. Such a selection is often ineffective, since the effect of each gene is small and is also influenced by the environment. Therefore, one would welcome a procedure for indirect selection, which is not influenced by the environment. For this purpose, linkage between quantitative traits and major marker genes should be known. Such marker genes involving morphological traits have been identified and mapped in Drosophilla and in several crop plants. Isozyme marker loci linked to quantitative traits of economic value had also been earlier identified in tomato and maize. However, in recent years, it became possible to identify and map quantitative trait loci (QTLs) using RFLP markers, particularly in some crop plants like tomato and maize. The RFLPs are recommended for this purpose, since these are numerous and codominant.

For the study of RFLP-QTL linkage relationships, the experimental procedures differ for self pollinated and crow pollinated species, although with the availability of inbred lines in cross pollinated species, similar methods can be used in the two cases. In self pollinated genotypes and among inbred lines (in cross pollinated species), parents are first selected which differ for a number of RFLPs as well as fur mean values of the quantitative traits. Once such parental lines are selected, they arc crossed and $F_2$ plants are derived. In the F, population, a chromosomal segment representing linkage between a RFLP marker and an OTL will be present in the background of random genetic variation due to independent assortment and recombination. By growing samples of' population in different environments, one may also study the linkage under different environments. In these cases, if mean values of a particular quantitative trait are determined in two groups of plants representing alternative alleles of RFLP, a significant difference in means (of two groups) for quantitative trait will indicate RFLP-QTL linkage. The observed marker associated difference in quantitative trait value will be a function of allelic effects at the QTL. Map distance between an RFLP and a QTL can also be determined.

Results of an experiment are shown in the table, wherein tomato linkage is evident between one of the two molecular markers ($A_1$, $A_2$) and a QTL for high soluble solids. Two lines (cullivars VF 36 and its near isogenic line LA 1503) used in the cross had molecular markers (RFLPs) that identified L.esculentum ( 5% soluble solids) and L.chrnielewskii(10% soluble solids), and also differed significantly for soluble solids (5% and 7-8%). In F, , allele 'e' of the marker $A_1$ showed association with the QTL for high mean value of soluble solids, while the allele 'e' of Al' and both the alleles of A, did not suggest that this QTL is found in the vicinity of A . RFLPs. associated with the expression of complex traits like insect resistance and W. U. E., i.e. water use efficiency have also been identified and mapped in tomato.

**Table**

The data on soluble solids content (mean per cent values) and two RFLPs in an $F_2$ derived from a crow between a tomato cultivur (5% soluble solids) ant its near isogenic line (NIL) carrying a QTL for high (7-8%) soluble solids derived from L chmielewskii with 10% soluble solids (data from Osborn et al., 1987).

| *RFLP locus* | *% Soluble solids* | |
|---|---|---|
| | *RFLP allele e/e from L. esculentun* | *RFLP allele c/c from L. chmielewskii* |
| $A_1$ | 5.82 | 6.24* |
| $A_2$ | 5.93 | 5.96 |

* significant difference at 5% level

In the experiments like the above, if parent differs for a sufficient number of molecular markers (RFLPs,) which are spaced out throughout the genome, a single cross can allow identification and mapping of al those QTLs for which the parent differs.

New analytical methods for the analysis of quantitative trait loci in plants using mapped RFLP markers were later suggested. These new methods included maximum likelihood method for interval mapping which was earlier used in humans. Interval mapping assesses the effects of each genome segment, located between a pair of marker loci, rather than the effect of a QTL associated with a single RFLP. This provides greater precision. Lod scores (log of the score of odds for the presence of a linkage against the odds of not having this linkage, e.g. if the score is 100: 1 in favour of linkage, the lod score is 2) are also used in this method to reduce the probability of false positives for linkage. The results are depicted as 'QTL likelihood maps'. The technique of interval mapping allowed the QTLs to be mapped to intervals of 20cM, which was a poor resolution. Therefore, the technique of interval mapping was supplemented through the use of substitution mapping using selected overlapping recombinant chromosomes. This technique is analogous to the use of overlapping deletions and allowed fine mapping of QTLs. The method has been successfully used for a study of tomato fruit characters, including mass per fruit, soluble solid concentration and fruit pH. The QTLs for these characters were mapped to intervals, as small as 3 cM.

In the above methods of QTL-RFLP linkage analysis, no distinction can be made between a group of tightly linked loci, each with small effect and a single locus of large effect for a QTL. A distinction between these two situations may be important for fundamental as well as for applied studies, since selection strategies may differ in two cases, and also because studies at the molecular level are often conducted at individual gene level. However, the success of genetic improvement based on selection of marker genes still remains to be tested, and will be effective only in specific uses. Further, it may not be necessary to use this method in cases where direct selection is more convenient and economic.

**Indirect selection in plant breeding using RFLPs**

If an RFLP locus is tightly linked with a gene of interest, indirect selection for the gene can be exercised by selecting for a particular RFLP phenotype, which is linked with the desired phenotype for the economic trait. Such an indirect selection has several advantages including the following: (i) during backcrossing, selection for a recessive gene can be exercised even in a heterozygous plant, because RFLPs are co-dominant; this will eliminate the need of selfing after each backcross as required in conventional backcross method of plant breeding and will thus help in speeding up the process; (ii) indirect selection may also 'be desirable for dominant genes in some cases during backcrossing because selection can be exercised at the seedling stage and the plants not carrying the desirable allele can be weeded out early, thus saving space and expense; (iii) during pyramiding of genes (when it is desired to incorporate more than one gene for the same trait) for traits like disease or insect resistance, it is not possible to select for the presence of an additional gene in the presence of an existing gene for the same trait; under such a situation, RFLP markers associated with different genes will differ and can be conveniently used to follow any number of genes during the breeding programme; (iv) selection for map QTLs representing several chromosome segments contributing to the quantitative trait, can be exercised simultaneously with the help of RFLPs associated with these chromosome segments.

**Identification of breeding lines and varieties using RFLPs**

It has been demonstrated that even closely related inidividuals of son species (e.g. humans) can be distinguished using DNA fingerprinting. Although probes equivalent in resoultion to minisatellites or VNTRs of human beings have not been reported in plants, genotype specific RFLP patterns have been detected and used for the identification of breeding lines and varieties and also for the characterization of germplasm resources. DNA fingerprinting, using simple sequence repeats oligonucleotides, eg. (CT)n or (AC)n, has also been utilized in plants.

Genotype specific RFLP patterns have already been detected. Their use for distinguishing varieties has been suggested in crops like main oilseed, lettuce, soybean, potato and rice. An exception, however, may be tomato, where intervarietal differences were rare. In rice, cultivar specific DNA fingerprints were obtained when a human minisatellite DNA probe was used.

Any character used for variety identification needs to fulfill the following basic criteria: (i) distinguishable intervarietal variation, (ii) minimum intravarietal variation, (iii) environmental stability and (iv) experiment reproducibility (in cases of biochemical or molecular markers). The requirements are met in most of the cases where molecular markers have been identified for the detection of varieties. In barley, cDNA clones of hardein proteins were used to demonstrate that polymorphism exists for RFLPs in several barley cultivars which can therefore be identified using RFLPs. In potato, two probe/enzyme combinations were found adequate for the identification of more than 100 varieties, and one probe/enzyme combination distinguished 19 out of 20 4x varieties and 33 out of 2x lines. However, the limitations for the practical use of RFLPs for variety identification include the following : (i) use of radioactive isotol is inconvenient, (ii) the technique is expensive, and (iii) the technique is labour intensive. As already discussed, non- radioactive labelling (biotin and digoxigenin) has already become possible for use of molecular probes. Similarly, automization (use of automatic equipments, reducing dependence on labour) and standardization of the technique will make less expensive and less labour intensive, so that the RFLP fingerprint technology will develop into a valuable and reliable varietal diagnosis tool.

## Characterization of genetic resources using RFLPs

Characterization of genetic resources becomes prohibitive due to limitation of time and resources, particularly when large collections are being evaluated. The collections may also have duplications which should be eliminated and a well characterized core collection should be prepared. Such core collections (5-10% of large collections) may be characterized using biochemical and molecular markers in addition to morphological and agronomic traits. These biochemical and molecular markers may also be used to assess genetic diversity within and among accessions and to monitor changes during maintenance and regeneration of genetic stocks.

In the form of biochemical markers, isozymes have been used, but their use is limited due to lack of polymorphism and a small number of loci and alleles available for analysis. The use of RFLPs for characterization overcomes this limitation, since they provide substantial polymorphism and they can be unlimited in number. The only disadvantage of using RFLPs for characterization of genetic resources is the cost of material and labour, which may gradually be minimized. Developments like PCR (polymerase chain reaction) and RAPD (random amplified polymorphic DNA, pronouced as 'rapid') will make the use of molecular markers more convenient and more economic.

## Study of phylogenetic relationships using RFLPs

In recent years, the techniques of molecular biology have been extensively utilized for revealing phylogenetic relationships. For instances sequences of small DNA segments (e.g. SS rRNA genes) have been used for suggesting phylogenetic relationships in Angiosperms. Similarly, nuclear RFLPs (not so much those for mitochondrial or chloroplast DNA) have proved useful for the study of phylogenetic relationships in several groups of plant species including the genera Brassica, Lens (lentils) and Solanum. In the genus Solanum, phenetic trees were obtained using wild and cultivated species and it was demonstrated that these trees do not change, when different computational methods or different genotypes of the same species or different sets of RFLPs were used.

In order to establish phylogenetic relationship among different species of a genus, data on restriction fragment patterns of these species using a large number (sometimes upto one hundred) of probe/enzyme combinations (several enzymes per probe) are first collected. Fragments on Southern blots are numbered 1 to n and a distance matrix is computed using presence or absence of these fragments in different species. It will be seen that in four genotypes, not all fragments are informative. In two genotypes of a pair, number of fragments for which two genotypes differ are worked out ($B_{ij}$) and divided by the total number of fragments present in at least one of the two genotypes ($M_{ij}$). From these estimates, distance ($D_{ij}$) between the two genotypes is calculated as follows: $D_{ij} = B_{ij} / M_{ij}$. This gives a distance

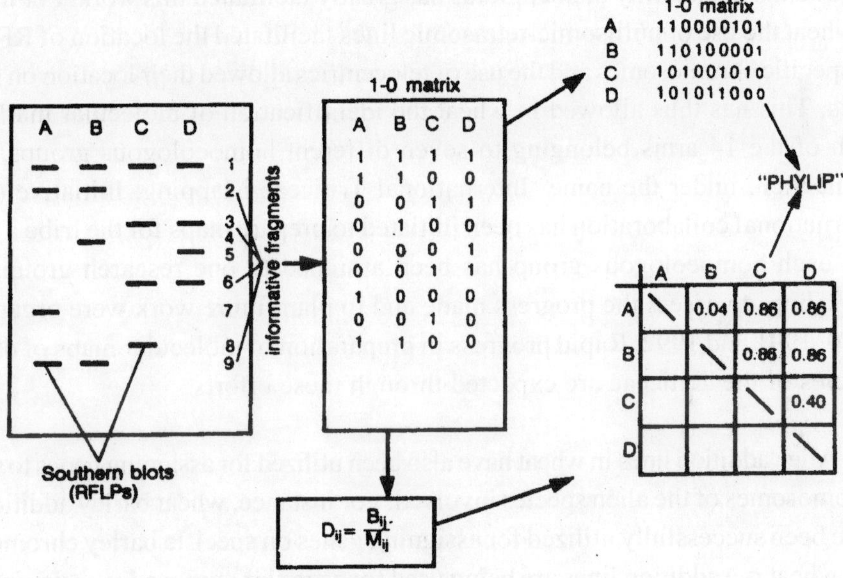


**Fig.** A distance matrix prepared for potato and its other Solanum species (based on fragments examined on Southern blots) and used for phylogenetic relationships.

matrix. This information is then used for computer analysis of the data and phylogenetic trees depicting phylogenetic relationships can be prepared.

## Cytogenetic RFLP Maps Using Aneuploids

Through a study of linkage of RFLP loci with morphological or biochemical (isozymes) markers having known chromosome location, it was possible to locate RFLP loci not only on specific chromosomes, but also on specific regions of these chromosomes. For this purpose, monosomics (particularly in maize), trisomics and various available translocation or deletion stocks were also utilized. This resulted in the preparation of RFLP chromosomes, maps for maize with more than 400 RFLP loci and those for tomato with more than 1000 loci.

The above technique also allowed mapping of duplicate loci, when more than one fragments homologous to a probe were detected on Southern blots. While in maize, 28.6% sequences detected duplicate loci in tomato, these duplicate loci were less than 2% of the loci tested. This gave new evidence about the presence of frequent duplicate chromosome segments in maize rather than in tomato.

Molecular chromosome maps' are also being prepared for wheat, barley, rice etc, and the availability of aneuploids has greatly facilitated this work. For instance, in wheat the use of nullisomic-tetrasomic lines facilitated the location of RFLP loci on specific chromosomes and the use of telocentrics allowed their location on specific arms. This has thus allowed in wheat the identification of molecular markers for each of the 14 arms belonging to seven different homoeologous groups. In this connection, under the name 'International Triticeae Mapping; Initiative (ITMI); international collaboration has been initiated to prepare maps for the tribe Triticeae and each homoeologous group has been assigned to one research group. Three workshops to assess the progress made and to plan future work were organised in 1990, 1991 and 1992. Rapid progress in preparation of molecular maps of different species of the Triticeae are expected through these efforts.

Alien addition lines in wheat have also been utilized for assigning genes to specific chromosomes of the alien species involved. For instance, wheat barley addition lines have been successfully utilized for assigning genes on specific barley chromosomes and wheat rye addition lines are being used for a similar purpose for assigning genes on specific rye chromosomes. Genome specific probes (isolated from DNA of Aegilops squarrosa) have also been isolated to identify the D genome chromosomes in wheat and several other polyploid species carrying the D genome e.g. Ae. cylindrica, CCDD;Ae. crassa, DDMM;Ae. ventricosn, DDUU. Recently, RFLP maps of the seven chromosomes of the D genome chromosomes in wheat have also been prepared through the use of genetic variability available in Aegilops squarrosa.

## RAPD Markers (using PCR) for Mapping and Crop Improvement

The technique of polymerase chain reaction (PCR), which has revolutionized a variety of techniques used in molecular biology, was earlier described. As discussed there, original PCR technique has been variously modified to suit a range of needs.

One such variation produces 'Randomly Amplified Polymorphic DNAs (RAPIDs); The use of RAPDs in preparation of linkage maps has been briefly describedanother in Chapter. The procedure is simple, quick to perform, requires relatively very small amounts of DNA and involves no radioactivity. A comparison of the properties of RFLPs and RAPDs is presented an in an early chapter. In early I990's, it was also shown that for RAPDs, instead of using a pair of oligonucleotides as primers, a single short oligonucleotide primer can be successfully used. This is based on the probability that a given DNA sequence (complementary to that of the primer) will occur in the genome, on opposite DNA strands, in opposite orientation, within a reasonable distance amplifiable by PCR. For most plant materials, primers that are 9-10 nucleotides long (random sequences with at least 50% G and C and lacking inverted repeats) are predicted to generate, on an average, 2-10 amplification poducts. Polymorphism results mainly due to changes in the primer binding site's presence or absence, and amplification products represent one allele per locus (dominant-null effect); The use of PCR technology for preparation of genetic maps and for a variety of crop improvement programmes will be discussed in this section.

**RAPDs and SSRs for genome mapping**

ı ne technique of molecular mapping using RAPD and SSR marker loci, using PCR has been described. These polymorphic loci can be mapped in much the same way as RFLPS. Although RAPD markers generate less information per locus examined than RFLP markers, they are considered more convenient and less time-consuming. For using these markers for preparing genetic maps, F2 and backcross segregating populations, recombinant inbred lines and doubled haploids can be used in the same manner as RFLPs.

A large number of plant materials have already been used for preparation of genetic maps using RAPD and SSR loci. The plant materials include the following: maize, rice, tomato, soybean, sugarcane, sunflower, flax and a number of trees (pine, eucalyptus, peach, citrus, etc.). These efforts will certainly be extended to other plant materials in future and will supplement the already available high resolution RFLP maps in several crops.

**Use of PCR for mapping STS and EST (or TST)**

Most of the molecular markers (RFLPs and RAPDs) in a variety of plant materials are now also being converted into sequence tagged sites (STS, where genomic DNA is involved) and expressed sequence tags or transcribed sequence tags (EST or TST, where cDNA probes are used). This has already been done in materials like rice and Arahidopsis and will be extended to other materials in future. These STS or EST maps will be useful, because each marker will be identified by a known sequence and not by a DNA or cDNA clone of unknown constitution.

## RAPDs for gene tagging

Earlier in this chapter, we discussed the use of RFLPs for selection of markers linked to monogenic and polygenic traits, so that these may be used for indirect selection in segregating populations during plant breeding programmes. RAPDs can be used for the same purpose with lesser effort. Following are some PCR approaches already utilized for gene tagging.

Tagging of major genes— At least three different approaches have already been utilized for the use of PCR in tagging major genes, and each has been successfully utilized in tomato or lettuce. (i) Use of near isogenic lines : In tomato, two NIls, which differed for the presence or absence of a gene (pto) conferring resistance to Pseudomonas, were screened using 144 random primers leading to identification of three RAPDs, which were tightly linked to pto gene (Martin et al., 1991). The entire work took four weeks, in contrast to two years required for the same objective using RFLPS. (ii) Use of homozygotes derived form $F_2$. In lettuce, from an $F_2$ population segregating for downy mildew resistance gene (Dm), homozygous dominant (Dm Dm) and homozygous recessive (dm dm) plants were identified (Dm Dm were distinguished from Dim dm through progeny testing). Two separate samples of DNA were prepared for RAPD analysis, one from ten homozygous dominant resistant plants and the other from ten homozygous recessive susceptible plants. In total, only 300 PCR reactions were required to obtain three markers linked to Dm, illustrating the efficiency of RAPDs for gene tagging (Michelmore et al., 1991) (iii) Use of RFLPs flanking a DNA segment. In tomato, high density linkage map was used to select two intervals (one 15 cM long and the other 6.5 cM long), each containing the gene responsible for pedicel abscission and fruit ripening. Each interval was flanked by RFLP markers. Individual plants having alternative alleles of flanking markers were classified in two groups, and DNA samples obtained from the two groups of bulked plants were screened with 200 random primers for PCR products. It was assumed that polymorphism in PCR products should be derived from either within the defined interval or from a segmant very close to it. Using this approach, three polymorphic products were identified linked to the selected intervals and to the gene of interest (Giovannoni et al., 1991).

Tagging of QTLs— The polygenic traits, controlled by quantitative trait loci, are characterized by a continuous range of variation. The plants at the two extremes of this range should differ at most loci controlling a particular trait. These two extremes may be classified in corresponding groups and two DNA samples from these bulked plants may be screened for RAPDs using a number of random primers for PCR. Thus, the markers linked to loci controlling a particular quantitative trait can be identified. For this approach, if recombinant inbred lines (RILs) or doubled haploids (DHs) arc used, this will greatly facilitate tagging of QTLs. RILs and DHs can also be grown in replicated trials at different locations for several years to partition heritable and non-heritable variation, so that markers linked to heritable loci showing minimum G X E (genotype X environment) interaction are identified. This approach, using RFLPs, has alerady been utilized in several crops and will be utilized in future with the help of RAPD.

## RAPDs for identification of somatic hybrids

One of the problems in production of somatic hybrids, often, is lack of appropriate methods for identification of these hybrid,, among the developing regenerants Some of the methods traditionally used, include the following: (i) combining light bleached protoplasts with green leaf mesophyll protoplasls, followed by manual isolation of heterokaryons; (ii) genetic complementation of auxotrophic mutants; (iii) fusion of different antibiotic and/or herbicide - resistant lines derived by plant transformation and (iv) use of hybrid vigour, isozymes, proteins and/or RFLPs. The methods, however, are time-consuming and costly. RAPDs can be utilizerd for the same purpose, where a small DNA sample from developed regenerant can be used to determine hybridity. Primers may be used which give polymorphic PCR products from each of the parents involved in fusty so that the hybrids can be identified by the presence of molecular profit which represented a combination of the profiles of the two parents. This has been achieved successfully in potato (Baird et al. 1992), and the approa will certainly be utilized in future for identification of somatic hybrids in a variety of plant materials.

## RAPDs for conservation of genetic resources

The subject of in situ and ex situ conservation of plant genetic resources (PGR) will be discussed in some detail later on. It will be seen that genetic resources are traditionally characterized by a combination of morphological and agronomic traits. The effectiveness of this approach is really limited and therefore RAPDs are recommended as a useful system for evaluation and characterization of genetic resources. Some success in this direction has already been acheived elsewhere, but efforts in this direction are also being made at NFPTCR (National Facility for PI Tissue Culture Repository), New Delhi. Wayne Powell and his coworker at Scottish Crop Research Institute (Dundee, U.K.) have also used RAPDs for evaluation and characterization of collections from genera Gliricidia (nitrogen-fixing trees within agrofureslry systems in the tropics) and Theobroma (T. cacao = cocoa).

**Physical Maps Using In Situ Hybridization (ISH)**

Utilizing molecular probes derived from cDNA or gDNA (even if they are not derived from the same species i.e. heteroluguus probes), in *situ* hybridization could be successfully utilized for mapping genes with multiple copies specific chromosomes. For instance, cDNA copies of wheat rDNA were used for in situ hybridization to mitotic metaphase chromosomes of bread wheat. It was found that 90% of the ribosomal genes were located on chromosomes 1B and 6B and the remaining repeat units loci on SD (Miller et al., 1980, Peacock et al., 1951). This had suggested that diploid genomes like A genome, which had two chromosomal sites for rRN genes (Gerlach et al., 1980) had undergone a change after incorporation in hexaploid (6x) bread wheat. Similarly,Appets et al. (1913-8) sucessfully collected information on physical location of the SS rRNA loci in wheat and barley. Using in sinut hybridization (ISH), they found that SS rRNA are located distant to the secondary constrictions on chromosome 1B of who and 1R of rye and on non-nucleolar chromosome in barley. Recently, another additional SS rRNA locus in rye was located on 5R (Reddy and Appels, 1989).

In another recent report, physical mapping of the SS rRNA multigene families in bread wheat was achieved by combining the technique of in situ hybridization (ISH) with deletion mapping (Mukai m al., 1990). Terminal deletion stocks, produced by using monosomic addition of a gametocidal chromosome from Aegilops cylindrion, were utilized for detection and location of twelve SS rRNA loci on short arms of chromosomes of homologous groups 1 (1AS, 1BS and IDS) and 5 (5AS, 5BS and 5DS). Specific chromosome arms and regions of the arms associated with only five different loci were determined using ditelocentrics (DT) and deletion stocks. The position of ISH sites on a telocentric chromosome was determined as a fraction of the total telsome length from centromere. The postion of five loci of SS rRNA designated as SS-Rrnn Al, SS-Rrfln Dl, SSRrna A2, SS-Rrna B2 and SS-Rrno D3 were 0.77, 0.96, 0.76, 0.63 and 0.64 on arms IAS, 1DS, 5AS 5BS and 5DS respectively. The SS rRNA locus on IAS was discovered for the first lime during this study.

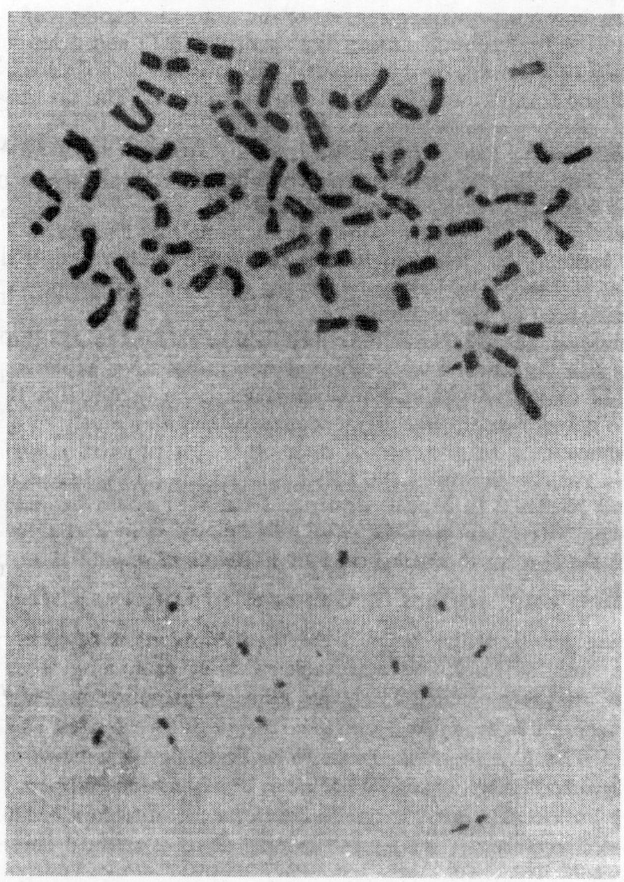

**Fig.** Terminal deletions in wheat chromosomes used for detection of twelve SS rRNA loci.

In sita hybridization has also been used for physically locating the molecular markes earlier mapped in rice. It has been shown that distances on linkage maps (in cM) are not proportionate to physical distances on chromosomes.

Techniques have also been described, where each of two or three DNA sequences can be labelled with different non-radioactive haptens, each of which can be visualized with an independent detection system. It is, therefore, possible to detect two or three DNA sequences simultaneously on animal or plant chromosomes, in presence of each other, for physical mapping. For instance, in rye chromosomes, the location of pTa71 (rDNA sequence from wheat) and pSc119.2 (a repeat sequence from rye) could be detected by labelling one with biotin and the other with digoxigenin and then using a mixture of the two for in situ hybridization (Leitch et al., 1991)

## Resolution Gap Differs in Genomes of Different Sizes

In preparing genetic RFLP maps, 1 cM (centiMorgan) is often considered to be the limit of resolution achievable without examining a very large population. At the molecular level, this limit of resoultion means different physical distances in organisms, with genomes of different sizes, as shown in Table. This gap, therefore, needs to be bridged, if a gene of interest is to be approached from a linked RFLP locus, that has been mapped. Bridging of this gap has become possible due to development of techniques like pulse field gel electrophoresis (PFGE) and cloning in yeast artifical chromosome (YAC) vectors, which will enable handling of large pieces of DNA. Similarly, jumping and linking libraries can be prepared, so that chromosome walking need not be followed throughout the genome to identify, locate and isolate a gene of interest.

**Table**

The resolution gap as worked out in four different genomes (Shields, 1989).

| *Organism* | *Genome size (kbp)* | *Map length (cm)* | *kbp equal to 1 cM* |
|---|---|---|---|
| Arabidopsis | $7 \times 10^4$ | 501 | 139 |
| Tomato | $7.15 \times 10^5$ | 1,400 | 510 |
| Maize | $3 \times 10^6$ | 1,400 | 2,140 |
| Human | $3 \times 10^6$ | 2,708 | 1,108 |

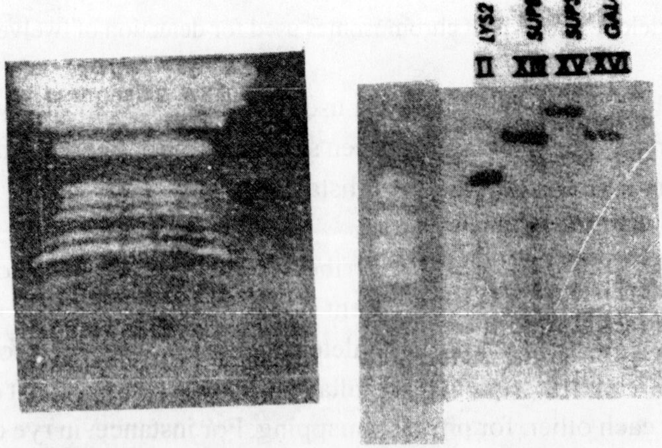


**Fig.** Separation of yeast genomic DNAs belonging to individual yeast, chromsomes on the gel by PFGE and stained with ethidium bromide

**Molecular Maps in Yeast and Other Fungi**

Since, E. coil can not serve the purpose of understanding eukaryotic genomes, yeast has been selected as the simplest eukaryote to understand the eukaryotic genome. Following are the reasons for the choice: (i) yeast has haploid DNA (14 × 103 kilobasepairs), which is only 3.5 times that in E. coil (4.0 × 103 kbp) ; Drosophila contains 25 times the DNA of E. coil, and higher vertebrates have 1000 times this DNA; (ii) work with yeast is not as expensive as with higher vertebrate cells, due to high cost of media required to grow them in culture; (iii) yeast cells grow and divide like bacteria, doubling their mass every 90 minutes; (iv) yeast genetic system has been very thoroughly understood where about 600 genes are already mapped, although they may still represent a fraction of the whole genome; (v) DNA molecules as long as 100kb can be routinely handled through PFGE. These can be produced by enzymes like Dpn 1 or Not 1, which recognize rare sites.

Yeast cell has 16 chromosomes, each with a single DNA molecule. Until recently, even the smallest of these DNA molecule was too large to be isolated in pure form, but due to the recent availability of pulsed field gel electrophoresis (PFGE), where short pulses of electricity are used in two different directions, separation of DNA belonging to each of the 76 individual chromosomes has become possible. Centromere sequence (CEM) about 130 bp and telomere sequence (TEL), about 100bp with repeat sequence, (CCCACACA)n have also been isolated. These sequences, when used as probes, can help in positioning the centromere and the telomeric ends on the chromosome map.

Using the technique of PFGE and its refinement known as contour clamped homogeneous electric field (CHEF) gel electrophoresis, genomes of several fungi could be resolved in chromosome bands, popularly described as electrophoretic karyotypes. These included Saccharomyces cerevisae (14 Mbp), Schizosaccharomyces pombe (15 Mbp), Neurospora crassa (47 Mbp), Aspergillus nidulans (31 Mbp) and Dictyostelium discoideum (52-56 Mbp). For electrophoretic karyotypes, intact chromosomal DNA has to be prepared in agarose plugs, which can also be purchased from commercial firms for organisms like yeast. These agarose plugs are subjected to PFGE or CHEF gel electrophoresis, which yields 15 bands for 16 chromosomes in case of yeast, 6 bands for 8 chromosomes in Aspergillus nidulans, and bands for 7 chromosomes in D. discoideum. In case of yeast, initally chromosome 12 did not clearly separate. Another strain of yeast (YPH 149 allowed identification of each of the 16 chromosomes as 17 distinct band using alternating CHEF gel electrophoresis (chromosome VII split into two).

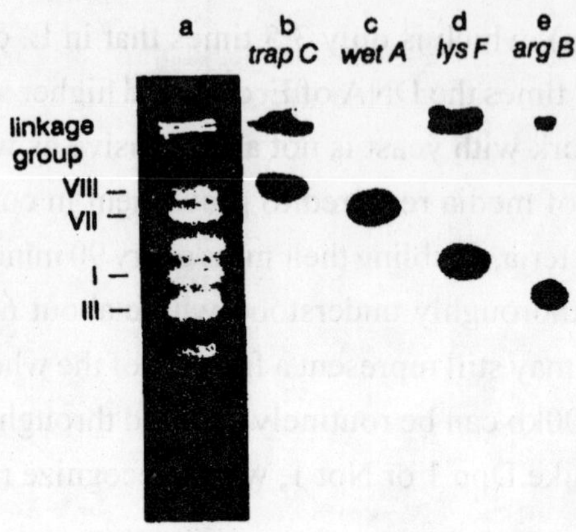


**Fig.** Separation of DNA of individual chromosomes from Aspergillus nidulans on the gel using CHEF. DNA was transferred to a nylon membrane which was cut into four strips and hybridized to DNA probes specific to four of the eight linkage groups (b, c, d, e); the four linkage groups are trp Co (VIII), Mw A (VII), lys F (I), arg B (III)

In Aspergillus, 2 of the 6 bands are doublets with 3.8 Mbp and 3.5 Mb respectively. Thus, these bands are in agreement with genetic and cytological information available in these cases The details of different DNA bands in yeast and Aspergillus nidulans are given in Table, along with the linkage groups to which these bands belong. The sizes of bands are measured by comparing with bands of known sizes obtained from agarose plugs having yeast intact chromosomes. The bands are assigned to specific linkage groups by using one of the several possible methods. For instance, in Aspergillus, this was achieved by combination of two methods: (i) hybridization with labelled DNA probes containing genes that are already mapped to their respective linkage groups; (ii) examination of CHEF gel band patterns of genomic DNAs from strains carrying known translocations. Utilizing these approaches, eight different linkage groups (I-Gs) in Aspargillus could be assigned to six electrophoretic DNA bands. The CHEF and genetic linkage data agreed that LGs VIII (5.0 Mb), VII (4.5 Mb) and II (4.2 Mb) are the largest, and` LG IV (2.0 Mb), the smallest.

**Table**

In Physical signs of chromosomes separated by PFGE, in yeast (Mortimer and Schild, 1985) and Aspergillus nidulans (Brody and Carbon, 1989).

| *Yeast (total DNA = l4 Mb)* | | | *Aspargillus nic(nlarrs (total DNA = 31M6)* | | |
|---|---|---|---|---|---|
| **Band** | **Chromosome (recombination length in cM in parentheses)** | **DNA (in kb)** | **Band and chromosome** | **Linkage group** | **DNA (in kb)** |
| 1 | 1 (98) | 198 | 1 | 1V | 2900 |
| 2 | 6 (137) | 225 | 2 | 111 | 3500 |
| 3 | 3 (137) | 311 | 3 | VI | 3500 |
| 4 | 9 (198) | 495 | 4 | 1 | 3800 |
| 5a | 8 (183) | 635 | 5 | V | 3800 |
| 5b | 5 (233) | 709 | 6 | 11 | 4200 |
| 6 | 11 (242) | 763 | 7 | VII | 4500 |
| 7 | 10 (200) | 835 | 8 | VIII | 5000 |
| 8 | 14 (283) | 864 | | | |
| 9 | 2 (244) | 948 | | | |
| l0a | 13 (229) | 991 | | | |
| l0b | 76 (176) | 1047 | | | |
| 11a | 15 (361) | 1146 | | | |
| 11b | 7 (391) | 1285 | | | |
| 12 | 4 (485) | 1571 | | | |
| 13 | 12 (294) | 2194 | | | |

# FOURTEEN

# Production of Transgenic Plants

Variety of methods can be used for transfer of foreign genes to plant cells, tissues or organs. This transformation has been achieved at the level of protoplasts or cells in many plant species although the ultimate objective should be the production of transgenic plants following regeneration of whole plants from transformed protoplasts/cells. Not in all cases, the success in transformation could be combined with success in regeneration. However, now there are more than 50 plant species, where transgenic plants have been successfully produced (Table). Initially, the production of transgenic plants was restricted to dicotyledons, but it has now been extended to several monocotyledons like wheat, maize, rice and oats.

Progress in this exciting area of research for production of transgenic plants has been so spectacular that by the turn of the century, we hope to be growing crops which have been tailored to market specifications by the addition, subtraction or modification of genes. Transgenes will also be important in increasing the efficiency of crop production systems. For instance, transgenic plants resistant to herbicides, insects, viruses and a host of other stresses have already been produced. Transgenic plants have also been produced, which are suitable for food processing (e.g. bruise resistance and delayed ripening in tomato). Another exciting example is the production of male sterile (due to barnase gene) and fertility restorer (due to barstar gene) plants in Brassica napus, so that hybrid seed in future will be conveniently produced without manual emasculation and controlled pollination as practiced in maize. This has also eliminated the need for a search of cytoplasmic male sterility (cms) and fertility restoration system in crops, where hybrids are intended to be produced for higher yields. Another major goal for production of transgenic plants, is their use as bioreactors or factories for production of speciality chemicals and pharmaceuticals. This area is described as 'molecular farming or molecular pharming'. The transgenic plants have also been produced for identification of regulatory sequences for many genes, using gene constructs with overlapping deletions. These achievements and prospects of their future use will be briefly described in this chapter.

## I.Transgenic Plants for Crop Improvement

### Transgenic plants in dicotyledons

Despite the totipotency of plant cells, thus obviating the need of transforming specifically the germ line cells as required in animals, the production of large number of transformed plants was, till recently (in 1980's), restricted mainly to tobacco, petunia or tomato of the family Solanaceae. There are, however, recent reports of transgenic plants from other dicotyledonous families like Cruciferae, Leguminosae, etc. and from monocotyledonous families like Liliaceae and Gramineae. Some of these examples, where transgenes of economic value have been utilized are included in Table, and will be described in this section.

**Table**

A list of higher plants where transgenic plants have been produced, using different methods.

**I. Herbaceous dicotyledons**

1. Nicotiana tabacum (tobacco)
2. N. phunbaginifolia (wild tobacco)
3. Petunia hybrida (petunia)
4. Lycopersicon esculerraan (tomato)
5. Solanurn tubersum (potato)
6. Solarium melongena (eggplant)
7. Arubidopsis thaliana
8. Lactuca saliva (lettuce)
9. Apium graveolens (celery)
10. Helianthus annuus (sunflower)
11. Linum usitatissimum (flax)
12. Brassica napes (oilseed rape; canola)
13. Brassica oleracea (cauliflower)
14. Brassica oleracea var. capitata (cabbage)
15. Brassica rapa (syn. B. campestris)
16. Gossypium hirsutum (cotton)
17. Beta vulgaris (sugarbeet)
18. Glycine max (soybean)
19. Pisum sativum (pea)
20. Medicago saliva (alfalfa)
21. M. varia
22. Lows conticulatum (lotus)
23. Vigna aconitifolia
24. Cocumis sativus (cucumber)
25. Cucumis melo (muskmelon)
26. Cichorium intybus (chicory)
27. Daucus carom (carrot)
28. Armoracia sp. (horse radish)
29. Glycorrhiza glabra (licorice)
30. Digitalis purpuria (foxglove)
31. Ipomoea batatas (sweet potato)
32. Ipomoea purpurea (morning glory)
33. Fragaria sp. (strawberry)
34. Actinidia sp. (Kiwi]
35. Corica papaya (papaya)
36. Vitis vinifera (grape)
37. Vaccinium rnacrncorpnn (cranberry)
38. Dianthus caryophyllus (carnation)
39. Chrysanthemum sp. (chrysanthemum)
40. Rosa sp. (rose)

**II. Woody dicotyledons**

41. Populus sy. (poplar)
42. Malus sylvestris (apple)
43. Pyrus communis (pear)
44. Azadirachta indica (neem)
45. Jiglans regia (walnut)

**III. Monocotyledons**

46. Asparagus sp. (asparagus)
47. Dachtylis glomerata (orchard grass)
48. Secale cercale (rye)
49. Oryza sativa (rice)
50. Triticum aestivurn (wheat)
51. Zea mays (corn)
52. Avena sativa (oats)
53. Festuca arundinacea (tall fescue)

**IV. Gymnosperms (a conifer)**

54. Picea glauca (white spruce)

**Herbicide resistance in transgenic plants**—Due to increasing concern about contamination of environment by herbicides, new herbicides are being developed that are safer and biodegradable. This has necessitated the development of resistance in crop plants against these new and safer herbicides. Herbicides, normally affect processes like photosynthesis or biosynthesis of essential amino acids (Table). Two approaches have, therefore, been used for the development of resistant plants: (i) In the first approach, we try that either the target molecules should become insensitive to herbicide or the target protein should be overproduced. (ii) In the second approach, a pathway is introduced that will detoxify the herbicide.

**Table**
Mechanisms of action of different herbicides and basis of achieving resistance against them in transgenic plants.

| | *Active principle of herbicide* | *Inhibited pathway* | *Target product* | *Use* | *Basis of resistance* |
|---|---|---|---|---|---|
| **I.** | **Amino acid biosynthesis inhibitors** | | | | |
| 1. | Glyphosate (Roundup) | Aromatic | EPSPS | Broad spectrum | Overexpression of EPS Gene |
| 2. | Sulphonylurea and Imidazolinones | Branched chain amino acids | ALS | Selected crops | Mutant ALS gene |
| 3. | Phosphinothricin (Basta) | Glutamine biosynthesis | GS | Broad spectrum | Gene amplification, bar gene: detoxification |
| **II.** | **Photosynthesis Inhibitors** | | | | |
| 4. | Atrazine (Lasso) | Photosystem II | QB (32 kDa protein) | Selected crops | Mutant Psba gene, GST gene; detoxification |
| 5. | Bromoxyni (Buctril) | Photosynt-hesis | — | Selected crops | bxn gene; detoxification |

**(a) Modification of flee target**—The target has been modified for developing resistance against at least three herbicides **(glyphosate, sulphonylureas and imidazolinones).** The transgenic petunia plants resistant to glyphosate (active ingredient of **Roundup** herbicide) were developed by transfer of a gene for EPSPS (5-enol-pyruvylshikimat-3-phosphate synthase), that overproduces this enzyme. This overexpressing gene was isolated from plants selected for herbicide resistance. In some other cases, a gene aroA was isolated from the bacteria Salmonella typhimurium or E. coli and was transferred to tomato and/or tobacco. Another class of herbicides includes sulphonylurea compounds (active ingredients of Glean & Qust herbicide) and imidazolinones that inhibit the enzyme acetolactate synthase (ALS), which is involved in the bisoynthesis of branched chain amino acids like leucine, isoleucine and valine. Transgenic tobacco plants expressing a mutant ALS gene from tobacco or Arabidopsis were produced that were tolerant to sulphonylurea herbicides. Similarly, transgenic tobacco plants were produced by incorporation of genes for resistance against (i) Lphosphinothricin PM, which is the active ingredient of herbicide 'Baste' and inhibits glutamine synthase (GS), and (ii) atrazine which inhibits photosynthesis. These genes were isolated from Medicago sativa and Amaranthus hybridus respectively. The transgenic tobacco plants expressed low level of resistance against the corresponding herbicides. Scientists at Plant Genetic Systems and at the German Company Hoechst, isolated a gene from 'Streptomyces hygroscopicus', which encodes an enzyme, capable of inactivating the herbicide 'Baste'. Transgenic plants with this gene have been produced and field tested, demonstrating effectiveness of this gene for protection against the herbicide 'Baste'.

***(b) Detoxification or degradation of herbicide***—Detoxification or degradation of herbicides is the basis of selective use of herbicide, so that the latter will kill the weeds and not the crop. A number of detoxifying enzymes have been identified in piants as well as in microbes. Some of these enzymes include (i) glutathione-S transferase or GST (in maize and other plants), which detoxifies the herbicide atrazine; (ii) nitrilase (coded by gene bxn in Klebsiella pneurnoniae, which detoxifies the herbicide bromoxynil, and (iii) phosphinothricin acetyl transferase or PAT (coded by bar gene in Streptomyces spp.), which detoxifies the herbicide PPT (L-phosphinothricin).

Transgenic tomato plants using the bxn gene from Klebsiella and bar gene from Streptomyces and transgenic plants in potato, oilseed rape (Brassica napes) and sugarbeet using her gene from Streptomyces have been obtained and were found to be herbicide resistant. Other target crops for engineered herbicide tolerance include soyabean, cotton and corn. Field trials with transgenic plants in some cases are either being conducted or will be conducted in the near future, so that the utility of transgenic herbicide resistant plants and the risk of growing them at the commercial scale will be known. It is estimated that the first transgenic plants will be commercially grown in mid 1990's and that a number of other transgenic plants will be grown commercially before the turn of the century.

**Insect resistance in transgenic plants**—(a) Genes for Bt toxins—The use of pesticides and insecticides is a common measure in plant protection programmes, since pests and insects cause appreciable damage to our crops. Most of these pesticides and insecticides are chemically synthesized. However, an exception is the Bt toxins produced by a bacterial species (Bacillus thuringiensis), so that a spore preparation of this bacterium has been used as a biological insecticide during the last 20 years. Insecticidal activity of this species depends on a protein (delta endotoxins) synthesized during sporulation. Since these toxins are very specific in their action, they are safe insecticides, but their use is limited due to high production cost and due to instability of crystal proteins when exposed in the field.

The above toxin gene (bt2) from B. thuringiensis has been isolated and used for Agrobacterium Ti plasmid mediated transformation of tobacco, cotton and tomato plants. The transgenic plants were resistant to the Manducta sexta, a pest of tobacco. Experiments of feeding the leaves of these plants to larvae of M. Sexta showed 75%-100% mortality of the larvae, while the control plants carrying no transgenes were severely damaged. The presence of the gene, bt2 as well as that of the toxin protein synthesized under its control was also demonstrated by appropriate experiments. When inheritance of insect resistance was studied using crosses with normal control plants, Fl showed resistance and F2 generation exhibited expected segregation.

Field tests using transgenic insect resistant plants were also conducted and the results were excellent with tobacco and tomato. In view of this, one can expect that transgenic crop plants for this trait may be released for commercial cultivation in the near future. There were also reports that India may acquire technology from USA for introducing Bt toxin gene in cotton for the development of resistance against pests in this major cash crop of India.

***(b) Genes for protease inhibitors (e.g. gene for cowpea trypsin inhibitor or CpTI).*** In cowpea (Vigna unguiculata), trypsin inhibitor (CpTI) level was shown to be responsible for its resistance to attack by the major storage: pest of its seeds (i.e. bruchid beetle = Callosobruchus maculatus). CpTI was later shown to be toxic to a variety of insects (Table) but cowpea seeds with high level of CpTI are not toxic to humans. It was, therefore, considered desirable to transfer genes) for CpTI for production of transgenic insect resistant plants. A number of binary vectors were developed using Ti plasmid, where CpTI gene was joined with CaMV 35S promoter, and one or more marker genes. The vector was mobilised into Agrohocterium, which was used to infect tobacco leaf discs, which led to the production of transgenic tobacco plants expressing high level of CpTi (as shown by western blotting) imparting resistance against a variety of insects (Table). The CpTI gene in transgenic plants is stably inherited and there is no serious 'yield penalty'. Thus, Like Bt toxin,CpTI can also be used as a protectant against insect attack in transgenic plants. However, extensive field trials will be necessary before these transgenic plants can be released to the farmers for cultivation.

**Table**

Insecticidal efficacy of CpTI in artificial diets and in transgenic plants (after Gatehouse et aL, 1991)

| *Insects killed by CpTI in artificial diets* | *Insects, against whom resistance noticed in transgenic tobacco plants (with CpTI)* |
|---|---|
| **I. Lepidoptera** | |
| Heliothis virescens | Helothis virescens |
| H. zea | H. zea |
| Spodoptera littoralis | Spodoptera littoralis |
| Chilo partellus | Manducta seta |
| Autographa gamma | |
| **II. Coleoptera** | |
| Callosobruchus maculatus | |
| Anthonomus grandes | |
| Diabrotica undecimpunctata | these insects do not attack control |
| Tribolium confusum | tobacco plants |
| Costelytra zealandica | |
| **III. Orthoptera** | |
| Locusts rnigratoria | |

**(c) Genes for other insecticidal secondary metabolites**—Secondary metabolites produced by plants have also been implicated in the resistance to insect attack (Table). However, biosynthesis of each of these metabolites involves a series of steps (sometimes even more than one biosynthetic pathway), each controlled by a separate gene. Furthermore, these genes are tissue specific in expression. These features make the production of transgenic plants difficult in this case. CpTI discussed in the previous section is also a secondary metabolite, but its transfer is easier, since it is single gene controlled.

The production of transgenic plants by transfer of genes for entire **multi-enzyme biosynthetic pathway** (or for its augmentation) is though not yet achieved, but seems possible in future. For this purpose, not only the transfer of genes is required, but the gene expression needs to be regulated, otherwise it leads to 'yield penalty' and also to toxic effect when consumed by humans and livestock. Defined promoter sequences are now available, which will allow in the transgenic plants a control over the temporal and spatial expression of genes for biosynthetic pathways of secondary metabolites. Utilizing these facilities, transgenic plants with insect resistance due to secondary metobolites will be available in the near future.

**Table**

Examples of secondary compounds from legume seeds with demonstrated insecticidal properties (after Gatehouse et eL, 1991).

| *Compounds* | *Bruchid (insect)* | *Legume* |
|---|---|---|
| **I. Non-protein antimetabolites** | | |
| **1. Alkaloids** | | |
| (i) 2,5-dihydtnxymethyl-3,4-dihydroxypytrolidine (DMDP) | CalloscbrucJurs rnaculaurs | Lonchocarlnrs sp. |
| (ii) Castanospermine | Callosobruchus rnaculanrs | Castanos Penne»n australe |
| **2. Non-protein amino acids** | | |
| p-Aminophenylalanine | Collosobnrchus rnaculantus | Vigrru sp. |
| | Zabrotes scrhJasciunrs | |
| 3. Rotenoids | Callosobnichus maculatus | Lonchocarlnrs salvadorcrtsis |
| 4. Saponins | Callosobruchus chinensis | Phascohrs vulgaris |
| | Callosobnrchus chirrensis | Glycine max |
| 5. Polysaccharides | | |
| (i) Peetosans | Callosobnrchrrs chinensis | Phascohrs vulgaris |
| (ii). Heteropolysaccharide | Callosobruchus chinensis | Phaseolus vulgaris |
| | Acanthoscelides obtecturs | Phuscohus vulgaris |
| (iii) Polysaccharide | Callosobnichus chinensis | Vigna spp. |
| **II. Protein anlimelabolites** | | |
| 1. Lectins (phytohaemagglutinins) | Callosobnichus »raculanrs | phaseolus vulgaris |
| 2. x-Amylase inhibitors | Zabrotes subfascianrs | Phascohrs vulgaris |
| | Callosobnichus chinensis | Phaseolus vulgaris |
| 3. Protease inhibitors | Callosobnrchus maculaurs | Vigrra unguiculata |
| 4. Areelin / LLP | Zabrotes subfasciaters | Phascolrrs vulgaris |
| | Zabrotes subfasciatus | Phaseolus vulgaris |

Resistance against viral infection—(a) Cross protection It has been: shown that if susceptible strain of a crop is inoculated with a mild strain I of a virus, then the susceptible strain develops resistance against more virulent strains. This phenomenon is known as cross protection and has been used to reduce yield losses in crops like tomatoes against tomato mosaic virus (TMV), in potato against potato spindle tuber viroid and inl citrus against citrus tristeza virus. In most cases of cross protection, the symptoms of infection are delayed, and even the replication of virus isl suppressed, although eventually the severe strain may be able to overcome the protection. There are various disadvantages of this practice of cross protection such as (i) possibility of mutation in inducing mild virus strain, (ii) possibility of synergism between inducing virus and another unrelated virus, (iii) possibility of unnecessary spread of mild virus causing threat for future yield losses and (iv) possibility of some yield losses due to mild strain also. These disadvantages can be overcome, if a single gene imparting this benefit is transferred and transgenic plants arc produced. Such transgenic plants have been produced in tobacco, tomato, and potato using a broad spectrum of plant viruses. Following three kinds of genes, have been used for this purpose.

***(b) Gene for virus coat or capsid protein (CP) from positive strand RNA viruses***—Coat protein gene from tobacco mosaic virus (TMV), classified as a positive strand RNA virus, has been transferred to tobacco and in the transgenic plant, expression of coat protein (CP) was observed. Further, when inoculated with TMV, the infection in the transgenic plants was very low and delayed relative to control plants that were not transformed. This provides a new approach for producing virus resistant plants complementing the efforts of classical plant breeding in producing disease resistant crop varieties. However, in such transgenic plants, coat protein gene should be constitutively expressed and may thus have effects on the nutritional value of plants. Subsequently, this approach has been applied to a range of crops (tomato, alfalfa, tobacco, potato, melons, rice) for developing resistance against a broad spectrum of positive RNA strand plant viruses (e.g. alfalfa mosaic virus, potato virus X = PVX, potato virus Y = PVY and potoato leaf roll virus). Potatoes have been produced which have coat protein genes and are tolerant in field tests to both PVX and PVY, DNA coding for a component of TMV replicase enzyme was also transferred to tobacco plants, conferring resistance to TMV.

***(c) Gene for nucleocapsid (N) protein from tomato spotted wilt virus (TSWV).*** In earlier reports of resistance against viruses in transgenic plants, a number of positive (+) strand RNA viruses were involved. No report of resistance against any negative (-) strand virus was available. An important negative strand virus is tomato spotted wilt virus (TSWV), which causes considerable yield losses in crops like tomato, tobacco, lettuce, groundnut, pepper and in ornamentals like Impatiens, Ageratum and Chrysanthemum. In this virus, genomic RNA is tightly associated with nucleocapsid (N) protein. This protein helps in (i) wrapping of viral RNA and also in (ii) regulation of transcription-to-replication switch during infection cycle.

For producing transgenic tobacco plants, N gene was associated with CaMV 35S promoter and a leader sequence (to enhance expression). The transgenic plants showed expression of N-gene and also significant resistance to TSWV. A correlation was also observed between the amount of N protein and the level of resistance. The mechamism, involved in imparting resistance against negative strand RNA virus like TSWV, may differ from that involved in positive strand RNA virus, where coat protein (CP) gene has been used.

***(d) Satellite RNA and its use for transformation***—Satellite RNAs are species of RNA associated with specific strains of some plant RNA viruses, although it is not necessary for their replication. Replication of this satellite RNA depends on the virus, so that it gets packaged with it to cause infection elsewhere. Therefore, satellite RNA depends on virus for its replication and transmission, even though it is unrelated to viral genome. Presence of sat-RNA leads to reduction in severity of disease symptoms, and therefore, has been used for developing resistance against specific viruses.

Using the above principle, transgenic tobacco plants have been produced which carried a DNA fragment, which when transcribed gives a species of satellite RNA which can reduce the severity of disease. Following two such attempts were made : (i) In one case, transgenic tobacco plants using DNA for RNA satellite of cucumber mosaic virus (CMV) were produced and when satellite free aggressive CMV strains were used for infection, the presence of satellite RNA (transcribed from DNA) in transgenic plants led to reduction in disease symptoms. These transgenic plants are also protected against another related virus i.e. tomato aspermy virus. (ii) In the other example involving tobacco ringspot virus (TobRV), transgenic tobacco plants could be produced, which carried DNA version of satellite RNA of tobacco ringspot virus (STobRV). These transgenic plants are protected against the deleterious effect of subsequent TobRV infection. The advantage of using satellite RNA strategy over coat protein is that satellite RNA is not constitutively expressed like coat protein, but is expressed only after virus infection.

**Resistance against bacterial and fungal pathogens**—Several examples are now available, where transgenic plants with resistance against bacterial and fungal pathogens have been produced. Some examples are given in Table.

**Table**

Some pathogens for which resistance has been transferred in some crop plants.

| *Pathogen* | *Disease* | *Resistance gene* | *Source of gene* | *Transgenic crop* |
|---|---|---|---|---|
| 1. Pseudomones syringae | Wild fire | Acetyl transferase gene | - | Tobacco |
| 2. Alternoria lonsipes | Brownspot | Chitinase gene | Serratia marcescens (soil bacterium) | Tobacco |
| 3. Rhizoctonia solani | | Chitinase gene | Bean | Tobacco |
| 4. Phytophthora infestatts | Late blight | Osmotin gene | Potato | Potato |

**Altered fatty acid composition in Brassica seed oil**—In the past, specialized fatty acid composition in the seed oil, desired for edible and industrial purposes, has been achieved by conventional plant breeding or by mutagenesis programmes. Following are some examples: (i) removal of erucic acid from rape-seed oil to create 'canola'; (ii) reducton in linolenic acid content in flax; (iii) increase in stearate content (six times that in wild type) in sanlower (upto 12% stearate) and soybean (upto 30% stearate).

Tissue specific antisense RNA expression has recently been used for reducing 'stearoyl-ACP desaturase' (ACP = acyl carrier protein) activity in seeds, leading to alteration in the ratio of saturated to unsaturated fatty acids. Due to tissue specific expression of anti-sense RNA, integrity of membrane lipids in leaf remained unaffected. 'Stearoyl-ACP desaturase' catalyzes the first desaturation step in seed oil biosynthesis, converting 'stearoyl-ACP' to 'oleoyl-ACP.' Seed specific anti-sense gene constructs of B. rapa's (syn-B.campestris) stearoyl-ACP desaturase were used for production of transgenic B. raga and B. napus plants. This led to reduction in desaturase activity, resulting into a dramatic increase in the level of stearate in the seeds. This example demonstrated the potential to engineer the composition of fatty acids in seed oil.

Resistance against stress—A number of genes responsible for providing resistance against stresses such as heat, cold, salt, heavy metals, phytohormones and nitrogen have been identified. Studies are also being conducted on metabolites like proline and betaines, that are implicated in stress tolerance. With this background, transgenic plants resistant to a variety of stresses will be produced in future.

In a recent report (Nature; 23 April, 1992) results were described, where resistance against chilling (1°C for 10 days) was introduced into tobacco plants by introducing a gene for 'glycerol-l-phosphate acyl transferase' enzyme from Arabidopsis (Arabidopsis is resistant to chilling). The enzyme encoded by nuclear genome and later transported to chloroplast, determines the level of unsaturation of fatty acids in the phosphatidyl-glycerol of chloroplast membranes. The plants with high proportion of cis-unsaturated fatty acids (e.g. spinach, Arabidopsis) are resistant to chilling, and so are the transgenic tobacco plants carrying the gene for the above enzyme.

Transgenic plants suitable for food processing—A number of examples are available, where transgenic plants suitable for food processing have been developed. (i) Bruise resistant tomatoes were developed which express antisense RNA against polygalacturonase (PG), which attacks pectin in the cell walls of ripening fruit and thus softens the skin (Fig.) (ii) Tomatoes exhibiting delayed ripening were developed, either by using antisense RNA against enzymes involved in ethylene, production (e.g. ACC synthase), or by using gene for ACC deaminase , which degrades 1 aminocyelopropane-i carboxylic acid (ACC), an immediate precursor to ethylene. This will increase the shelf life of tomato. These tomatoes can also stay on the plant longer, giving more time for accumulation of sugars and acids for improving flavour. Therefore, they are described as 'Fiavr Savr, (iii) Tomatoes with elevated sucrose and reduced starch could also be produced using sucrose phosphate synthase gene. (iv) Starch content in potatoes could be increased by 20 - 40% by using a bacterial ADP glucose pyrophosphorylase gene (ADP GPPase)

Male sterility and fertility restoration in transgenic plants (suitable for hybrid seed production). During 1990-1992, an exciting example of producing transgenic plants with male sterility and fertility restoration genes has become available in Brassica napes. This should facilitate production of hybrid seed without manual emasculation and controlled pollination as often practiced in maize.

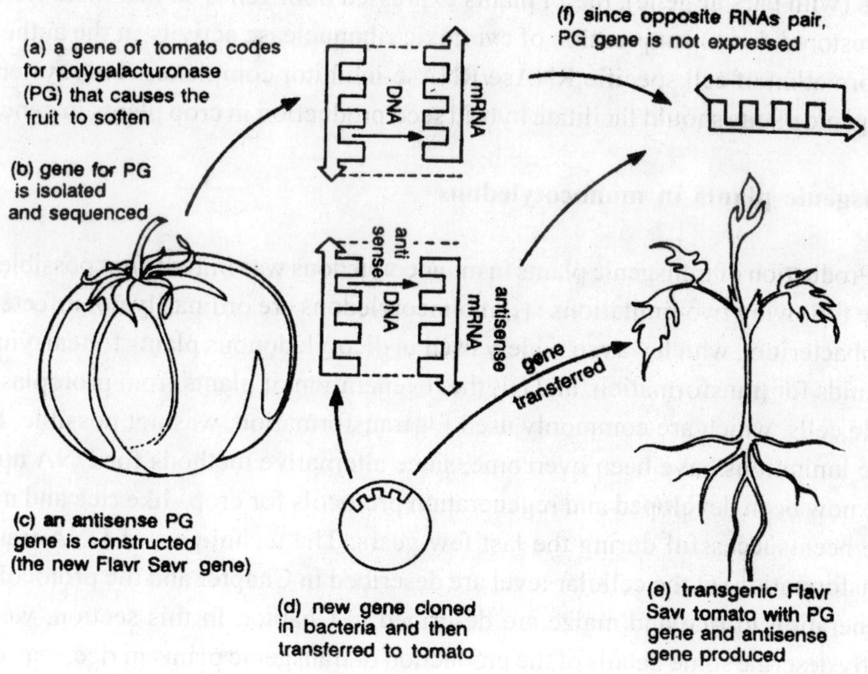


**Fig.** Production of transgenic Flay Save tomato using an antisense RNA gene

In 1990, C. Mariani and others from Belgium, successfully used a gene construct having an anther specific promoter (from TA29 gene of tobacco) and bacterial coding sequence for a ribonuclease (barnase gene from Bacillus amyloliquefaciens for production of transgenic plants in B. napus. The results were spectacular in the sense that the transferred gene prevented normal pollen development leading to male sterility. The product of barnase gene is cytotoxic, killing the tapetal cells, thus preventing pollen development. Utilizing this male sterility barnase gene construct (TA-29- RNase), it was possible to introduce male sterility in other crops also. These crops include tobacco, lettuce, cauliflower, cotton, tomato, corn, etc.

The same group of workers (*Mariani et al.*, 1992) used another gene construct later involving the same anther specific promoter i.e. TA 2% and the barstar gene from B. amyloliquefaciens, for production of transgenic plants in B. napus. The product of barstar gene is a ribonuclease-inhibitor. It forms a complex with ribonuclease and neutralises its cytotoxic properties.

In Bacillus, the ribonuclease is active extracellularly and the bacterium itself is protected by a ribonuclease inhibitor protein coded by hamstar gene. When transgenic male sterile plants (with barnase gene) were crossed with transgenic male fertile plants (with barstar gene), the Fl plants expressed both genes so that male fertility was restored due to suppression of cytotoxic ribonuclease activity in the anther by the formation of cell specific RNAse/RNAse-inhibitor complexes. This system of transgenic plants should facilitate hybrid seed production in crop plants in general.

**Transgenic plants in monocotyledons**

Production of transgenic plants in monocotyledons was initially not possible due to the following two limitations : (i) monocotyledons are ordinarily not infected by Agrobacterium, which was so widely used in dicotyledonous plants for carrying Ti plasmids for transformation, and (ii) the regeneration of plants from protoplasts or single cells, which are commonly used for transformation, was not possible. Both these limitations have been overcome, since alternative methods for DNA uptake have now been developed and regeneration protocols for crops like rice and maize have been successful during the last few years. The techniques of DNA transfer (transformation) at the cellular level are described in Chapter and the protocols for regeneration in rice and maize are described in Chapter. In this section, we will briefly describe some details of the production of transgenic plants in rice, maize and wheat.

**Transgenic plants in rice**—In rice (both in japonica and indica varieties), there are reports (1988-1991) of successful production of transgenic plants. In one case, transgenic plants were produced, which carried a functional gene for aminoglycoside phosphotransferase II(APT (3') 11) along with the reporter gene for neomycin phosphotransferase (NPT II). In the other case, the transgenic rice plants possessed a bacterial hph gene, encoding hygromycin B resistance ( Hmt ) along with the reporter gene encoding ß glucuronidase (GUS).

Transgenic plants in maize—In maize, a reporter gene for neomycin phosphotransferase (NPT II) associated with 35S promotor region of the cauliflower mosaic virus (CMV) was used for production of transgenic plants. At the enzyme level, these plants exhibited NPT II activity, and at the DNA level, the presence of NPT-II gene was demonstrated by Southern blot analysis.

In USA, the biotechnology company, Monsanto recently produced insect resistant transgenic maize plants, using the Kurstaki gene derived from B. thuringiensis. The gene was attached to CaMV 35 S promoter, which enhanced the production of toxin 1000 times. When each transgenic plant was inoculated with 100 larvae of European corn borer, complete control of insect was observed. Transgenic rice plants, resistant to a certain virus, were also produced using coat protein (CP) gene from the virus.

**Transgenic plants in wheat**—In wheat, transgenic plants were produced (in 1992) by a non resident Indian (NRI) scientist (Indra K. Vasil) and his coworkers. In these transgenic plants, resistance gene (bar gene) against the herbicide PPT (commercial name 'Basta' = 20% PPT) along with 'gus' marker gene associated with CaMV 35S promoter + Adh I intron (Adh intron was shown to increase the activity of 35S promoter leading to enhanced expression of the associated gene) was introduced into wheat plants. The transgenic wheat plants showed expression of gus activity and also the resistance against the herbicide PPT. The development of these transgenic wheat plants has been considered a major breakthrough, since wheat is a major cereal crop and its improvement through transfer of foreign genes will be a great advantage to wheat breeders.

In the above successful reports for the production of transgenic plants in monocots, initially no useful genes were used for transformation and transformed plants were often sterile, although reports of fertile transgenic plant are now available. Therefore, they had mainly demonstrated that genes could be transferred into monocotyledons also. Efforts are being made now to produce transgenic monocotyledonous plants using unmodified desirable genes or suitably modified genes to meet the desired needs. Virus resistant rice plants using a coat protein gene have been porduced. Similarly, herbicide resistant wheat (also oats) plants have also been produced as disucssed above. It is, thus, apparent that transgenic cereals will be produced and field tested in future to be released for cultivation. However, before this is achieved, there is publicity against these transgenic plants questioning their safety for human consumptions. Therefore, fears of their safety will need to be dispelled before one can be successful in growing these transgenic plants at any large scale.

### Transgenic Plants for Molecular Farming

An additional major goal of biotechnology industry is also the use of transgenic plants as 'factories' for manufacturing speciality chemicals and pharmaceuticals. Sugars, fatty acids, starches, celluloses, rubber and wax are traditionally obtained from plants and genetic engineering can be used to increase their production. Following are some of the examples: (1) Transfer of a gene for mannitol dehydrogenase from E.coli to tobacco was achieved, which led to increase in the level of mannitol in transgenic tobacco plants. (2) Transfer of bacterial gene (from Klebsiella) from cyclodextrin glueosyltransferase (CGTase) to potato was achieved successfully, leading to the production of a and cyclodextrins (CDs) in potato tubers. Cyclodextrins are cyclic oligosaccharides containing six, seven or eight glucose molecules in a , P or y CDs respectively. CDs can be obtained from starch by the action of cyclodextrin glucosyl transferase (CGTase) enzyme, and are useful for pharmaceutical delivery systems, flavour and odour enhancement and for removal of undesired compounds (such as caffeine) from foods. The major use of CDs is due to their ability to form inclusion complexes with a wide variety of organic compounds. Tissue specific expression of CGTase in potato tuber and its targeting to plastids for action on stored starch was achieved through the use of a chimeric gene construct consisting of the following : (i) a patatin promoter (for tuber specific expression of CGTase gene) ; (ii) a DNA sequence encoding small subunit of roublose bis phosphate carboxylase (SSU) transit peptide, (iii) the CGTase structural gene from Klebsiella and (iv) nopaline synthase (nos) 3' region. More transgenic plants with CGTase gene may be produced in view of the present high cost and world demand of CDs (Table) and due to availability of a number of CGTase genes (Table). (3) Transfer of genes for acetoacetyl CoA reductase (phh B) and polyhydroxybutyrate (PHB) synthase (pltb C), which catalyze two seeps in the production of polyhydroxybutyrate or PHB (a biodegradahic thermoplastic polymer). These genes were successfully transferred and their expression was demonstrated in transgenic plants of Arahidopsis thalin. (4) Chimeric genes having CaMV promoter and encoding human serum albumin (HSA) were transferred successfully and transgenic potato and tobacco plants obtained. The secretion of protein was achieved by using either the human preprosequence or the signal sequence from extracellular PR -S protein from tobacco. HSA was secreted in transgenic leaf tissue. (5) The production of pharmaceutically active compounds like enkephalins was achieved in transgenic oil seed rape.

**Table**

Present and anticipated world market for cyclodextrins
(source : Consortium for elect rochemische Industries GmbH)

| *Application* | *Market (tonnes year$^{-1}$)* present | *Expected for 1995* |
|---|---|---|
| Pharmaceuticals | 50 | 2000 |
| Food | 700 | 2500 |
| Cosmetics | 50 | 500 |
| Agriculture | 10 | 100 |
| Chemical industry (biotransformations, separation, catalysis) | 30 | 300 |
| Other purposes (e.g. diagnostics) | 10 | 200 |

**Table**

Bacteria which produce CGTases

| *Organism* | *CGTase type* | *Gene cloned* |
|---|---|---|
| Klebsiella axytoca M5al | α | + |
| Bacillus macarans | α | + |
| Bacillus stearothennophilus | α | + |
| Bacillus circulans | β | + |
| Bacillus megaterium | β | – |
| Bacillus ohbensis | β | – |
| Micrococcus sp. | β | – |
| Alkalophilic Bacillus 38-2 | β | + |
| Alkalophilic Bacillus 17-1 | β | + |
| Alkalophilic Bacillus 1011 | β | + |
| Alkalophilic Bacillus 1-1 | β | + |
| Bacillus subtilis No. 313 | γ | + |
| Alkalophilic Bacillus 290-3 | γ | + |

## Transgenic Plants to Study Regulated Gene Expression

Transgenic plants have particularly been found useful in the study of development, since with their help gene expression can be studied in time and space, i.e. in different environmental conditions or at different developmental stages, causing induction or suppression of gene expression. Transgenic plants thus facilitated the recognition of regulatory sequences involved in differential expression of gene activity. For this purpose, sequences upstream to a number of structural genes were cloned either with their own structural genes or with other unrelated structural genes (e.g. reporter genes) to study the specific regulatory role of these sequences in time and spare. A number of structural genes, whose regulatory sequences were examined in this manner, are listed in Table and the details of some of these are described in the following text.

**Gene for small subunit of roublose bisphosphate carboxylase (rbcS).** Dozens of genes in plants are affected by light and many of these regulated by phytochrome. The expression of phytochrome gene is depressed and that of RUBISCO (roublose, 1,5 bisphosphate carboxylase/oxygenase) gene is stimulated by red light. RUBISCO is involved in CO, fixation and is the most abundant enzyme on the earth. It consists of two subunits, each encoded by a separate gene, the smaller subunit 'rbcS' encoded by a nuclear gene and the larger subunit 'rbcL' encoded by a chloroplast gene. Although rbcS from different plants are fairly similar, small differences that exist, can be used to distinguish mRNA of a transgene from that of the host plant using specific probes. The pea rbcS gene was transferred to

**Table**

Transgenes that are inducible (by a stimulus) and/or exhibit tissue specific expression in transgenic plants

| | *Gene (product) whose regulartory sequence examined* | *Donor plant* | *Transgenic host* | *Inducer stimulus* | *Specific host tissue* |
|---|---|---|---|---|---|
| 1. | rbc S | Pea | Tobacco, petunia | Light | Green tissues |
| | rbc S | N. plumba | N.plumba, petunia, tobacco | Light | Green tissues |
| | rbc S | Soybean | Petunia | Light | Green tissues |
| 2. | Cab | Pea, wheat | Tobacco | Light | Green tissues |
| | Cab | N. plumba | Tobacco | Light | |
| | Cab | Wheat | Tobacco | Circadian clock | — |
| 3. | ST-L,SI | Potato | Tobacco | Light | Green tissue |
| 4. | Chalcone synthase | A. majus | Tobacco | Light | — |
| 5. | Hsp | Maize | Petunia | Heat | — |
| | Hsp | Soybean | Tobacco | heat | — |
| 6. | Protease inhibitor | Potato | Tobacco | Wounding | — |
| 7. | Patatin | Potato | Potato | — | Tuber |
| 8. | Leghaemog-lobin | Soybean | Lotus | — | Nodule |

*(Table Contd.)*

| | Gene (product) whose regulartory sequence examined | Donor plant | Transgenic host | Inducer stimulus | Specific host tissue |
|---|---|---|---|---|---|
| 9. | Nodulin | Soybean | Lours | — | Nodule |
| 10. | β-phaseolin | French bean | Tobacco | — | Seed |
| 11. | β-conglycin | Soybean | Petunia, tobacco | — | Seed |
| 12. | Lectin | Bean | Tobacco | — | Seed |
| | Lectin | Soybean | Tobacco | — | Seed + root |
| 13. | Glutenin | Wheat | Tobacco | — | Seed |
| 14. | Hordein | Barley | Tobacco | — | Seed |
| 15. | Legumin | Pea | N-Plumba | | Seed |
| 16. | EPSP synthase | Petunia | Tobacco | | Pollen |
| 17. | RNAase | 1. Aspergillus oryzae | Tobacco | Anther development | Anthers |
| | RNAase | 2. Bacillus arryloliqrrefaciens | Tobacco | ” | Anthers |
| 18. | Sucrose synthase-1 | Corn | Tobacco | — | Phloem cells |

*N. plumba = Nicotiana phrrrrhaginifolia*

petunia and tobacco plants and the soybean rbcS gene was transferred to petunia. The gene expressed itself in transgenic plants, rbcS subunit from the introduced gene sometimes even combining with rbcL subunit of the host plant to yield the enzyme. 'Deletion mutants' having sequences deleted upstream of the transgene rhcS led to the identification of a sequence upstream (-1G6 to -149 bp) of TATA box, which was light responsive and was, therefore, conveniently called 'light responsive element (LRE)'. This LRE sequence could also regulate genes other than rbcS genes so that these unrelated foreign genes became light, organ and age specific in a manner similar to rbcS gene (high expression in leaves, low expression in stem and no expression in roots). LRE was shown to have three regions, box I, box II and box III and it was shown that box II region also had a negative regulatory element. This LRE also seems to be repetitive (with two repeats) and each repeat unit can mediate light responsiveness and organ specificity. However, it has also been shown that both repeats of LRE are required for full expression of the rbcS gene in youngest leaves of tobacco plants.

It has also been shown using transgenic plants that a different sequence upstream to rbcS is responsible for age responsiveness. For example, it was shown that the region from –410 to –166 bp is essential for high level of expression in mature leaves, but not important for light responsive expression in leaves, for which the region -16f to -149 by was most important.

Gene for chlorophyll a/b binding protein (Cab)—Examples of the transfer of Cab gene from pea, wheat and Nicotiana plunrbaginifolia are listed in Table. In this case, it was shown that a 247 by sequence for pea Cab AB80 (–347 to –100) and a 268 bp sequence (–357 to –89) for wheat Cab-1, when separately fused with heterologous TATA sequence fused to neomycin phosphotransferasa (NPT Il) coding sequence, conferred on this coding sequence a light responsive expression. This regulatory sequence when present caused induction of expression of NPT II by red light and the expression could be reversed by flash of far-red light as observed in phytochrome mediated expression. In another experiment, negative elements suppressing expression were also identified, when Cab-E gene of Nicotiana plumbaginifolia was studied in a similar manner in transgenic plants.

Heat shock genes—Thermal stress induces the transcription of a small set of heat shock genes in diverse organisms, and also reduces the expression of many other vital genes. The upstream region of heat shock genes regulating their expression is highly conserved. Thus, when a gene construct, carrying NPT II reporter gene, fused with upstream region of heat shock gene hsp70 from Drosophila, was introduced into tobacco, induction of NPT II expression due to heat shock was comparable to that of endogenous plant heat shock genes (inducible in all organs except pollen). Similarly, from maize a homologue of above gene, hsp70 with 1.1 kolobases of upstream sequence, when introduced in petunia exhibited heat inducibility.

A heat shock gene hsp6871 from soybean was also introduced in tobacco and the deletion mutants allowed identification of a 36 by element responsible to confer heat inducibility in all organs independent of the presence of light. Interestingly, when this 30 by sequence was placed upstream from rbcS promotor, a light dependent and organ specific heat shock response was observed (gene expressed due to heat shock only in the presence of light and only in leaves). This suggested that a negative control element of rbcS did not allow expression in roots and in the dark even in leaves.

Alcohol dehydrogenase gene (ADH)—Alcohol dehydrogenase gene expresses under anaerobic conditions leading to oxygen deprivation. Using transgenic plants, it could be shown that regulatory sequences upstream to ADH gene from maize lead to suppression (down regulation) of gene activity under aerobic conditions, this activity being normal and uninducible under anaerobic conditions. The inability of tha ADH upstream regulatory sequences to induce (these sequences only suppress) gene expression became evident when these sequence failed to induce chloramphenicol acetyltransferase (CAT) expression under aerobic as well as anaerobic conditions in transgenic tobacco plants. Other enhancer sequences (from octopine synthase or CaMV35S gene), when transferred at a position upstream of ADH gene, led to induction under anaerobic conditions. Therefore, with normal conditions, ADH activity is inhibited under aerobic conditions due to negative control, and expression is promoted under anaerobic conditions simply due to the absence of a negative control rather than the presence of positive induction.

Genes for seed storage proteins in legumes and cereals—The genes for seed storage proteins provide an excellent example of a cell and tissue specific expression. Using transgenic plants, sequences could be identified which were responsible for tissue or cell specific expression of these genes. For instance, when wheat glutenin gene or barley hordein gene (each with its regulatory sequence) was fused to chloramphionicol acelyltransferase (CAT) gene and introduced into tobacco, CA-1 activity was found only in the endosperm. Another reporter gene utilized in the same purpose is β glueuronidase (GUS) gene, which also showed endosperm specific expression when fused with a wheat giutonin gene and its regulatory sequence. The regulatory sequence, thus, identified was called endosperm box of the storage protein genes.

EPSP. synthase gene—5-enolpyruvyl shikinlate 3-phosphate synthase (EPSPS) is an enzyme involved in aromatic amino acid biosynthetic pathway and is the target of glyphosaic based herbicides. In petunia, enzyme synthesized in flowers at a high level and is hardly found in leaves, althou in tomato it is synthesized both in flowers and leaves. Transgenic tobacco plants carrying CAT gene or GUS gene fused with deletion mutant for of EPSPS gene were analysed which suggested that sequences between-1800 to -800 of petunia EPSPS gene were responsible for this gene's tissue sped expression.

## II Chloroplast and Mitochondrion Engineering

In Transgenic plants produced in most cases involve insertion of foreign genes in the nuclear genome and only in rare cases, foreign genes could be successfully inserted in the genomes of plastids. However, perhaps no examples are known (till 1992), where mitochondria genome could be modified using recombinant DNA technology.

Chloroplast engineering or manipulation really means changes at the level of plastid proteins. Genes for these plastid proteins are located either in the nuclear genome or in the chloroplast genome itself. In the former case, the proteins are synthesized in the cytoplasm and targeted to the plastids, while in the latter, the proteins are synthesized within the chloroplasts themselves. It is also known that plastid genome is small in size and a vast majority of the plastid proteins are really coded by nuclear genes. In view of this, introduction of novel proteins, or enzymes into plastids can be achieved by either of the two following routes : (i) insertion of a foreign gene directly into the chloroplast for expression by the plastid's own protein synthesis machinery or (ii) insertion of a chimeric foreign gene construct (containing the gene of interest fused to sequences coding for a plastid transit peptide plus appropriate transcriptional control signals) into the nuclear genome. In the latter case, the gene product after its synthesis in the cytoplasm, is targeted to the chloroplast. Successful efforts following both these routes have been made and will be discussed in this chapter.

**Engineering of Chloroplast Genome in Chlamydomonas**

Although cases of stable transformation of plastids in higher plants also are now known; Chlamydomonas has been a particularly favourable organism for chloroplast transformation for the following reasons : (i) It has a single large chloroplast. (ii) Photosynthetic mutants are viable, if reduced carbon is provided in the form of acetate. In view of this genetic transformation of chloroplasts in Chlamydomonas has been successfully achieved with high efficiency. Atleast two different methods have be en successfully utilized: (i) bombardment of cells with DNA-coated tungsten particles; (ii) agitation of cells in the presence of glass beads and DNA. Both these methods were earlier proved to be very efficient for the transfer of nuclear genes. Chloroplast transformation, like nuclear transformation requires suitable selectable markers. Several such markers have recently been developed and include the following : (i) spectinomycin resistance (spec$^R$) and streptomycin resistance (str$^R$), encoded by 16S rDNA (ii) erythromycin resistance encoded by 23S rDNA; these two markers are separated by 4.2 kb in the chloroplast genome; (iii) atpB gene (coding for a subunit of chloroplast ATPaes essential for photosynthesis. During 1988-1991, at least four reports of chloroplast transformation in Chlamydomonas were published, three of them utilizing bombardment of cells with DNA-coated tungsten particles, and the fourth utilizing agitation of cell wall deficient cells in the presence of glass beads and DNA.

## Transformation of Chloroplast (Cp) Genome in Higher Plants

In higher plants like Nicotiana tabacum also, transformation of the chloroplast genome has been successfully achieved, but with a 100-fold lower efficiency. The term transplastomic lines has also been proposed for lines carrying transgenic plastome. Since several useful genes of agronomic value reside in chloroplast genome, efforts are being made to transfer foreign genes to plastids. These efforts will certainly be intensified in future. The methods used for chloroplast transformation, the achievements made, and the future prospects of using this technology for improvement of crop plants will be discussed in this section. At least two cases are known, where specific foreign genes have been introduced in the chloroplast genome of Nicotinna tabacum. In one case, the transformation was achieved using a vector based on Agrobacterium: tumefaciens -Ti plasmid, and in the other case, it was achieved through bombardment of leaves with tungsten particles coated with DNA. These two experiments are briefly described

## Agrobacterium mediated transformation of Cp genome

Ti plasmid vector was constructed using a chimeric gene construct consisting of the following : (i) promoter region tit nopaline synthase gene (Pnos); (ii) coding

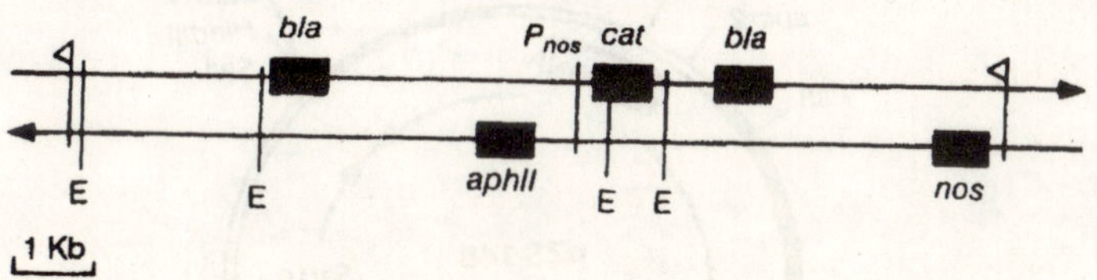


Structure of T-DNA construct used for chloroplast transformation; Pvos-cat = chimeric gene for CAT with nos promoter; bla = # lactamase gene; aph II = aminoglycoside phosphotransferase gene; nos = nopaline synthase gene.

sequence of chloramphenicol acetyltransferase (CAT) gene (cat); (iii) complete gene for nopaline synthase (nos); (iv) gene for amino - glycoside phosphotransferase type II or APH II (aph II) and (v) the gene for lactamase (bla). The structure of the TDNA construct carrying the above sequences is shown in Figure. The gene construct was introduced into tobacco leaf proptoplasts by cocultivation with A. tumefaciens containing the hybrid Ti plasmid. Calli were derived from cocultivated protoplasts on a medium containing chloramphenicol, so that chloramphenicol resistant ($Cm^R$) calli were selected. Of these $Cm^R$ calli, only 8% had nopaline, suggesting that nos gene was lost or inactivated in majority of cases. Of 20 plantlets regenerated from $Cm^R$nos- calli, only eight (40%) were able to root on medium containing chloramphenicol.

One of these plants exhibited the presence of CAT activity and APH II activity in chloroplast fraction obtained after cell fractionation. In reciprocal crosses with wild plants, it was found that cat gene was transmitted through the ovule and not through the pollen (maternal inheritance), confirming location of cat gene in the Cp genome. The presence of cat gene in chloroplast genome was further confirmed, when Southern blots derived from XhoI restriction digests of nuclear mitochondria and chloroplast DNAs were hybridized with a probe having cat gene (only CpDNA showed hybridization). Evidence was also available showing integration of cat and *aphII* genes with the chloroplast DNA.

**Stable transformation of Cp genome using particle gun method**

A plastid transformation vector pZS148 was constructed, whose structure is shown in Fig. It contained the following DNA sequences : (i) 16S rRNA gene cloned as a 3.7 kb Sac 1 - EcoRV fragment derived from tobacco genotype (SPC 2); this fragment carried resistance to both streptomycin and spectinomycin due to mutations in 16S rDNA; and also had a *Pst* I site. (ii) A 2.9 kb Sal I fragment carrying cp on region needed for CpDNA replication; (iii) A 3.0 kb segment of pBluescript I KS + phagemid vector.

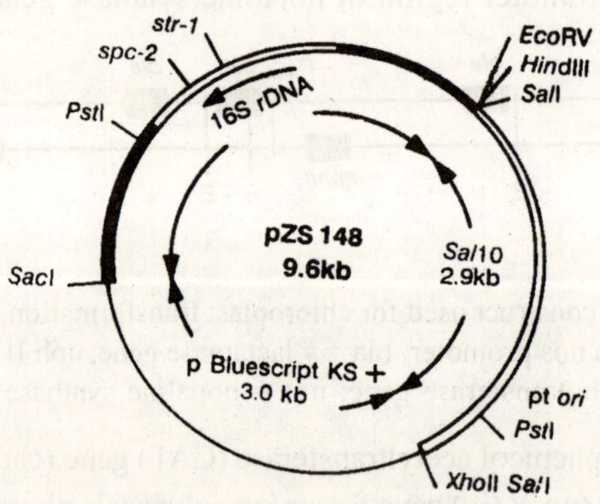


**Fig.** Transformation vector pZS148 (for details, see text).

For transformation, following two lines of Nicotina tabacum were used: (i) Nt (tbc) having original cytoplasm and (ii) Nt (hhg) carrying the cytoplasm of Nicotiana plumbaginifolia; this was functionally male sterile due to short filaments, but seeds could be obtained by manual pollination. Whole leaves of both, Nt (the) and Nt (phg) plants, were bombarded with tungsten particles coated with pZS74R DNA (carrying resistance against spectinomycin as well as streptomycin). pZS 134 carrying resistance against only spectinomycin (not against streptomycin) was used as control. Three transplastomic clones were isolated from calli obtained from leaves bombarded with pZSl48 and none from those bombarded with pZSl34. Screening for Pst1 site confirmed the replacement of l6S rDNA region in leaves by that of engineered 16S rDNA using SPC 2. The transplastomic plants obtained as above, were selfed and crossed to wild type plants. Its was shown that in plants raised from seeds obtained due to shelling, there was no segregation, while in crosses the resistance was maternally inherited.

In higher plants, the transformation of plastids by bombardment is 100 fold less efficient (1 transformant per bombardment) than transformation of nucleus (2-6 transformants per 5 bombardments), despite the high number of CpDNA copies in a cell (a cell may have upto 100 plastids with 3000-12,000 copies of CpDNA molecules). In Chlamydomonas (80 Cp DNA copies per cell), however, the efficiency is as high as that for nuclear genes (2-100 transplastomic clones per bombardment). Further, the selection for a nonlethal marker (e.g. neomycin, kanamycin, spectinomycin resistance) is critical for obtaining the transplastomic lines, because it allows sufficient time for the resistant plastid genome copies to increase in number to allow phenotypic expression. This explains the low efficiency of transformation of CpDNA, since all transformed plastids may be derived from a solitary plastid or from a solitary DNA molecule.

## Other possible methods for introduction of DNA into plastids

Several other methods have been suggested for DNA delivery into chloroplasts, although none of them has been successfully utilized so far. These methods include the following:

**DNA uptake by chloroplasts in vivo**— Since chloroplasts are compared with prokaryotes, they should be able to take up DNA in a manner analogous to bacterial transformation using naked DNA. If DNA is injected into the protoplasts either by microinjection or by electroporation, this DNA will be available in cytoplasm to be taken up by the chloroplasts. This may be facilitated, if a DNA sequence is identified, which may work as a signal and is recognized by a receptor in the chloroplast, thus facilitating internalisation of DNA.

**Microinjection of DNA into chloroplasts in vivo**— Microinjection of DNA directly into chloroplasts should also be possible in future, although there may be technical problems which will have to be overcome. For instance, in higher plants, each cell has many plastids and each plastid has several genomes. Therefore, number of plastids per cell may have to be artificially reduced and the optimal number of chloroplasts to be injected per cell will have to be determined. This will be difficult, because the fate of chloroplasts and their genomes during regeneration of protoplasts is not fully understood. However, microinjection as a method for transformation of Chlamydomonas chloroplasts should be a distinct possibility due to its large size and the presence of a single chloroplast per cell.

**DNA uptake by chloroplasts in vitro**— Introduction of DNA in isolated chloroplasts from higher plants will help in overcoming the following problems faced in DNA uptake in vivo: (i) a large number of chloroplasts per cell may not take up the DNA; (ii) the concentration of DNA in the cytoplasm, in the vicinity of chloroplasts, will be 'slow' thus reducing the chances of DNA uptake by chloroplasts. The major problem in this method, however, will be the return of manipulated chloroplasts back to plant cells. There are reports of successful uptake of chloroplasts by plant protoplasts, but stable incorporation of chloroplasts and subsequent regeneration of transformed plants from protoplasts has not been achieved.

## Targeting of Foreign Proteins into Chloroplasts

Many important metabolic processes including the following are carried on in chloroplasts: (i) synthesis of amino acids, proteins and lipids; (ii) light capture and carbon fixation and (iii) photorespiration. For successful operation of these metabolic processes and resulting from them, there are proteins (either synthesized outside or within the chloroplasts), which are targeted to the following six different compartments : outer membrane (OM), inner membrane (IM), intermembrane space (IMS), stroma, thylakoid membrane (TM) and thylakoid lumen (TL). The nuclear encoded chloroplast

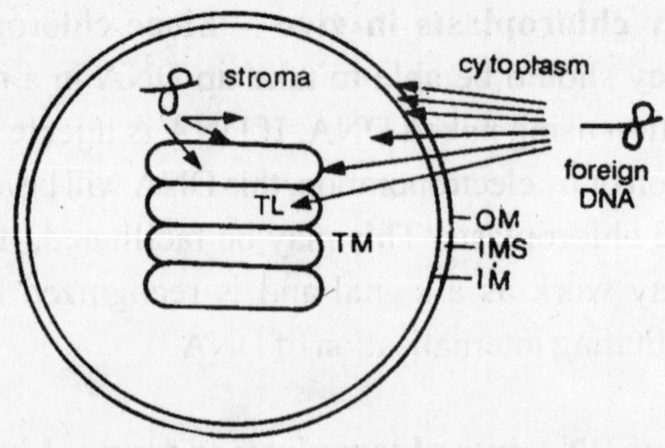


**Fig.** Transport of proteins Into and within the chloroplast.

proteins are synthesized in the cytoplasm and targeted to all these six compartments, but the chloroplast - encoded proteins (after synthesis in stroma), either remain in stroma or are targeted perhaps only to TM or TL.

We have already discussed above the techniques, achievements and future possibilities of transferring genes to chloroplast genome, but chloroplasts can also be engineered by transfer of foreign nuclear genes (fused with signal peptide DNA sequence) meant for chloroplast proteins. Some success in this direction has already been achieved, although majority of studies were aimed at (i) understanding the mechanism of targeting, (ii) identification of signal peptides and the receptor molecules and (iii) understanding the interaction between signal peptides and receptor molecules.

Several studies have suggested that the presequences of several stromal proteins contain all the information needed to direct the import and processing of foreign proteins. In one report, a gene for NPT (neomycin phosphotransferase- a bacterial protein), fused with DNA sequence coding for presequence of small subunit of rubisco (ribulose bisphosphate carboxylase - oxygenase) i.e. pre=SSU was used for production of transgenic plants. The enzyme (NPT) was targeted to chloroplasts in these transgenic plants. Other similar examples, where successful targeting has been demonstrated either in vitro or in vivo, include the following : brume mosaic virus coat protein. b- glucuronidase, yeast mitrochondrial superoxide dismutase, light harvesting chlorophyll-binding protein (LHCP), etc. Even a chloroplast encoded protein (e.g. larger subunit of rubisco = LSU) could be successfully targeted using a signal sequence like pre-SSU. Therefore, theoretically it should be possible to produce transgenic plants by incorporating chimeric foreign genes containing sequences meant for chloroplast proteins, whether coded by nuclear genome or by chloroplast genome. Proteins can also be targeted efficiently to plastids other than chloroplasts (e.g. leucoplasts).

The above results hold enormous implications for the genetic engineering of plants, since novel or mutated proteins can be efficiently targeted into chloroplasts, even though their corresponding foreign genes would be inserted into the nuclear genome. However, only very few examples are yet available, where properties of chloroplasts have been altered in transgenic plants by retargeting technology. One popular example is the targeting of a mutant bacterial EpSP synthase imparting resistance to herbicide glyphosale. This herbicide inhibits the chloroplast EpSP synthase resulting in inhibition of plant growth (due to inhibition of the synthesis of aromatic amino acids). The bacterial gene was fused to a 5' cDNA sequence encoding the presequence of the corresponding petunia chloroplast enzyme, and was used for producing transgenic plants. The translation product from the inserted gene construct (or chimeric gene, as it is often called) was imported by chloroplasts in vitro and in vivo, imparting stable resistance to glyphosate.

**Targeting of Foreign Proteins into Mitochondria**

Targeting of foreign proteins into plant mitochondria has also been achieved in vivo. Following are three examples (all involving transgenic tobacco plants; listed in Chapter 6 of 'Plant Genetic Engineering' edited by Don Grierson,1991). (i) Using amino - terminal 90 residues of the tobacco ß-$F_1$ ATPase precursor, bacterial CAT (chloramphenicol acetyltransferase), could be targeted into the mitochondrial matrix. (ii) Using the presequence of a yeast mitochondrial protein, GUS (ß-glucuronidase) could be targeted to mitochondria in plants. (iii) Using again the presequence of cytoplasmic plant glutamine synthetase enzyme, tobacco ß-F1 ATPase could be retargeted to the mitochondrial matrix.

# FIFTEEN

# Methods of Gene Transfer

To produce transgenic animals, DNA is usually microinjected into pronuclei of embryonic cells at a very early stage after fertilization, or alternatively gene targeting of embryo stem (ES) cells is employed. This is possible in animals due to the availability of specialized in vitro fertilization technology, which allows manipulation of ovule, zygote or early embryo. Such techniques are not available in plants. In contrast to this in higher plants, cells or protoplasts can be cultured and used for regeneration of whole plants. Therefore, these protoplasts can be used for gene transfer followed by regeneration leading to the production of transgenic plants. Besides cultured cells and protoplasts, other meristem cells (immature embryos or organs), pollen or zygotes can also be used for gene transfer in plants. The enormous diversity of plant species and the availability of diverse genotypes in a species, made it necessary to develop a variety of techniques, suiting different situations. These different methods of gene transfer in plants will be discussed in this chapter.

### Target Cells for Transformation

The first step in gene transfer technology is to select cells that are capable of giving rise to whole transformed plants. Transformation without regeneration and regeneration without transformation are of limited value. In many species, identification of these cell types is difficult. Some of these cell types and the methods used for regeneration of whole plants from them are listed in Table.

As can be noticed in Table, cultured cells or protoplasts are often used for gene transfer. This is unlike the situation in animals, because the plant cells are totipotent and can be stimulated to regenerate into whole plants in vitro via organogenesis or embryogenesis. However, in vitro plant regeneration imposes a degree of 'genome stress', especially if plants are regenerated via a callus phase.

**Table**

Theoretical routes for production of transgenic plants.

| | *Target cell type* | *Method used for regeneration* |
|---|---|---|
| 1. | Cultured cells or protoplasts | Organogenesis or embryogenesis vie callus phase |
| 2. | Meristem cells from immature embryo or organ | In vitro plant regeneration from transformed cells |
| 3. | Cells in immature embryos, shoot and flower meristems | Normal development (in vivo) of embryo, shoot or flower followed by use of pollen (transformed) from chimaeric plant to produce transformed seed |
| 4. | Pollen | Pollen (treated with DNA) used for pollination leading to the production of transgenic plants |
| 5. | Zygote | Direct development (in vivo) of transgenic plants |

This may lead to chromosomal or genetic abnormalities in regenerated plants—a phenomenon referred to as somaclonal variation. In contrast to this, gene transfer into pollen (or possibly egg cells) may give rise to genetically transformed gametes, which if used for fertilization (in vivo) may give rise to transformed whole plants. Similarly, insertion of DNA into zygote (in vivo or in vitro) followed by embryo rescue, may also be used to produce transgenic plants. Another alternative approach is the use of individual cells in embryos or meristems, which may be grown in vitro or may be allowed to develop normally for the production of transgenic plants.

## Vectors of Gene Transfer

### General Features

The common features of vectors used for transformation of eukaryotic cells are shown in Figure. Most vectors carry marker genes, which allow recognition of transformed cells (other cells die due to the action of an antibiotic or herbicide) and are described as selectable markers. A number of selectable markers are now known and are listed in Table. Among these marker genes, the most common selectable marker is II, providing kanamycin resistance.

Other common features of suitable transformation vector include the following : (i) multiple unique restriction sites (a synthetic polylinker); (ii) bacterial origins of replication (e.g. ColE1) ). The vectors having these properties may not necessrily have features, which facilitate their transfer to plant cells or integration into the plant nuclear genome. Therefore, Agrobecterium Ti plasmid is preferred over all other vectors, because of wide host range of this bacterial system and the capacity to transfer genes due to the presence of T - DNA border sequences (sec below).

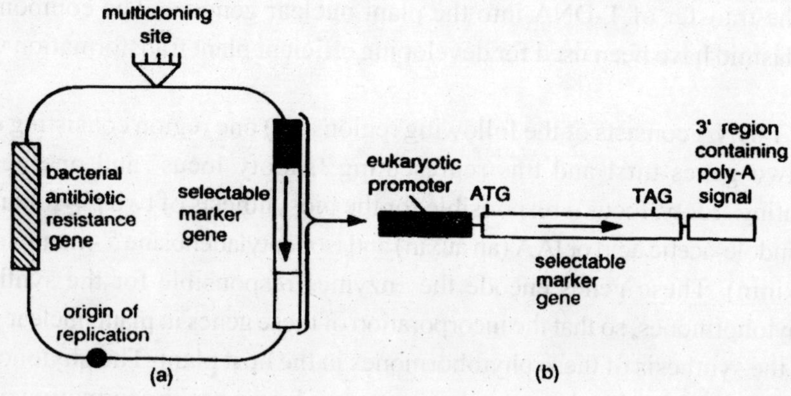


**Fig.** (a) Generalised vector for the transformation of eukaryotic cells; (b) details of the selectable marker genes. (Redrawn from D. Grierson, 'Plant Genetic Engineering', 1991)

## Vectors based on Ti and Ri plasmids of Agrobacterium

Structure and function of Ti and Ri plasmids—The most commonly used vectors for gene transfer in higher plants are based on tumour inducing mechanism of the soil bacterium. Agrobacterium tuntefuciens which is the causal organism for crown gall disease. A closely related species A. rhizogenes causes hairy root disease. An understanding of the molecular basis of these diseases led to the utilization of these bacteria for developing gene transfer systems. It has been shown that the disease is caused due to the transfer of a DNA segment from the bacterium to the plant nuclear genome. The DNA segment, which is transferred is called T - DNA and is part of a large Ti (tumour inducing) plasmid found in virulent strains of Agrobacterirrm tuntefuciens. Similarly Ri (root inducing) megaplasmids are found in the virulent strains of A. rhizogenes. The Ti and Ri plasmids, inducing 'crown gall disease' and 'hairy root disease' respectively have been studied in great detail during the last decade. However, we will discuss only those aspects of these plasmids which are relevant to the design of vectors for gene transfer in higher plants.

Most Ti plasmids have four regions in common, which are shown in Figure. (i) Region A, comprising T-DNA is responsible for tumour induction, so that mutations in this region lead to the production of tumours with altered morphology (shoot' or root' mutant galls). Sequences homologous to this region are always transferred to plant nuclear genome, so that the region is described as T-DNA (transferred DNA). (ii) Region B is responsible for replication. (iii) Region C is responsible for conjugation. (iv) Region D is responsible for virulence, so that mutation in this region abolishes virulence. This region is therefore called virulence (vir) region and plays a crucial role in the transfer of T-DNA into the plant nuclear genome. The components of this Ti plasmid have been used for developing efficient plant transformation vectors.

The T-DNA consists of the following regions : (i) one region consisting of three genes (two genes tntsl and tms representing 'shooty locus' and one gene tnn-representing 'rooty, locus') responsible for the biosynthesis of two phytohormones, namely indole-acetic acid or IAA (an auxin) and isopentyladenosine 5'-monophosphate (a cytokinin). These genes encode the enzymes responsible for the synthesis of these phytohormones, so that the incorporation of these genes in plant nuclear genome leads to the synthesis of these phytohormones in the host plant. The phytohormones in their turn alter the developmental programme, leading to the formation of crown gall. (ii) An os region responsible for the synthesis of unusual amino acid or sugar derivatives, which are collectively given the name opines. Opines are derived from a variety of compounds (e.g. arginine + pyruvate), that are found in plant cells. Two most common opines are octopine and napoline. For the synthesis of octopine and napoline, the corresponding enzymes octopine synthase and napoline synthase are coded by T- DNA. Depending upon whether the Ti plasmid encodes octopine or napoline, it is described as octopine-type Ti plasmid or nopaline-type Ti plasmid. Many organisms including higher plants are incapable of utilizing opines, which can be effectively utilized by Agrobac-outside the TDNA region, Ti plasmid carries genes that catabolize the opines, which are utilized as a source of carbon and nitrogen.

The T-DNA regions on all Ti and Ri plasmids are flanked by almost perfect 25bp direct repeat sequences, which are essential for T-DNA transfer, acting only in its orientation. It has also been shown that any DNA sequence, flanked by these 256p repeat sequences in the correct orientation, can be transferred to plant cells, an attribute that has been successfully utilized for Agrobacteritium mediated gene transfer in higher plants leading to the production of trangenic plants.

Besides 25bp flanking border sequences (with T DNA), vir region is also essential for T-DNA transfer. While harder sequences function in this orientation with respect to TDNA, vir region is capable of functioning even in traps orientation. Consequently physical separation of T-DNA and vir region onto two different plasmids does not affect T-DNA transfer, provided both the plasmids are present in the same

Aprobacterium cell. This property played an important role in designing the vectors for gene transfer in higher plants.

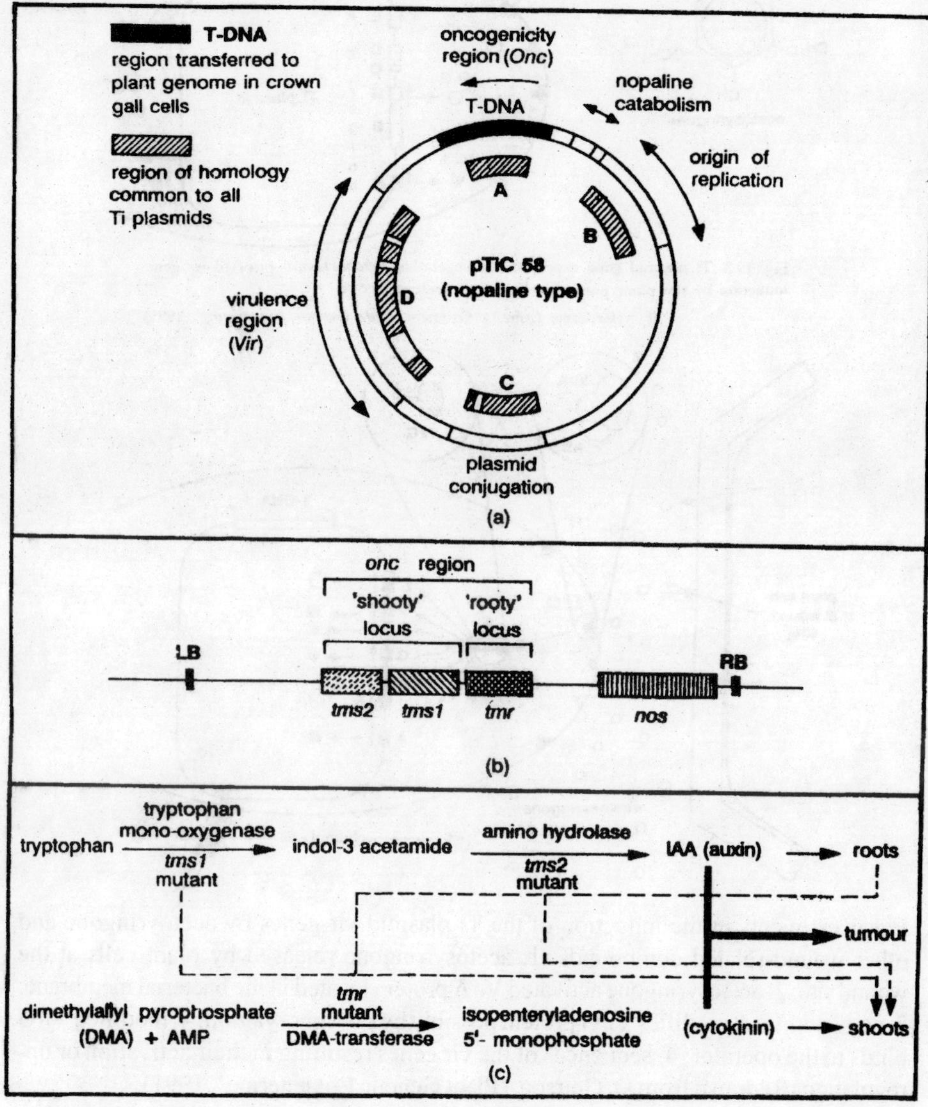


**Fig.** (a) General features of Ti plasmids; (b) general structure of the T-DNA; (c) phytohormone biosynthesis in crown gall tissue. (Redrawn from D. Grierson 'Plant Genetic Engineering' 1991)

The vir region (approx 35 kbp) is organized into six operons, namely vir A, vir B, vir C, vir D, or E, and vir G, (Fig.), of which four operons (except vir A and vir G) are polycistronic. Genes vir A, B, D, and G are absolutely required for virulence; the remaining two genes vir C and E are required for tumour formation. The vir A

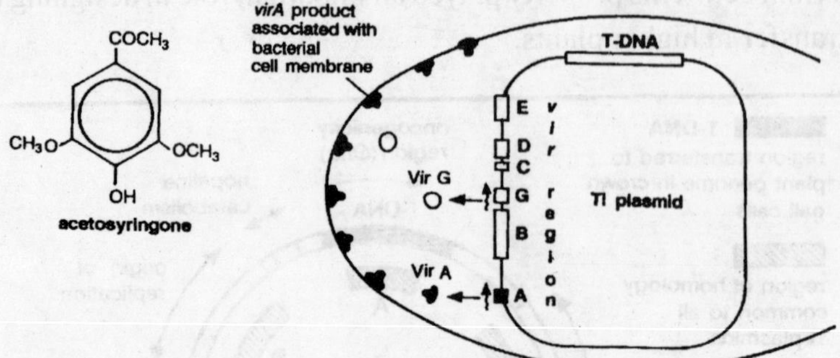


Fig. 19.3. Ti plasmid gene expression in vegetative *Agrobacterium* prior to *vir* gene induction by the plant phenolic compound acetosyringone.

(Redrawn from D. Grierson *'Plant Genetic Engineering'*, 1991)

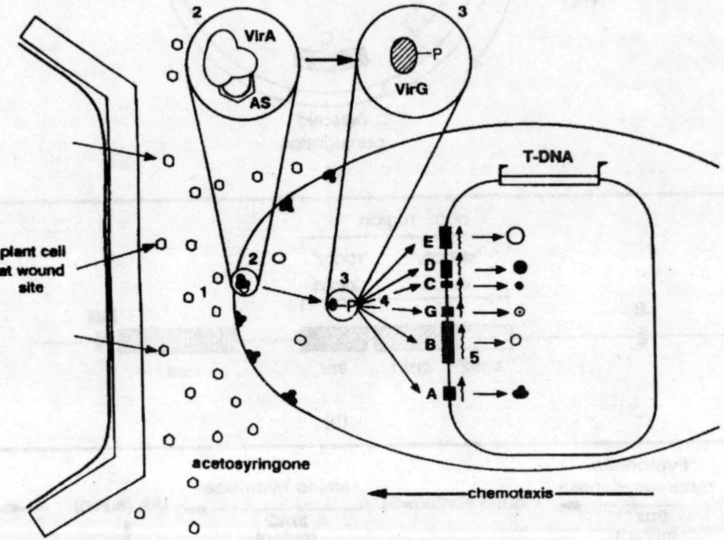


Fig. Key events in the induction of the Ti plasmid vir genes by acetoyringone and other wound-related compounds: 1. acetosytringone released by plant cells at the wound site; 2. acetosyringone activated VirA protein located in the bacterial membrane; 3. activated VirA modifies VirG protein possibly by phosphorylation; 4. modified VirG binds to the operator; 4. sequences of the vir genes resulting in their activation or up-regulation (Redrawn from D. Geirson. 'Plant Genetic Engienering', 1991).

locus is expressed constitutively under all conditions. The vir G locus is expressed at low levels in vegetative cells, but is repidly induced to higher expression levels by exudates from wounded plant tissue. The vir A and vir G gene products regulate the expression of other vir loci (Fig.). The vir A product (Vir A) is located on the inner membrane of Agrobacterium cells and is probably a chemoreceptor, which senses the presence of phenolic compounds (found in exudates of wounded plant tissue), such as aeetosyringone and β-hydroxyaceto- syringone. Signal transduction proceeds via activation (possibly phosphorylation) of Vir G (product of gene virG), which in its turn induces expression of other vir genes (Fig.).

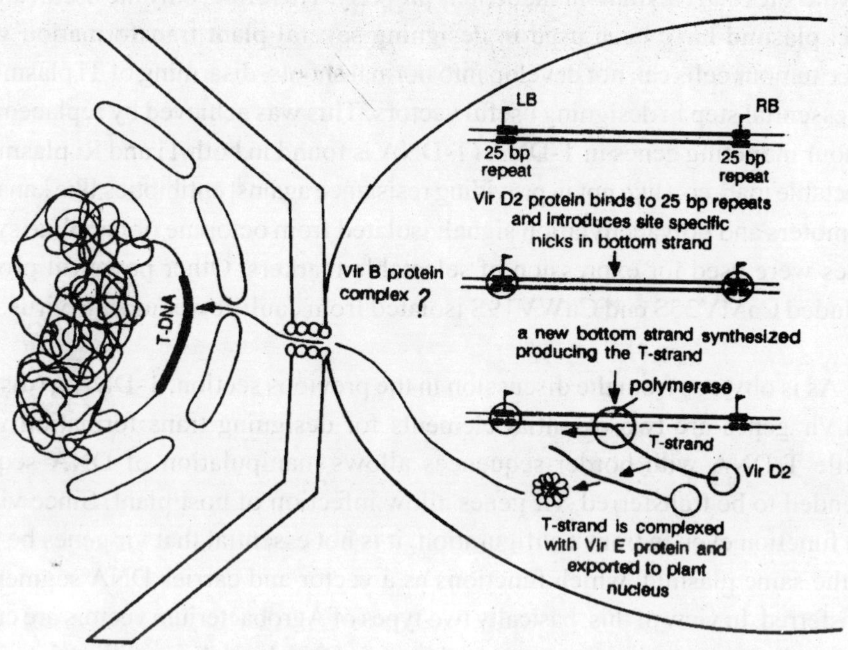


**Fig.** Molecular events in the transfer of the T-DNA to plant cells. (Redrawn from D. Grierson. 'Plant Genetic Engineering' 1991).

T-DNA transfer process—An early event in the T-DNA transfer process is the nicking of Ti plasmid at two specific sites, each between the third and fourth base of the bottom strand of each 25bp repeat. This initiates DNA synthesis from the nick in the right hand 25bp repeat sequence in 5' - 3' direction, thus displacing a single T-DNA strand. This T-DNA single strand forms a complex with protein Vir E and gets transported to the plant nucleus (Fig.). The vir D operon encodes an endonuclease that produces the nicks in the border sequences. Several gene products of the or B operon have been identified in the bacterial envelope, a location, which suggests that they may play a role in directing T-DNA transfer extracellularly. The functions of several other vir gene products are largely unknown.

Apart from the role of Ti plasmid, the genes located on Agrobacterium chromosome also help in virulence. These genes are involved in the synthesis and secretion of glucons, cellulose fibrils and cell surface proteins. These loci are constitutively expressed and are also found in other soil bacteria associated with higher plants. Thus, these loci play a more general role in the virulence of Agrobacterium, and thus also in the Agrobacterium mediated gene transfer.

Vectors based on Ti and Ri plasmids—The following properties of Ti plasmids did not allow their direct use as vectors : (i) large size, (ii) absence of unique restriction enzyme sites and (iii) tumour inductions property. Therefore, only the useful attributes of Ti plasmid have been used in designing several plant transformation vectors. Since tumour cells can not develop into normal shoots, disarming of Ti plasmids was an' essential step in designing useful vectors. This was achieved by replacements of tumour inducting genes in T-DNA (T.-DNA is found in both Ti and Ri plasmids) by selectable markers (like npt in providing resistance against antibiotics like kanamycin. Promoters and polyadenylation signal; isolated from octopine or napoline synthase genes were used for expression of selectable markers. Other powerful promoters included CaMV35S end CaWV19S isolated from cauliflower mosaic virus.

As is obvious from the discussion in the previous section, T-DNA g (disarmed) and vir genes are two essential elements for designing trans formation vectors. While T-DNA with border sequences allows manipulation of DNA sequences intended to be transferred, vir genes allow infection of host plant. Since vir genes can function even in trans configuration, it is not essential that vir genes be present on the same plasmid, which functions as a vector and carries DNA segment to be transferred. In view of this, basically two types of Agrobacterium vectors are currently in use : (i) Cointegrative vectors recombine, via DNA homology, with an intermediate cloning vector, which is used for manipulation and cloning of the gene in E. coli (Fig.). Agrobacterium i containing co:ntegrative vector and E. coli containing intermediate cloning vector are allowed to undergo conjugation, but the intermediate vector can not replicate in Agrobacterium so that it has to transfer the marker F genes as well as the DNA segment to the resident Ti plasmid (cointegrative f vector) through recombination in the region of DNA homology. One of the first cointegrative vectors, pGV3850 was developed from a nopaline-type ti plasmid (C58), where almost all T-DNA has been deleted and replaced by pBR322, a common small E. coli cloning vector. The intermediate vector (e.g. pGV1103) based on pBR322 is conjugated into pGV3850 at the region of pBR322 homology. After cointegration, pGV3850, 1103 T-DNA contains the whole of intermediate vector including the unwanted DNA, which is also transferred to host plant along with the gene intended to be transferred. To overcome this difficulty, the right border repeats or both borders are introduced to flank a cloning site in the intermediate Vector, so that only a part of intermediate vector cointegrates. The utility of this vector system is however limited. (ii) Binary vectors are based on the principle that vir genes may be located on a 'helper' Ti plasmid having the whole of T-DNA deleted (e.g. pAL4404), because vir genes can function even in trans configuration. In this case, T-DNA is found on a separate vector (binary vector) designed to replicate in both E. coli and

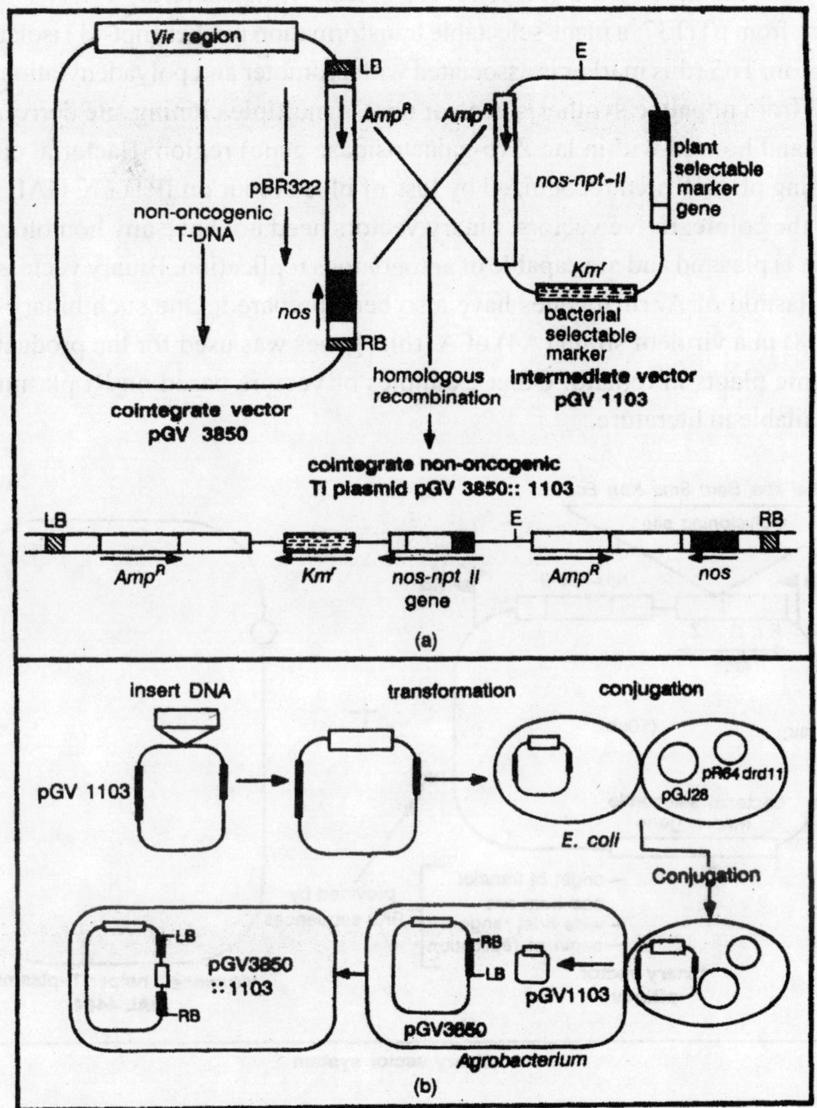


**Fig.** (a) Use of the cointegrate vector pGV3850 with the intermediate pGV1103.
(b) Preparation of a cointegrative vector for the transfer of a desired DNA segment.
(Redrawn from D Grierson 'Plant Genetic Engineering, 1991).

Agrobaclerium and capable of conjugal transfer between these two bacterial species (Fig.). Many binary vectors have been developed, which differ in size and the source of 25bp repeat sequences, plant selection marker, bacterial selection marker and cloning sites. pBin19, which was designed in 1984, is still popular and is based on wide host-range, replicon pRK252. This vector contains the following elements :

kanamycir resistance gene (APfI-I)- for selection in bacteria, T-DNA borders derivec= from pTiT37, a plant-selectable transformation marker (npt-11) isolate from transposon, Tn5 (this marker is associated with promoter and polyadenylation signal derived from nopaline synthesis gene or nos), a multiple cloning site derived from pUC19 and housed within lac Z (b-galactosidase gene) region. Bacterial colonies containing pBin19 area recognized by loss of blue colour on IPTG/X-GAL plates. Unlike the cointegrative vectors, binary vectors need not have any homology with resident Ti plasmid and are capable of autonomous replication. Binary vectors based on Ri plasmid of A. rhizogenes have also been prepared. One such binary vector (pARS 8) in a virulent strain (A4) of A. rhizogenes was used for the production of transgenic plants in tomato. Other examples of vectors based on Ri plasmids are also available in literature.

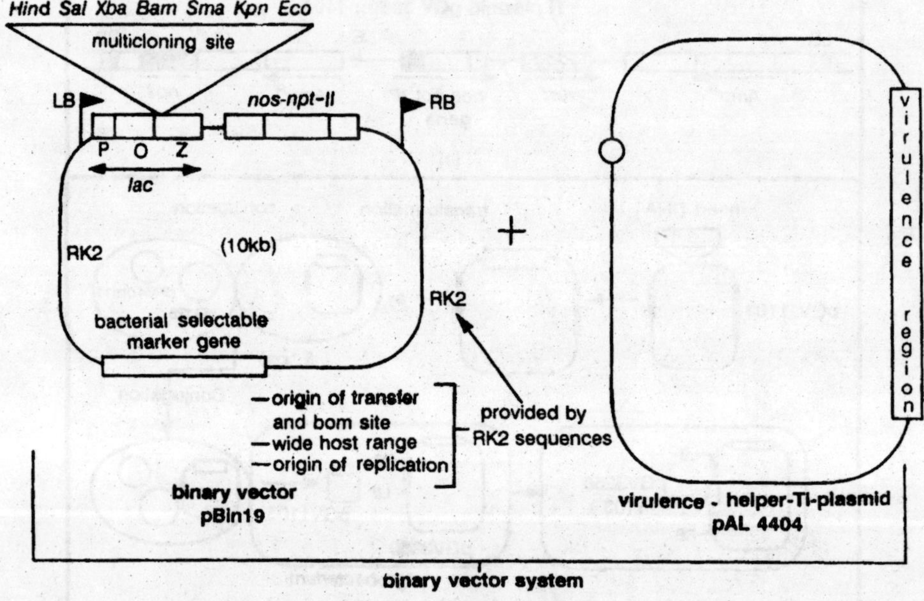


**Fig.** Typical binary vector strategy (Redrawn from D. Grierson. 'Plant Genetic Engeering, 1991).

## Virus vectors

Many viruses or their isolated genomes are capable of infecting intact plant tissue. This made them suitable for use as plant transformation vectors. But since viral genomes can not integrate with plant genome, they can be used only for a study of transient expression of transferred genes. All the three kind of viruses including caulimoviruses (double stranded DNA), gemini viruses (single stranded DNA) and tobacco mosaic virus (RNA) are capable of delivering genes into intact plant tissues, where they are expressed. However, virus vectors can not be used for the production of stably transformed plants, popularly described as transgenic plants.

**Transformation Techniques using Agrobacterium**

Agrobacteriurn infection (utilizing its plasmids as vectors) has been extensively utilized for transfer of foreign DNA into a number of dicotyledonous species. The only important species that have not responded well are major seed legumes, even though transgenic soybean (Glycine max) plants have been obtained. The success in this approach for gene transfer has resulted from improvement in tissue culture technology. However, monocotyledons could not be successfully utilized for Agrobacterium mediated gene transfer except a solitary example of Asparagus. The reasons for this are not fully understood, because T-DNA transfer does occur at the cellular level. It is possible that the failure in monocots lies in the lack of wound response of monocotyledonous cells.

**Prerequisites for production of transgenic plants**

The important requirements for Agrobacterium mediated gene transfer in higher plants include the following : (i) The plant explants must produce acetosyringone or other active compounds in order to induce or genes for virulence; alternatively Agrobacterianm may be preinduced with synthetic acetosyringone. (ii) The induced agrobacteria should have access to cells that are competent for transformation ; for gene transfer to occur, cells must be replicating DNA or undergoing mitosis (wounded end dedifferentiated cells, fresh explants or protoplasts have these properties). (iii) Often transformed tissues or explants do not regenerate and it is difficult to combine transformation competence with totipotency (regeneration ability) ; therefore, the transformation competent cells should be able to regenerate in whole plants, a combination that can be easily achieved only in some species, such as tobacco. In some cases, undifferentiated cells of embryos may undergo transformation, so that the embryos may develop into chimeric plants.

**Explants used for transformation**

The explants used for inoculation or cocultivation with Agrobacterium carrying the vector, include protoplasts, suspension cultured cells, callus cell clumps (undifferentiated and proembryogenic), thin cell layers (epidermis), tissue slices, whole organ sections (e. g. leaf discs, sections of roots, stems or floral tissues), etc. Wounding and inoculation of whole plants may also be used.

**Marker genes for selection find scoring of transformed cells/Calluses or shoots**

After explants are inoculated with Agrobacterium carrying the requisite vector having the gene of interest, we need to select the transformed cells/tissues. This is facilitated by the presence of selectable marker genes available in the vector (see Table). The selectable marker genes enable the transformed cells to survive in media containing toxic levels of the selection agent, which is usually an antibiotic or a herbicide. Tobacco is used as a model transformation system, where explants (e.g. leaf discs) are placed on regeneration medium containing an antibiotic like kanamycin and transformed shoots can be obtained directly. Any cells, which are not transformed, die due to the presence of kanamycin. Other antibiotics and herbicides may require more judicious use, since even low concentrations can cause rapid cell death. In some cases, selection is exercised only after the regeneration is achieved, because adventitious root formation is sensitive to antibiotics.

**Table**

Selectable marker genes in vectors used for gene transfer

| *Selectable marker gene* | *Substrate used for selection* |
|---|---|
| | **I. Antibiotics** |
| 1. Neomycin phosphotransferase (*npt II*) | G418, kanamycin, neomycin |
| 2. Hygromycin phosphotransferase (*hpt*) | Hygromycin B |
| 3. Dihydrofoate reductase (*dhfr*) | Methotrexate terimethoprim |
| 4. Gene ble (enzyme not known) | Bleomycin |
| 5. Gentamycin acetyltransferase | Gentamycin |
| 6. Streptomycin phosphotransferase | Streptomycin |
| | **II. Herbicides** |
| 7. Mutant form of acetolactate synthase (*als*) | Chlorosulfuron imidazolinones |
| 8. Bromoxynil nitrilase | Bromoxynil |
| 9. Phosphinothricin acetyltransferase (*bar*) | L - phosphinothricin (PPT; also known as bialaphos) |
| 10. 5 - enolpyruvylshikimate -3 -phosphate (EPSP) synthase (*aro A*) | Glyphosate (Roundup) |

Following successful selection, the next obstacle is the regeneration of shoots from transformed calli. Since the explants may be heterogeneous; and non-transformed tissue may escape initial selection, several measures are used to ensure that very few nontransformed shoots escape the selection procedure. The use of leaf discs ensures this, because most cells at the wounded edges, which undergo regeneration, are in contact with the selective agent.

Screening techniques offer an alternative in which no selection pressur is imposed on ceps or shoots developing from inoculated explants. In this case, samples of tissues are taken from all regenerated shoots and tested for the expression of a marker gene. A number of marker genes, described as screenable genes or scoreable genes or reporter genes are listed in Table. The most commonly used reporter genes are nptll, cat, gets and lux, which will be briefly described.

**Table**

Reporter genes used as scoreable markers, and their enzyme assays

| *Reporter gene for* | *Substrate* and assay* | *Identification* |
|---|---|---|
| 1. Chloramphenicol acetyl-transferase (CAT) | ($^{14}$C) chloramphenicol + acetyl CoA; TLC separation | Detection of acetyl chloremphenicol by autoradiography |
| 2. Neomycin phosphotrans-ferase (NPT II) | Kanamycin+ ($^{32}$P) ATP (in situ assay) | Radioactivity detection (also dot blots) |
| 3. β-glucuronidase (GUS) | Glucuronides (PNPG, x-GLUC, NAG, REG) | Fluorescence detection; colourimetric, fluorometric and histochemical |
| 4. β-galactosidase (lac Z) | β-galactoside (X-gal) | Colour of the colony |
| 5. Luciferase | (a) Decimal and FMNH2<br>(b) ATP +O2 +luciferin | Bioluminiscence (exposure of x-rays film) |
| 6. Octopine synthase (OCS) | Ariginine + pyruvate + NADH | Electrophoresis |
| 7. Nopaline synthase (NOS) | Arginine + ketoglutaric acid + NADI | Electrophoresis |

* PNPG = p-nitrophenyl glucuronide; x-GLUC = 5-bromo, 4-chloro. 3-indolyl glucuronide; NAG = naphthol AS-BI glucuronide; REG = resorufin glucuronide.

Neomycin phosphotransferase gene (*npt* 11)—This gene is used both as a selectable and a scoreable marker in experiments involving transfer of genes leading to the production of transgenic plants. It imparts kanamycin resistance, so that the transformed tissue can be selected on kanamycin. An assay for NPT II enzyme is also used to detect its presence in transformed tissue or transgenic plants. The gene for NPT II enzyme is often used with nos promoter, which drives its synthesis. In some cases, npt II gene had adverse effect on the expression of the desirable gene introduced (eg. bt2 gene for insect resistance; see next chapter), so that alternative approaches for improving its expression had to be used.

Chloramphenicol acetyl transferase gene (cat)—This gene is primarily used as a reporter gene or a scoreable marker and not as a selectable marker. The gene first isolated from E. coli codes for an enzyme (CAT), which is absent in mammals and higher plants, so that whenever transferred in a gene construct, its presence can be detected by enzyme assay. Rarely screening for this enzyme may also be used for selection of transformed regenerants, although no selection pressure can be applied. The enzyme uses acetyl CoA + chloramphenicol (32P) as substrates and helps in the transfer of acetyl group to chloramphenicol. The presence of acetyl chloramphenicol is detected through autoradiography. The gene cat has also been used for identification of a number of regulatory sequences.

β-glucuronidase (GUS) gene (gus)—The enzyme P-glucuronidase, popularly described as GUS, breaks down glucuronides giving a coloured reaction, so that its' presence can be detected in sin., i.e. inside the plant tissue, used either as thin sections or in any other form. Several glucuronides, which can be used as substrates, include p-nitro phenyl glucuronide (PNPG), 5-bromo, 4-chloro, 3-indolyl glucuronide (BCIG), naphthol AS-B1 glucuronide (NAG) and resorufin glucuronide (REG). The enzyme GUS is coded by a gene gus, first isolated from E. coli. The major advantage of using this reporter gene (scoreable marker) lies in its assay, which requires no DNA extraction, electrophonesis or autoradiography.

Luciferase gene (lux)—In the presence of a suitable substrate, the enzyme luciferase confers on the organism carrying it, the ability to glow in the dark. In other words, the organism carrying this enzyme exhibits luminiscence on an assay for the enzyme luciferase. The enzyme luciferase differs in bacteria and firefly. While in bacteria, it consists of two peptide subunits coded by genes lux A and lux B. In firefly, it consists of a single polypeptide coded by a gene lux. The enzyme luciferase in these two systems uses different substrates for luminiscence. Therefore, depending upon the source of luciferase gene used as a scoreable marker gene, the assay for detection of its presence also differs. When the source of the gene is bacteria, the substrate used is an aldehyde (decanal), which is supplied exogenously with FMNH2. Alternatively, when the source is firefly, luciferin (a heterocyclic carboxylic acid) is instead used as a substrate, and is supplied with ATP and oxygen (Table).

## Agroinfection and Gene Transfer

Agoinfection is a phenomenon, in which a virus infects a host as a part of T-DNA of Ti plasmid carried by Agrobacterium Viral. DNA can be integrated into the T-DNA and can be delivered into plant cells with the normal Agrobacterium T-DNA transfer process. After infection, viral DNA is released to form functional virus that replicates and spreads systemetically. Agroinfection may also lead to the integration of viral DNA so that transgenic plant containing integrated viral DNA can be produced.

In maize, agroinfection with maize streak virus has been demonstrated. This suggested that Agrobacterium based vector system can be used for genetic engineering in cereals, although ordinarily Agrobacterium does not infect monocotyledons. Thus, agroinfection can lead to the production of transgenic plants, even though it has no better chances of yielding transgenic cereals than does Agrobacterium infection alone. However, agroinfection has great potential for studies in virus biology, because it can transfer deletion mutations or even single viral genes.

## Physical Delivery Methods or

## DNA Mediated Gene Transfer (DMGT)

Agrobacterium mediated gene transfer has been the most commonly used method of gene transfer in plants but the cereals, comprising the most important food crops, are not amenable to this method of gene transfer. Further, in many crops including cereals and legumes, tissue culture techniques for regeneration are not very successful. These two limitations forced intensive search for alternative methods for gene transfer. Physical delivery of DNA or DNA mediated gene transfer (DMGT), as it is often described, employs methods, which can be grouped according to the type of target cell. For instance, chemically stimulated endocytosis of plasmids or DNA-loaded liposomes and electroporation are employed for delivery of DNA to protoplasts only. On the other hand, techniques like microinjection, macroinjection and shooting with microprojectiles can be used with a variety of explants (e.g. immature embryos, organ meritems, gametes, zygotes, etc.). These latter techniques achieve significance, because regeneration of transformed plants from protoplast derived tissues is still difficult in many species particularly in cereals (although encouraging results have been recently obtained in case of maize rice, wheat, oafs. etc.).

### Chemically stimulated DNA uptake by protoplasts

Direct DNA uptake by protoplasts can be stimulated by chemicals like polyethylene glycol (PEG). The technique is so efficient that virtually every protoplast system has proven transformable. PEG is also used to stimulate the uptake of liposomes and to improve the efficiency of electroporation (see below for details). PEG at high concentration (15-25%) will precipitate ionic macromolecules such as DNA and stimulate their uptake by endocytosis without any gross damage to protoplasts. This is followed by cell wall formation and initiation of cell division. These cells can now be plated at low density on selection medium.

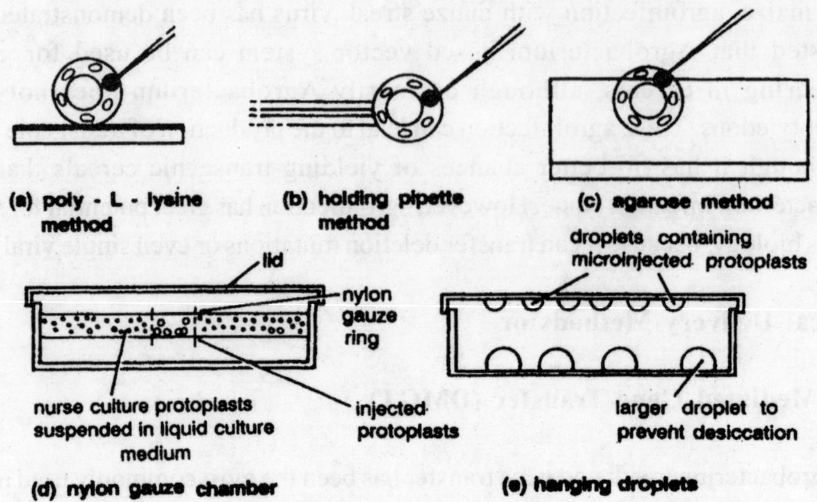


**Fig.** Methods for the microinjection and culture of plant protoplasts

Initial studies using the above method were restricted to Petunia and Nicotiana. However, other plant systems (rice, maize, etc.) were also successfully used later. In these methods, PEG was used in combination with pure Ti plasmid, or calcuim phosphate precipitated Ti plasmid mixed with a carrier DNA. Transformation frequencies upto 1 in 100 have been achieved by this method. Nevertheless, there are serious problems in using this method for getting transgenic plants and all these problems relate to plant regeneration from protoplasts.

## Microinjection and macroinjection

Plant regeneration from transformed protoplasts still remains a problem. Therefore, cultured tissues, that encourage the continued development of immature structures, provide alternate cellular targets for transformation. These immature structures may include immature embryos, meristems, immature pollen, germinating pollen, isolated ovules, embryogenic suspension cultured cells, etc. The main disadvantage of this technique is the production of chimeric plants with only a part of the plant transformed. However, from this chimeric plant, transformed plants of single cell origin can be subsequently obtained. Utilizing this approach, transgenic chimeras have actually been obtained in oilseed rape (Brassica napus).

When cells or protoplasts are used as targets in the technique of microinjection, glass micropipettes with 0.5-10,mm diameter tip are used for transfer of macromolecules into the cytoplasm or the nucleus of a recipient cell or protoplast.

The recipient cells are immobilized on a solid support (cover slip or slide, etc.) or artificially bound to a substrate or held by a pipette (Fig.) under suction (as done in animal systems). Often a specially designed micromanipulator is employed for microinjecting the DNA. Although, this technique gives high rate of success, the process is slow, expensive and requires highly skilled and experienced personnel.

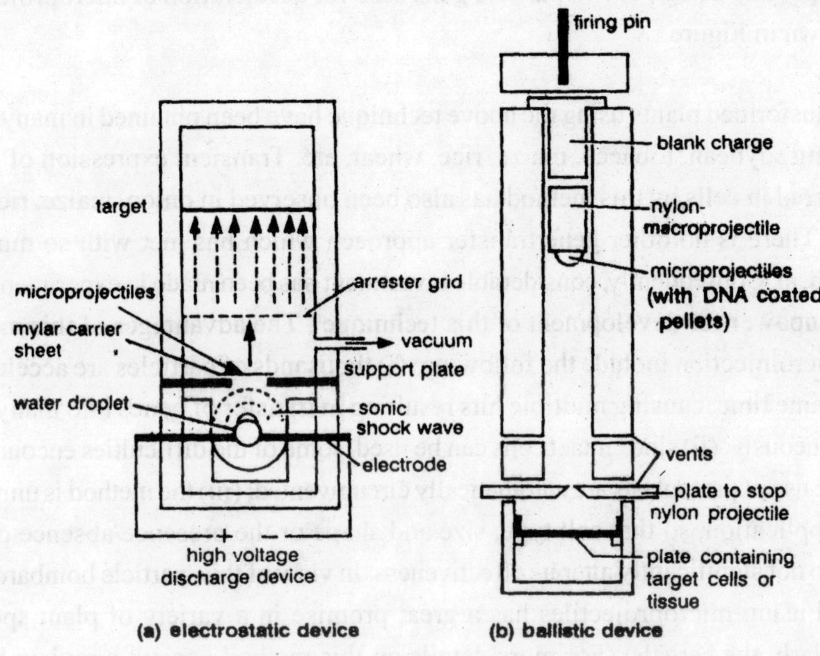


**Fig.** Microprojectile acceleration devices

DNA macroinjection employing needles with diameters greater than cell diameter has also been tried. In rye (Secale cereale), a marker gene was macroinjeced into the stem below the immature floral meristem, so is to reach the sporogenous tissue (De la Pena et al., 1987) leading to successful production of transgenic plants. Unfortunately, this technique could not be successfully repeated with any other cereal, when tried in several laboratories. Therefore, doubt is expressed about the validity of earlier experiments conducted with rye (Potrykus, 1991).

**Microprojectiles (biolistics or particle gun) for gene transfer**

In recent years, it has been shown that DNA delivery to plant cells is also possible, when heavy microparticles (tungsten or gold) coated with the DNA of interest are accelerated to a very high initial velocity (1,400 ft per, sec). These microprojectiles, normally 1-3μm in diameter are carried by a 'macroprojectile' or the 'bullet' and are accelerated into living plant cells (target cells can be pollen, cultured cells, cells in differentiated tissues and meristems) so that they can penetrate cell walls of intact tissue. The acceleration is achieved either by an explosive charge (cordite explosion) or by using shock waves initiated by a high voltage electric discharge. The design of two particle guns used for acceleration of microprojectiles are shown in Figure.

Transformed plants using the above technique have been obtained in many cases including soybean, tobacco, maize, rice, wheat, etc. Transient expression of genes transferred in cells by this method has also been observed in onion, maize, rice and wheat. There is no other gene transfer approach which has met with so much of enthusiasm. Consequently, considerable investment has been made in experimentation and manpower for development of this technique . The advantages of this method over microinjection include the following: (i) thousands of particles are accelerated at the same time, causing multiple hits resulting in transfer of genes into many cells simultaneously; (ii) since intact cells can be used, some of the difficulties encountered with the use of protoplasts are automatically circumvented; (iii) the method is universal in its application, so that cell type, size and shape or the presence/absence of cell walls do not significantly alter its effectiveness. In vie v of this, particle bombardment method using microprojectiles has a great promise in a variety of plant species, particularly the cereals. (For more details on this method consult a review 'Gene Transfer in Higher Plants through Biolistic Delivery' by Amita Pal, published in Proc. Indian natn. Sci. Acad. B59, pp. 87-94, 1993).

**Electroporation method for gene transfer**

This method is based on the use of short electrical impulses of high field strength. These impulses increase the permeability of protoplast membrane and facilitates entry of DNA molecules into the cells, if the DNA is in direct contact with the membrane. In view of this, for delivery of DNA to protoplasts, electroporation is one of the several routine techniques for efficient transformation. However, since regeneration from protoplasts is not always possible, cultured cells or tissue explants are often used. Consequently, it is important to test whether electroporation could transfer genes into walled cells. In most of these cases no proof of transformation was available.

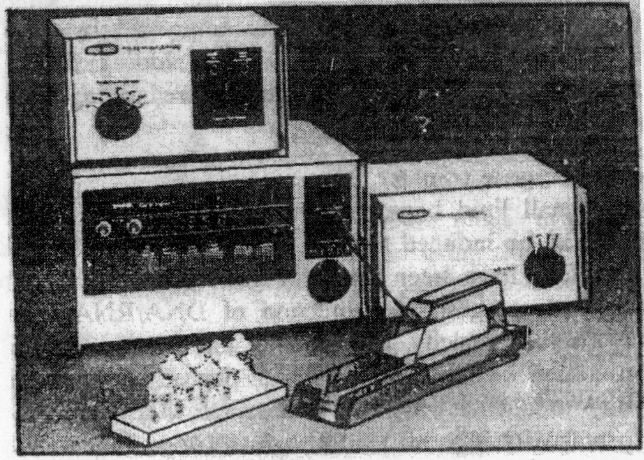

**Fig.** An electroporation system, suitable for transformation of bacţeria, mammalian cells and plant cells

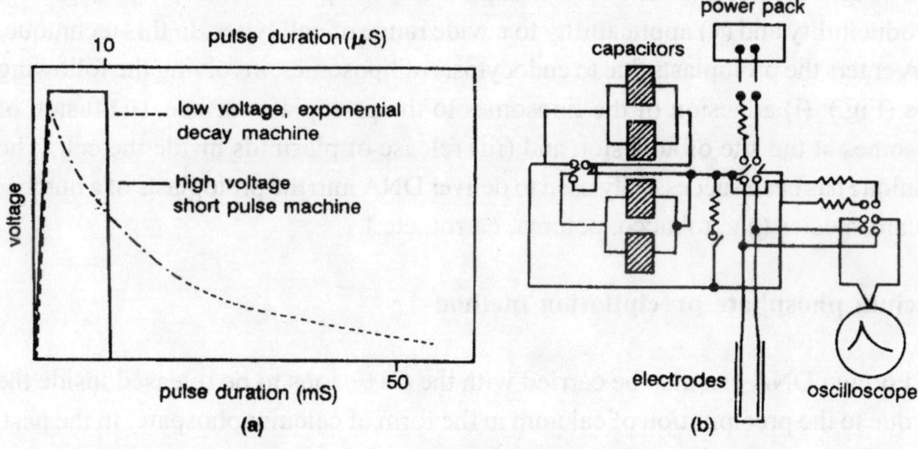


**Fig.** (a) Details of the output characteristics of the two principle types of electroporation devices in use: (b) circuit diagram of low voltage, exponential decay machine. (Redrawn from D. Geierson. 'Plant Genetic Engineering', 1991)

The electroporation pulse is generated by discharging a capacitor across the electrodes in a specially designed electroporation chamber (Fig.). Either a high voltage (1.5 kV) rectangular wave pulse of short duration or a low voltage (350V) pulse of long duration is used (Fig.). The latter can be generated by a home made machine. Protoplasts in an ionic solution containing the vector DNA, are suspended between the electrodes, electroporated and then plated as usual. Transformed colonies are selected as described earlier. Using electroporation method, successful transfer of genes was achieved with the protoplasts of tobacco, petunia, maize, rice, wheat and sorghum. In most of these cases, cat gene associated with a suitable promoter sequence was transferred.

Transformation frequencies can be further improved by (i) using field strength of 1.25kV/cm, (ii) adding PEG after adding DNA, (iii) heat shocking protoplasts at 45°C for 5 minutes before adding DNA and (iv) by using linear instead of circular DNA.

**Liposome mediated gene transfer**

Liposomes are small lipid bags, in which large number of plasmids are enclosed. They can be induced to fuse with protoplasts using devices like PEG, and therefore, have been used for gene transfer. The technique offers following advantages: (i) protection of DNA/RNA from nuclease digestion, (ii) low cell toxicity, (iii) stability and storage of nucleic acids due to encapsulation in liposomes, (iv) high degree of reproducibility and (v) applicability to a wide range of cell types. In this technique, DNA enters the protoplasts due to endocytosis of liposomes, involving the following steps (Fig.): (i) adhesion of the liposomes to the protoplast surface, (ii) fusion of liposomes at the site of adhesion and (iii) release of plasmids inside the cell. The technique has been successfully used to deiiver DNA into the protoplasts of a number of plant species (e.g. tobacco, petunia, carrot, etc.).

**Calcium phosphate precipitation method**

Foreign DNA can also be carried with the Ca++ ions to be released inside the cell due to the precipitation of calcium in the form of calcuim phospate. In the past, this method was considered to be very important for gene transfer in plants.

**Transformation using pollen or pollen tube**

There has been a hope that DNA can be taken up by the germinating pollen and can either integrate into sperm nuclei or reach the zygote through the pollen tube pathway. Both these approaches have been tried and interesting phenotypic alterations suggesting gene transfer have been obtained. In no case, however, unequivocal proof of gene transfer has been available. In a number of experiments, when marker genes were used for transfer, only negative results were obtained. Several problems exist in this method and these include the presence of cell wall, nucleases, heterochromatic state of acceptor DNA, allow plugs in pollen tube, etc. Transgenic plants have never been recovered using this approach and this method, though very attractive, seems to have little potential for gene transfer.

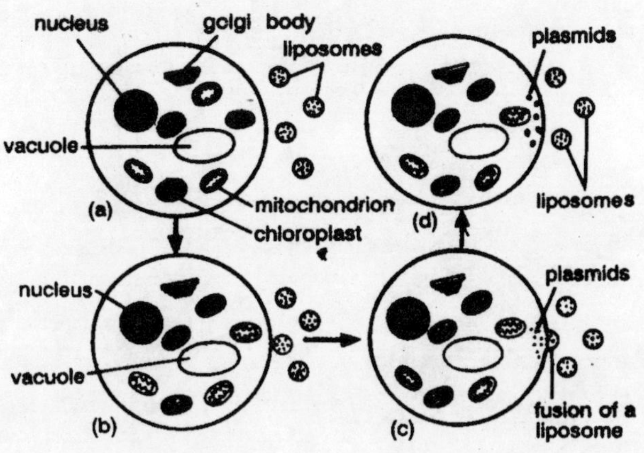


**Fig.** Fusion of plasmid-filled liposomes with isolated protoplasts as a method of plasmid uptake

## Incubation of dry seeds, embryos, tissues or cells in DNA

Incubation of dry seeds, embryos, tissues or cells in known DNA (viral or non-viral having defined marker genes) has been tried in many cases and expression of defined genes has been witnessed. However, in no case proof of integrative transformation could be available. In all these cases, plant cell walls not only work as efficient barriers, but are also efficient traps for DNA molecules. It would be very surprising if DNA can cross cell walls efficiently without permeabilizing them either by PEG, or by electroporation or by any other device.

## Transformation by ultrasonication

In wheat, tobacco and sugarbeet, explants after being cultured for a few days, were sonicated with plasmid DNA (carrying marker genes like cat, npt II and gus). When sonicated calli were transferred to selective medium, shoots were obtained, although all control calli (not sonicated with plasmid DNA) died. In tobacco transgenic plants were obtained at a frequency of 22 per cent. Biotechnology Reserch Center at Beijing (China) carried out this work.

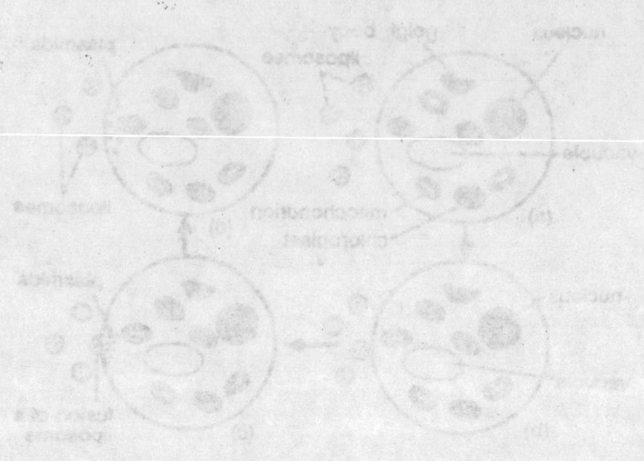

# Index

G



H



I



J



L



M

N



O



P

**W**



**Y**



□□□